The Modifications of Metallic and Inorganic Materials by Using Energetic Ion/Electron Beams

The Modifications of Metallic and Inorganic Materials by Using Energetic Ion/Electron Beams

Editor

Akihiro Iwase

MDPI • Basel • Beijing • Wuhan • Barcelona • Belgrade • Manchester • Tokyo • Cluj • Tianjin

Editor
Akihiro Iwase
Department of Research and
Development
The Wakasa Wan Energy
Research Center
Tsuruga
Japan

Editorial Office
MDPI
St. Alban-Anlage 66
4052 Basel, Switzerland

This is a reprint of articles from the Special Issue published online in the open access journal *Quantum Beam Science* (ISSN 2412-382X) (available at: www.mdpi.com/journal/qubs/special_issues/metallic_inorganic).

For citation purposes, cite each article independently as indicated on the article page online and as indicated below:

LastName, A.A.; LastName, B.B.; LastName, C.C. Article Title. *Journal Name* **Year**, *Volume Number*, Page Range.

ISBN 978-3-0365-3095-6 (Hbk)
ISBN 978-3-0365-3094-9 (PDF)

Contents

About the Editor

Akihiro Iwase

Akihiro Iwase obtained a Doctorate of Science from the University of Tokyo. He worked as a researcher at the Japan Atomic Energy Research Institute (JAERI; current name, Japan Atomic Energy Agency) from 1978 to 2003. In 2003, he moved to Osaka Prefecture University and worked as a professor for 15 years. After retirement from the University in 2018, he became an executive director of the Osaka Nuclear Science Association (ONSA) for 1 year. He is currently a director general of the Wakasa Wan Energy Research Center (WERC). He is also a visiting professor at Osaka Prefecture University, a visiting researcher at National Institutes for Quantum Science and Technology (QST), and a member of the Editorial Board of *Quantum Beam Science*, MDPI. His main research interests cover the fundamentals of the interaction between energetic ion/electron beams and materials, the modification of metallic and inorganic materials by using ion/electron beams, material analysis using ion beam and synchrotron radiation, and the radiation damage of materials related to nuclear energy. He has authored or co-authored over 300 research articles.

Preface to "The Modifications of Metallic and Inorganic Materials by Using Energetic Ion/Electron Beams"

It is with great pleasure that I served as a Guest Editor for this Special Issue entitled "The Modifications of Metallic and Inorganic Materials by Using Energetic Ion/Electron Beams". This book consists of 14 original articles, 5 review papers, and 1 editorial, which show the experimental and computer simulation results for the modifications of several physical properties and lattice structures of pure metals, metallic alloys, oxides, and ceramics by using energetic ion/electron beams. In the book, the developments of ion accelerators and transmission electron microscopes used for the study of material modifications are also described.

I thank all of the authors, the reviewers, and the QuBS Editorial Board Members for their valuable contributions. I believe that many researchers will enjoy this book.

Akihiro Iwase
Editor

Editorial

Modifications of Metallic and Inorganic Materials by Using Ion/Electron Beams

Akihiro Iwase 

The Wakasa Wan Energy Research Center (WERC), 64-52-1 Nagatani, Tsuruga 914-0192, Fukui, Japan; aiwase@werc.or.jp

Citation: Iwase, A. Modifications of Metallic and Inorganic Materials by Using Ion/Electron Beams. *Quantum Beam Sci.* **2022**, 6, 1. https://doi.org/10.3390/qubs6010001

Received: 22 December 2021
Accepted: 27 December 2021
Published: 29 December 2021

Publisher's Note: MDPI stays neutral with regard to jurisdictional claims in published maps and institutional affiliations.

Welcome to the Special Issue of *Quantum Beam Science* entitled "Modifications of Metallic and Inorganic Materials by Using Ion/Electron Beams". This Special Issue has collected original and review papers using energetic ion/electron beams in basic and applied research for new and novel metallic and inorganic materials' modifications. When materials are irradiated with energetic particles (ions or electrons), their energies are transferred to electrons and atoms in materials, and the lattice structures of the materials are largely changed to metastable or non-thermal-equilibrium states, causing modifications of several physical properties. Such phenomena will engage the interest of researchers as a basic science and can also be used as promising tools for adding new functionalities to existing materials and developing novel materials. Compared with organic materials such as polymers, however, not many studies on the modifications of metallic or inorganic materials by electron or ion irradiations have been performed so far.

The present Special issue of *Quantum Beam Science*, therefore, focuses on experimental investigations and computer simulations related to the modification of lattice structures and various physical properties (mechanical, electronic, magnetic, optical, and so on) of metallic and inorganic materials. The developments of accelerators and ion or electron beam equipment for the materials' modification are also included in the scope of this Special Issue.

The original and review articles of this Special issue cover the electron/ion beam induced modifications of optical properties of oxides, such as colors [1] and refractive indices [2]; the electronic properties of solar cells [3] and superconductors [4,5]; the mechanical property (hardness) of metallic alloys [6]; and chemical properties, such as the catalyst function of a metal surface [7], the corrosion of a metallic alloy [8], and the hydrogen desorption and retention of a metallic alloy [9]. In addition, several articles show ion/electron-beam-induced modifications of crystal structures, such as the lattice disordering of non-metallic materials [10], the phase transformation of an oxide [11], the hillock and ion-track formation in ceramics [12], the formation of nanostructured materials in oxides [13] and the generation of self-organized nanostructures on the surfaces of pure metals [14]. One of the review papers reports the details of the swift heavy-ion irradiation effects in CeO_2 with many references [15]. Moreover, the ion-beam-induced modifications of lattice structures and/or magnetic properties in oxides are discussed by using computer simulations (Monte Carlo method [16] and molecular dynamics [17]). As for the facility and the equipment used for the study of electron or ion beam irradiation effects, the ion accelerators at the Wakasa Wan Energy Research Center for the study of irradiation effects on space electronics [18] and the pulsed transmission microscope for high-speed observation and material nanofabrication [19] are introduced.

The original and review articles reported here include a lot of interesting results. I hope many readers will enjoy this Special Issue.

Funding: This research received no external funding.

Conflicts of Interest: The author declares no conflict of interest.

References

1. Kobayashi, T.; Nishiyama, F.; Takahiro, K. Chromatic Change in Copper Oxide Layers Irradiated with Low Energy Ions. *Quantum Beam Sci.* **2021**, *5*, 7. [CrossRef]
2. Amekura, H.; Li, R.; Okubo, N.; Ishikawa, N.; Chen, F. Irradiation Effects of Swift Heavy Ions Detected by Refractive Index Depth Profiling. *Quantum Beam Sci.* **2020**, *4*, 39. [CrossRef]
3. Imaizumi, M.; Ohshima, T.; Yuri, Y.; Suzuki, K.; Ito, Y. Effects of Beam Conditions in Ground Irradiation Tests on degradation of Photovoltaic Characteristics of Space Solar Cells. *Quantum Beam Sci.* **2021**, *5*, 15. [CrossRef]
4. Sueyoshi, T. Modification of Critical Current Density Anisotropy in High-T_c Superconductors by Using Heavy-Ion Irradiations. *Quantum Beam Sci.* **2021**, *5*, 16. [CrossRef]
5. Ozaki, T.; Kashihara, T.; Kakeya, I.; Ishigami, R. Effect of 1.5 MeV Proton Irradiation on Superconductivity in $FeSe_{0.5}Te_{0.5}$ Thin Films. *Quantum Beam Sci.* **2021**, *5*, 18. [CrossRef]
6. Fukumoto, K.; Kitamura, Y.; Miura, S.; Fujita, K.; Ishigami, R.; Nagasaka, T. Irradiation Hardening Behavior of He-Irradiated V-Cr-Ti Alloys with Low Ti Addition. *Quantum Beam Sci.* **2021**, *5*, 1. [CrossRef]
7. Sato, Y.; Koshikawa, H.; Yamamoto, S.; Sugimoto, M.; Sawada, S.; Yamaki, T. Fabrication of Size- and Shape-Controlled Platinum Cones by Ion-Track Etching and Electrodeposition Techniques for Electrocatalytic Applications. *Quantum Beam Sci.* **2021**, *5*, 21. [CrossRef]
8. Okubo, N.; Fujimura, Y.; Tomobe, M. Effect of Irradiation on Corrosion Behavior of 316L Steel in Lead-Bismuth Eutectic with Different Oxygen Concentrations. *Quantum Beam Sci.* **2021**, *5*, 27. [CrossRef]
9. Watanabe, H.; Saita, Y.; Takahashi, K.; Yasunaga, K. Desorption of Implanted Deutrium in Heavy Ion Irradiated Zry-2. *Quantum Beam Sci.* **2021**, *5*, 9. [CrossRef]
10. Matsunami, N.; Sataka, M.; Okayasu, S.; Tsuchiya, B. Modification of SiO_2, ZnO, Fe_2O_3 and TiN Films by Electronic Excitation under High Energy Ion Impact. *Quantum Beam Sci.* **2021**, *5*, 30. [CrossRef]
11. Okuno, Y.; Okubo, N. Phase Transformation by 100 keV Electron Irradiation in Partially Stabilized Zirconia. *Quantum Beam Sci.* **2021**, *5*, 20. [CrossRef]
12. Ishikawa, N.; Taguchi, T.; Ogawa, H. Comprehensive Understanding of Hillocks and Ion Tracks in Ceramics Irradiated with Swift Heavy Ions. *Quantum Beam Sci.* **2020**, *4*, 43. [CrossRef]
13. Tanaka, S. Control and Modification of Nanostructured Materials by Electron Beam Irradiation. *Quantum Beam Sci.* **2021**, *5*, 23. [CrossRef]
14. Niwase, K. Self-Organized Nanostructures Generated on Metal Surfaces under Electron Irradiation. *Quantum Beam Sci.* **2021**, *5*, 4. [CrossRef]
15. Cureton, W.F.; Tracy, C.L.; Lang, M. Review of Swift Heavy Ion Irradiation Effects in CeO_2. *Quantum Beam Sci.* **2021**, *5*, 19. [CrossRef]
16. Iwase, A.; Nishio, S. Simulation of Two-Dimensional Images for Ion-Irradiation Induced Change in Lattice Structures and Magnetic States in Oxides by Using Monte Carlo Method. *Quantum Beam Sci.* **2021**, *5*, 13. [CrossRef]
17. Sasajima, Y.; Kaminaga, R.; Ishikawa, N.; Iwase, A. Nanopore Formation in CeO_2 Single Crystal by Ion Irradiation: A Molecular Dynamics Study. *Quantum Beam Sci.* **2021**, *5*, 32. [CrossRef]
18. Hatori, S.; Ishigami, R.; Kume, K.; Suzuki, K. Ion Accelerator Facility of the Wakasa Wan Energy Research Center for the Study of Irradiation Effects on Space Electronics. *Quantum Beam Sci.* **2021**, *5*, 14. [CrossRef]
19. Yasuda, H.; Nishitani, T.; Ichikawa, S.; Hatanaka, S.; Honda, Y.; Amano, H. Development of Pulsed TEM Equipped with Nitride Semiconductor Photocathode for High-Speed Observation and Material Nanofabrication. *Quantum Beam Sci.* **2021**, *5*, 5. [CrossRef]

quantum beam science

Article

Chromatic Change in Copper Oxide Layers Irradiated with Low Energy Ions

Takuya Kobayashi [1], Fumitaka Nishiyama [2] and Katsumi Takahiro [1,*]

[1] Faculty of Materials Science and Engineering, Kyoto Institute of Technology, Matsugasaki, Sakyo, Kyoto 606-8585, Japan; m0672011@edu.kit.ac.jp

[2] Research Institute of Nanodevice and Bio Systems, Hiroshima University, Kagamiyama 1-4-1, Higashi-Hiroshima, Hiroshima 739-8527, Japan; fnishi@hiroshima-u.ac.jp

* Correspondence: takahiro@kit.ac.jp

Abstract: The color of a thin copper oxide layer formed on a copper plate was transformed from reddish-brown into dark blue-purple by irradiation with 5 keV Ar^+ ions to a fluence as low as 1×10^{15} Ar^+ cm^{-2}. In the unirradiated copper oxide layer, the copper valence state of Cu^{2+} and Cu^+ and/or Cu^0 was included as indicated by the presence of a shake-up satellite line in a photoemission spectrum. While for the irradiated one, the satellite line decreased in intensity, indicating that irradiation resulted in the reduction from Cu^{2+} to Cu^+ and/or Cu^0. Furthermore, nuclear reaction analysis using a $^{16}O(d, p)^{17}O$ reaction with 0.85 MeV deuterons revealed a significant loss of oxygen (5×10^{15} O atoms cm^{-2}) in the irradiated layer. Thus, the chromatic change observed in the present work originated in the irradiation-induced reduction of a copper oxide.

Keywords: ion beam; copper oxide; chromatic change; photoemission spectrum; beam viewer

check for
updates

Citation: Kobayashi, T.; Nishiyama, F.; Takahiro, K. Chromatic Change in Copper Oxide Layers Irradiated with Low Energy Ions. *Quantum Beam Sci.* **2021**, *5*, 7. https://doi.org/10.3390/qubs5010007

Academic Editor: Akihiro Iwase

Received: 2 February 2021
Accepted: 9 March 2021
Published: 10 March 2021

Publisher's Note: MDPI stays neutral with regard to jurisdictional claims in published maps and institutional affiliations.

1. Introduction

On ion beam experiments including materials analysis and modification with ion beams, a beam monitoring system is installed in the sample chamber to monitor the beam position and uniformity in the beam spot. Most of the beam monitor consists of a fluorescent plate, enabling real time visualization of a beam spot on the plate. A SiO_2 plate, for example, has been used for beam monitoring because of strong emission [1–5] in the visible range when irradiated with MeV-ion beams. A Cr-doped Al_2O_3 (e.g., AF995R, Desmarquest) is also suitable for beam profiling [6,7] for ion beams with energies larger than several hundred keV. The aforementioned materials are insulators and therefore electric charging takes places on the fluorescent plate irradiated with ion beams, resulting in deflection of ion beams in the vicinity of the fluorescent plate if their acceleration voltage is comparable to the charged potential of a few tens kilovolts [8]. This means that the fluorescent point would be different from the real position, and further the fluorescent point would not appear at all in the case of low energy ion beams with <10 keV. In addition to the fluorescent materials, phosphors such as ZnS based materials [9–11], which have been widely used for screens of a cathode-ray tube, are applicable to beam monitoring. Other candidates of inorganic luminescent materials can be found in the literature [12]. Powders of such materials, mixed with a conducting paste and deposited onto a conducting plate, are a candidate for ion beam monitoring materials. However, such powders cost very high, but their lifetimes are very short because radiation damage causes degradation of light-emitting efficiency. It is, therefore, not easy to view a beam spot of a low energy ion beam with energy of several keV on real-time.

On the other hand, non-real time beam monitoring can be conducted by the color change of materials irradiated with ion beams. A polyimide film is, for example, widely used to check both the position and uniformity of an ion beam, because blackening due to graphitization [13–15] occurs when the film is irradiated with ion beams. A polyimide film is, however, non-conducting and is inapplicable to the beam viewer for low energy ion beams with energies of several keV. The favorable beam viewer for low energy ion beams should be composed of electrically conducting materials. A metallic copper plate, even if a thin upper layer of copper oxide is present, is a good conducting material. The color of the copper plate covered with thin oxide is reddish-brown, largely different from polished metallic copper. The present authors made an attempt to fabricate a beam viewer in which the appearance of a beam spot turns shiny due to removal of the oxide layer by physical sputtering. In the irradiation apparatus with base pressure of 2×10^{-4} Pa, the shiny beam spot could be clearly recognized after irradiation with 5 keV Ar$^+$ ions to a fluence of 1×10^{15} Ar$^+$ cm^{-2}. Surprisingly, in the other irradiation equipment with base pressure of 2×10^{-6} Pa, the color of the beam spot turned dark blue-purple after irradiation under the same conditions above. In the present work, the chromatic change observed in the irradiation equipment with such a high vacuum is examined to fabricate a new type of beam viewer for low energy ion beams.

2. Materials and Methods

Oxygen-free copper plates of $10 \times 10 \times 1$ mm^3 (purity 99.99%) were supplied from NILACO, Tokyo, Japan. The Cu plate was put on a laboratory hot plate setting to 473 K in ambient air for 2 min to form the Cu oxide layer. The maximum temperature measured on the surface with chromel-alumel thermocouple was 440 K, somewhat lower than the setting temperature of 473 K, as recorded in Figure 1. The color of heat-treated Cu plate was reddish-brown. The samples were irradiated with 5 keV-Ar$^+$ ions up to a fluence of 1×10^{15} cm^{-2} using a 5 kV-ion gun installed in a vacuum chamber with a base pressure of 2×10^{-6} Pa. The ion incidence was normal with respect to the sample surface.

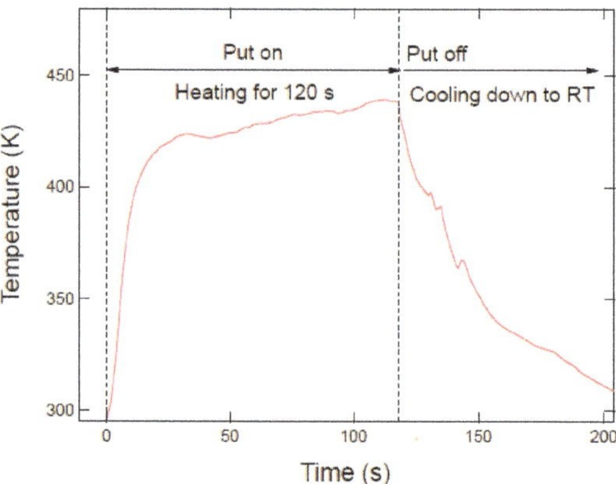

Figure 1. Temperature measured on the sample surface with chromel-alumel thermocouple during oxidation.

X-ray photoelectron spectroscopy (XPS) using Mg Kα radiation ($h\nu$ = 1253.6 eV) was performed with a JEOL 9010 X-ray photoelectron spectrometer (JEOL, Tokyo, Japan) to analyze chemical states of Cu before and after the irradiation. The XPS analysis was carried out immediately after the irradiation to avoid the change in chemical states upon the ambient air exposure. Rutherford backscattering spectrometry (RBS) and nuclear reaction analysis (NRA) using the ^{16}O(d, p)^{17}O reaction were conducted for chemical composition analysis with 2 MeV He ions and 0.85 MeV deuterons, respectively, produced from the Van de Graaff accelerator of Hiroshima University. The standard sample for the NRA was a SiO$_2$ layer formed on a Si wafer (SiO$_2$/Si), which contains 5.8×10^{17} O atoms·cm^{-2} determined by RBS.

3. Results and Discussion

Figure 2a shows a photograph of the sample surface irradiated with Ar^+ ions to a fluence of 1×10^{15} cm^{-2} using a 5 kV ion gun. The color of the sample changed from reddish brown to dark blue-purple at an irradiation spot. This darkening was visible through a viewing port, and started at the fluence as low as 10^{14} cm^{-2}. On the sample surface irradiated to a fluence of 10^{14} cm^{-2}, the color of the edge at the beam spot was not clear as can be seen in Figure 2b, indicating that the border between irradiated and unirradiated areas was not so abrupt. Thus, the uniformity of a beam intensity inside the beam spot could be estimated by the uniformity of color. The minimum fluence to recognize the beam spot with naked eyes will be examined for further discussion of the sensitivity and applicability of the chromatic change for a beam viewer.

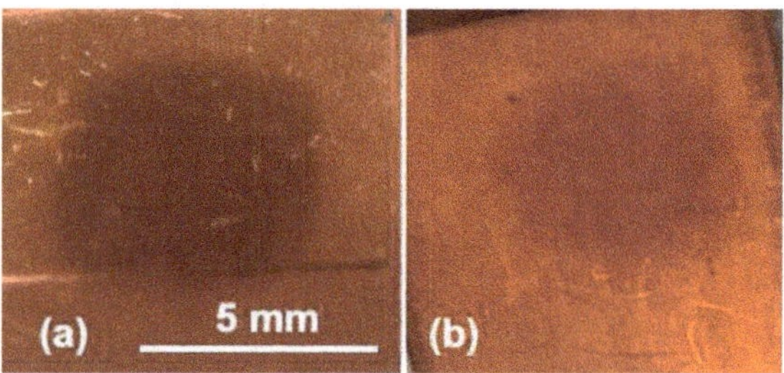

Figure 2. Photographs of the surface of samples irradiated with 5 keV-Ar^+ ions to fluences of 1×10^{15} cm^{-2} (**a**) and 1×10^{14} cm^{-2} (**b**). The photographs were taken after removing the samples from the vacuum chamber.

The mechanism of the observed chromatic change was discussed below, along with RBS, NRA, and XPS data. Figure 3 shows the backscattering spectrum of the Cu oxide layer formed on a Cu plate before irradiation. The chemical composition and thickness of the oxide layer was estimated to be $CuO_{0.4}$ and 1.9×10^{17} $CuO_{0.4}$ cm^{-2}, respectively, by fitting a simulation spectrum to experimental data, where the program SIMNRA 6.06 [16] was used to obtain the simulation spectrum. The thickness of 1.9×10^{17} $CuO_{0.4}$ cm^{-2} can be converted into 38 nm, much larger than the ion projected range of 5 nm predicted by the SRIM simulation [17], assuming the atomic density to be 5×10^{22} $CuO_{0.4}$ cm^{-3}. The chemical composition $CuO_{0.4}$ indicates that the oxide layer contains the mixture of metallic copper and copper oxides.

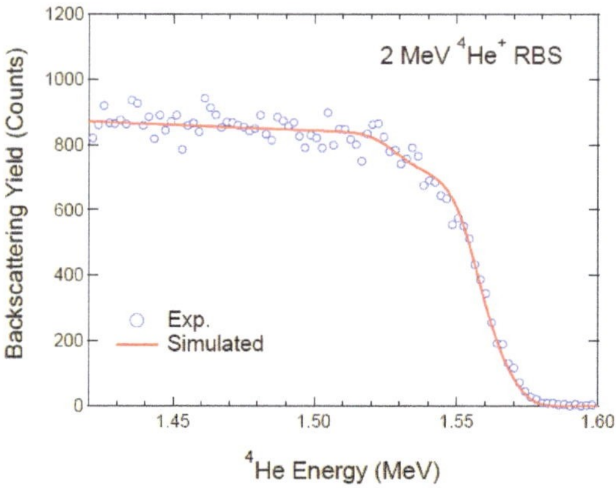

Figure 3. RBS spectrum of the sample before irradiation (blue open circles). A simulated spectrum (red solid line) is also shown.

Figure 4 presents the NRA spectra of the sample with and without irradiation. Peaks located around channel number of 260 correspond to protons emitted by the $^{16}O(d, p)^{17}O$ reaction. The peak intensities are $(2.15 \pm 0.05) \times 10^3$ and $(2.01 \pm 0.04) \times 10^3$ counts for the unirradiated and irradiated samples, respectively. The oxygen content in the irradiated sample was determined to be $(7.1 \pm 0.2) \times 10^{16}$ O atoms·cm^{-2} by the SiO$_2$/Si standard sample, smaller by 6% than that in the unirradiated sample ($(7.6 \pm 0.2) \times 10^{16}$ O atoms·cm^{-2}). Thus, oxygen atoms were found to be released from the CuO$_{0.4}$ layer by the irradiation with 5 keV-Ar$^+$ ions to a fluence of 1×10^{15} cm^{-2}. The SRIM simulation [17] predicts the sputtering yield of O in CuO$_{0.4}$ to be 4.9 O atoms·ion^{-1}, which means that the sputtered O atoms will be approximately 5×10^{15} O atoms·cm^{-2}, corresponding to a difference in the oxygen contents between the unirradiated and irradiated samples. The NRA results and the SRIM calculation suggest that the release of O atoms originates in physical sputtering. In the sputtering process of CuO$_{0.4}$ bombarded with 5 keV-Ar$^+$ ions, approximately 1.4×10^{15} Cu atoms·cm^{-2} as calculated by the SRIM will be lost, leading to the change in chemical composition from CuO$_{0.4}$ to CuO$_{0.38}$ in the layer.

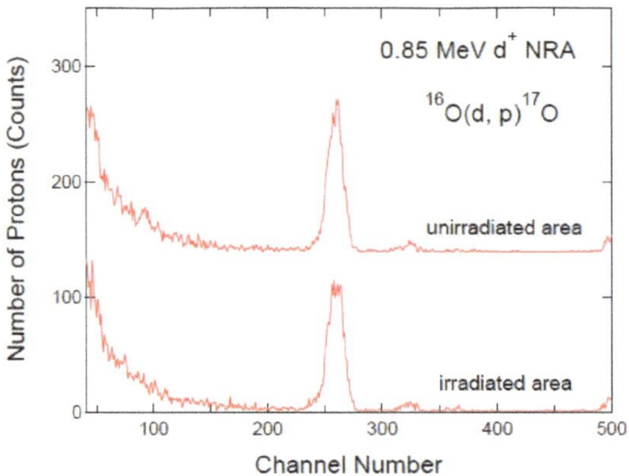

Figure 4. Nuclear reaction analysis (NRA) spectra of the unirradiated and irradiated areas in the sample irradiated with 5 keV-Ar+ ions to a fluence of 1×10^{15} cm^{-2}.

Figure 5 shows XPS Cu 2p core level photoemission spectra (PS) of the CuO$_{0.4}$ layer with and without irradiation. Considering the inelastic mean free path of photoelectrons whose kinetic energy is approximately 300 eV, information about copper valence states can be obtained within 0.8 nm [18] below the surface by the photoemission spectra. As can be seen in Figure 5, a broad shake-up satellite line due to the charge transfer [19,20] appeared in the binding energy (BE) region from 937 to 941 eV of PS for the unirradiated sample and decreased in intensity after irradiation. More quantitatively, the intensity ratio of the satellite to the main line at BE of 932.4 eV was 0.16 and 0.10 for the unirradiated and irradiated samples, respectively. These results indicate that the unirradiated sample included the copper valence state Cu^{2+} and it was transformed into Cu$^+$ and/or Cu^0 (Cu$^+$/Cu^0) by irradiation. Thus the 5 keV-Ar$^+$ irradiation reduced Cu^{2+} to Cu$^+$/Cu^0. The irradiation-induced reduction observed in the present work could be seen in the change in the shape of the Cu 2p$_{3/2}$ lines before and after irradiation.

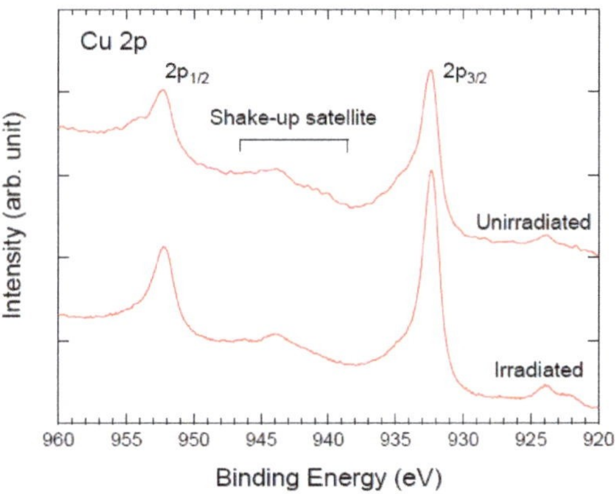

Figure 5. XPS Cu 2p core level photoemission spectra of the unirradiated and irradiated areas in the sample irradiated with 5 keV-Ar$^+$ ions to a fluence of 1×10^{15} cm^{-2}.

Figure 6a,b depicts Cu 2p$_{3/2}$ lines for the samples before and after irradiation, respectively. Each spectral line consists of three components denoted by I, II, and III. The lines, after background subtraction by the Shirly method [21], were decomposed by three Voigt functions using a curve fitting procedure. The component I located at BE of 932.4 eV was assigned to be Cu and/or Cu$_2$O. The BE of Cu (932.6 eV) [22–24] is very close to that of Cu$_2$O (932.5 eV) [25–28], therefore the component I cannot be further decomposed by a curve fitting. The components II and III located at BE of 933.6 eV and 934.8 correspond to CuO [28–31] and Cu(OH)$_2$ [28,32], respectively. The fractions of each component obtained by the curve fitting are summarized in Table 1 for the samples with and without irradiation. The fractions corresponding to copper valence state Cu^{2+} (CuO and Cu(OH)$_2$) decrease, while the fraction of Cu$^+$/Cu^0 (Cu$_2$O/Cu) increased by irradiation, indicating that the Ar$^+$ irradiation reduced Cu^{2+} to Cu$^+$/Cu^0. This was consistent with the result deduced from the decrease in intensity of shake-up satellite line as described above.

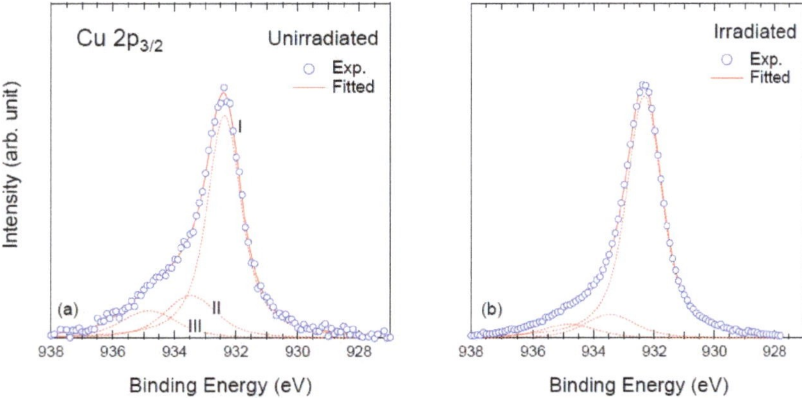

Figure 6. Detailed XPS Cu 2p3/2 core level photoemission spectra of the unirradiated (**a**) and irradiated (**b**) areas in the sample irradiated with 5 keV-Ar$^+$ ions to a fluence of 1×10^{15} cm^{-2}. Each spectrum was decomposed into three components denoted by I, II, and III.

Table 1. Compositions of Cu, Cu2O, CuO and $Cu(OH)_2$ determined by Cu $2p_{3/2}$ photoemission spectra (PS) lines for the unirradiated and irradiated samples.

Samples	Compositions (%)			
	Cu	Cu_2O	CuO	$Cu(OH)_2$
Unirradiated	73.8 (65.7) [1]	(8.1) [1]	16.5	9.7
Irradiated	94.1 (65.7) [1]	(28.4) [1]	5.5	0.4

[1] These values were obtained by the assumption that the composition of the analyzing layer was $CuO_{0.4}$ before irradiation and the fraction of Cu was unchanged after irradiation.

The concentration ratio of Cu_2O to Cu can be indirectly determined by the atomic ratio O/Cu of analyzing layer using the fractions of three components. Of course, the O/Cu can be calculated by the intensity ratio of O 1s to Cu 2p PS lines. However, it is impossible to accurately determine the atomic ratio O/Cu because of the presence of adventitious carbon contamination that includes oxygen atoms in the outermost layer. Therefore, the atomic ratio O/Cu in the analyzing layer was assumed to be 0.4 that was determined by RBS, resulting in the concentration ratio Cu_2O/Cu of 8.2/65.7 for the sample without irradiation. These values are presented with parenthesis in Table 1. For the sample with irradiation, the atomic ratio O/Cu in the analyzing layer was definitely different from that in the oxide layer, and thus, the composition $CuO_{0.38}$ determined by the combination of RBS and NRA could not be used to calculate the fraction of Cu_2O/Cu. Therefore, the fraction of metallic copper was assumed to be unchanged after irradiation. In fact, Panzner et al. [33] demonstrated that the oxide CuO was reduced to Cu_2O, while the oxide Cu_2O was much more stable and no more reduction to Cu occurred under sputtering with 3–5 keV Ar^+ to a low fluence. Then the concentration ratio Cu_2O/Cu was calculated to be 28.4/65.7 for the sample with irradiation. These values were also presented inside parenthesis in Table 1.

As described above, XPS analysis revealed that the 5 keV-Ar^+ irradiation reduced Cu^{2+} to Cu^+/Cu^0 in the cuprate. This result is consistent with the previous studies that CuO thin films were transformed into Cu_2O by ion irradiation [33,34]. Next, the relationship between the reduction and the chromatic change was discussed briefly.

It is well known that the color of cuprous oxide (Cu_2O) powder is red. In addition, Cu_2O in the form of both the thin film [35,36] and nanoparticle [37,38] would be reddish considering their optical absorption spectra. On the other hand, the color of the irradiated layer was found to be dark blue-purple, different from the color of pure Cu_2O. Fredj and Burleigh [39] showed that the color of a copper oxide layer in which Cu_2O is primarily included varied from bare copper color to green depending on its thickness. Thus, they demonstrated the possibility that the Cu_2O layer exhibited the color other than red. One possibility for the chromatic change is the presence of Cu_2O phase in a metallic Cu phase. Another possibility is the change in a refractive index of the oxide layer accompanied by the change in chemical states of Cu and/or radiation damage by Ar ions. Unfortunately, the dominant effect on the chromatic change is unclear at present.

As mentioned in the Introduction, the beam spot in the sample irradiated with 5 keV-Ar^+ ions using the other machine turned bare copper color due to the removal of an oxide layer by sputtering, different from the present work. The reason for this difference is unknown but may result from the initial thickness and composition of the oxide layer. Further studies with various thicknesses and compositions are needed to clarify the mechanism underlying the irradiation-induced chromatic change in copper oxide layers. In addition, the minimum fluence at which the chromatic change can be recognized is necessary to examine for the applicability of the new type of a beam viewer. Such investigations are now in progress.

4. Conclusions

The color of a thin copper oxide layer formed on a copper plate turned from reddish-brown to dark blue-purple by irradiation with 5 keV Ar^+ ions to a fluence of 1×10^{15} Ar^+ cm^{-2}. Nuclear reaction analysis revealed that a significant amount of oxygen (5×10^{15} O atoms$\cdot cm^{-2}$) was released from the irradiated layer. The reduction of cupric oxide (CuO) to cuprous oxide (Cu_2O) occurred in the layer after the irradiation as confirmed by the decrease in intensity of a shake-up satellite line and the change in the shape of a Cu $2p_{3/2}$ line in photoemission spectra. The reduction led to the compositional change in the mixture of $Cu/Cu_2O/CuO$, which would result in the chromatic change.

Author Contributions: Conceptualization, K.T. and T.K.; methodology, K.T.; RBS and NRA, F.N.; data curation, T.K. and K.T.; writing—original draft preparation, K.T.; writing—review and editing, K.T. and T.K. All authors have read and agreed to the published version of the manuscript.

Funding: This research received no external funding.

Conflicts of Interest: The authors declare no conflict of interest.

References

1. Peña-Rodríguez, O.; Crespillo, M.L.; Díaz-Nuñez, P.; Perlado, J.M.; Rivera, A.; Olivares, J. In situ monitoring the optical properties of dielectric materials during ion irradiation. *Opt. Mater. Express* **2016**, *6*, 734–742. [CrossRef]
2. Rivera, A.; Olivares, J.; Prada, A.; Crespillo, M.L.; Caturla, M.J.; Bringa, E.M.; Perlado, J.M.; Peña-Rodríguez, O. Permanent modifications in silica produced by ion-induced high electronic excitation: Experiments and atomistic simulations. *Sci. Rep.* **2017**, *7*, 10641. [CrossRef]
3. Zhou, J.; Li, B. Origins of a damage-induced green photoluminescence band in fused silica revealed by time-resolved photoluminescence spectroscopy. *Opt. Mater. Express* **2017**, *7*, 2888–2898. [CrossRef]
4. Crespillo, M.L.; Graham, J.T.; Zhang, Y.; Weber, W.J. In-situ luminescence monitoring of ion-induced damage evolution in SiO_2 and Al_2O_3. *J. Lumin.* **2016**, *172*, 208–218. [CrossRef]
5. Bandourko, V.; Umeda, N.; Plaksin, O.; Kishimoto, N. Heavy-ion-induced luminescence of amorphous SiO_2 during nanoparticle formation. *Nucl. Instrum. Meth. B* **2005**, *230*, 471–475. [CrossRef]
6. Forck, P.; Andre, C.; Becker, F.; Haseitl, R.; Reiter, A.; Walasek-Höhne, B.; Krishnakumar, R.; Ensinger, W. Scintillation Screen Investigations for High Energy Heavy Ion Beams at GSI. In Proceedings of the DIPAC2011, Hamburg, Germany, 16–18 May 2011; pp. 170–172.
7. Johnson, C.D. *The Development and Use of Alumina Ceramic Fluorescent*; CERN/PS/90-12(AR); European Laboratory for Particle Physics: Geneva, Switzerland, 1990.
8. Takahiro, K.; Terai, A.; Kawatsura, K.; Naramoto, H.; Yamam, S.; Tsuchiya, B.; Nagata, S.; Nishiyama, F. Rutherford backscattering spectrometry of electrically charged targets: Elegant technique for measuring charge-state distribution of backscattered Ions. *Jpn. J. Appl. Phys.* **2006**, *45*, 1823–1825. [CrossRef]
9. Parajuli, R.K.; Kada, W.; Kawabata, S.; Matsubara, Y.; Sakai, M.; Miura, K.; Satoh, T.; Koka, M.; Yamada, N.; Kamiya, T.; et al. Ion-Beam-Induced Luminescence Analysis of β-SiAlON:Eu Scintillator under Focused Microbeam Irradiation. *Sens. Mater.* **2016**, *28*, 837–844.
10. Warren, A.J.; Thomas, C.B.; Reehal, H.S.; Stevens, P.R.C. A study of the luminescent and electrical characteristics of films of ZnS doped with Mn. *J. Lumin.* **1983**, *28*, 147–162. [CrossRef]
11. Calusi, S.; Colombo, E.; Giuntini, L.; Giudice, A.L.; Manfredotti, C.; Massi, M.; Pratesi, G.; Vittone, E. The ionoluminescence apparatus at the LABEC external microbeam facility. *Nucl. Instrum. Meth. B* **2008**, *266*, 2306–2310. [CrossRef]
12. Feldmann, C.; Jüstel, T.; Ronda, C.R.; Schmidt, P.J. Inorganic Luminescent Materials: 100 Years of Research and Application. *Adv. Funct. Mater.* **2003**, *13*, 511–516. [CrossRef]
13. Kleiman, J.; Iskanderova, Z.; Best, C.; Dennison, J.R.; Wood, B. Long-term stability of ion-beam treated space polymers in geo-simulated environment. In Proceeding of the 14th ISMSE $ 12th ICPMSE, Biarritz, France, 1–5 October 2018.
14. Plis, E.A.; Engelhart, D.P.; Cooper, R.; Johnston, W.R.; Ferguson, D.; Hoffmann, R. Review of Radiation-Induced Effects in Polyimide. *Appl. Sci.* **2019**, *9*, 1999. [CrossRef]
15. Matienzo, L.J.; Emmi, F.; Van Hart, D.C.; Gall, T.P. Interactions of high-energy ion beams with polyimide films. *J. Vac. Sci. Technol. A* **1989**, *7*, 1784–1789. [CrossRef]
16. Computer Simulation of RBS, ERDA, NRA, MEIS and PIGE by Matej Mayer. Available online: https://home.mpcdf.mpg.de/~{}mam/Version6.html (accessed on 5 December 2020).
17. Ziegler, J.F.; Ziegler, M.D.; Biersack, J.P. SRIM—The stopping and range of ions in matter (2010). *Nucl. Instrum. Meth. B* **2010**, *268*, 1818–1823. [CrossRef]
18. Tanuma, S.; Powell, C.J.; Penn, D.R. Calculations of electron inelastic mean free paths. II. Data for 27 elements over the 50–2000 eV range. *Surf. Interface Anal.* **1991**, *17*, 911–926. [CrossRef]

19. Kawai, J.; Tsuboyama, S.; Ishizu, K.; Miyamura, K.; Saburi, M. Covalency of copper complex determined by Cu 2p X-ray photoelectron spectrometry. *Anal. Sci.* **1991**, *7*, 337–340. [CrossRef]
20. Pawar, S.M.; Kim, J.; Inamdar, A.I.; Woo, H.; Jo, Y.; Pawar, B.S.; Cho, S.; Kim, H.; Im, H. Multi-functional reactively-sputtered copper oxide electrodes for supercapacitor and electro-catalyst in direct methanol fuel cell applications. *Sci. Rep.* **2016**, *6*, 21310. [CrossRef] [PubMed]
21. Végh, J. The Shirley background revised. *J. Electron Spectrosc. Relat. Phenom.* **2006**, *151*, 159–164. [CrossRef]
22. Anderson, C.R.; Lee, R.N. Comparison of APS and FRESCA core level binding energy measurements. *J. Vac. Sci. Technol.* **1982**, *20*, 617–621. [CrossRef]
23. Seah, M.P.; Gilmore, I.S.; Beamson, G. XPS: Binding energy calibration of electron spectrometers 5—re-evaluation of the reference energies. *Surf. Interface Anal.* **1998**, *26*, 642–649. [CrossRef]
24. Bird, R.J.; Swift, P. Energy calibration in electron spectroscopy and the re-determination of some reference electron binding energies. *J. Electron Spectrosc. Relat. Phenom.* **1980**, *21*, 227–240. [CrossRef]
25. McIntyre, N.S.; Cook, M.G. X-ray photoelectron studies on some oxides and hydroxides of cobalt, nickel, and copper. *Anal. Chem.* **1975**, *47*, 2208–2213. [CrossRef]
26. Schön, G. ESCA studies of Cu, Cu2O and CuO. *Surf. Sci.* **1973**, *35*, 96–108. [CrossRef]
27. Losev, A.; Rostov, K.; Tyuliev, G. Electron beam induced reduction of CuO in the presence of a surface carbonaceous layer: An XPS/HREELS study. *Surf. Sci.* **1989**, *213*, 564–579. [CrossRef]
28. Deroubaix, G.; Marcus, P. X-ray photoelectron spectroscopy analysis of copper and zinc oxides and sulphides. *Surf. Interface Anal.* **1992**, *18*, 39–46. [CrossRef]
29. Haber, J.; Machej, T.; Ungier, L.; Ziółkowski, J. ESCA studies of copper oxides and copper molybdates. *J. Solid State Chem.* **1978**, *25*, 207–218. [CrossRef]
30. Gaarenstroom, S.M.; Winograd, N. ESCA spectra of cadmium and silver oxides. *J. Chem. Phys.* **1977**, *67*, 3500–3506. [CrossRef]
31. Tobin, J.P.; Hirschwald, W.; Cunningham, J. XPS and XAES studies of transient enhancement of Cu[1] at CuO surfaces during vacuum outgassing. *Appl. Surf. Sci.* **1983**, *16*, 441–452. [CrossRef]
32. Biesinger, M.C. Advanced analysis of copper X-ray photoelectron spectra. *Surf. Interface Anal.* **2017**, *49*, 1325–1334. [CrossRef]
33. Panzner, G.; Egert, B.; Schmidt, H.P. The stability of CuO and Cu2O surfaces during argon sputtering studied by XPS and AES. *Surf. Sci.* **1985**, *151*, 400–408. [CrossRef]
34. Dementyeva, M.M.; Prikhodko, K.E.; Gurovich, B.A.; Bukina, Z.V.; Komarov, D.A.; Kutuzov, L.V. Phase transitions in copper oxide thin films under proton irradiation. *IOP Conf. Ser. Mater. Sci. Tecnol.* **2017**, *256*, 012020. [CrossRef]
35. Pan, J.; Yang, C.; Gao, Y. Investigations of Cuprous Oxide and Cupric Oxide Thin Films by Controlling the Deposition Atmosphere in the Reactive Sputtering Method. *Sens. Mater.* **2016**, *28*, 817–824.
36. Gevorkyan, V.A.; Reymers, A.E.; Nersesyan, M.N.; Arzakantsyan, A.M. Characterization of Cu2O thin films prepared by evaporation of CuO powder. *J. Phys. Conf. Ser.* **2012**, *350*, 012027. [CrossRef]
37. Goua, L.; Murphy, C.J. Controlling the size of Cu2O nanocubes from 200 to 25 nm. *J. Mater. Chem.* **2004**, *14*, 735–738. [CrossRef]
38. Butte, S.M.; Waghuley, S.A. Optical properties of Cu2O and CuO. *AIP Conf. Proc.* **2020**, *2220*, 020093.
39. Fredj, N.; Burleigh, T.D. Transpassive Dissolution of Copper and Rapid Formation of Brilliant Colored Copper Oxide Films. *J. Electrochem. Soc.* **2011**, *158*, C104–C110. [CrossRef]

quantum beam science

Article

Irradiation Effects of Swift Heavy Ions Detected by Refractive Index Depth Profiling

Hiroshi Amekura [1],* , Rang Li [2] , Nariaki Okubo [3], Norito Ishikawa [3] and Feng Chen [2]

1   National Institute for Materials Science (NIMS), Tsukuba, Ibaraki 305-0003, Japan
2   School of Physics, State Key Laboratory of Crystal Materials, Shandong University, Jinan 250100, China; sdurangli@163.com (R.L.); drfchen@sdu.edu.cn (F.C.)
3   Japan Atomic Energy Agency (JAEA), Tokai, Ibaraki 319-1195, Japan; okubo.nariaki@jaea.go.jp (N.O.); ishikawa.norito@jaea.go.jp (N.I.)
*   Correspondence: amekura.hiroshi@nims.go.jp; Tel.: +81-29-863-5479

Received: 25 September 2020; Accepted: 10 November 2020; Published: 16 November 2020

Abstract: Evolution of depth profiles of the refractive index in $Y_3Al_5O_{12}$ (YAG) crystals were studied under 200 MeV $^{136}Xe^{14+}$ ion irradiation, since the index can be related with the stress change and/or the defect formation by the irradiation. Using the prism-coupling and the end-surface coupling methods, various waveguide (WG) modes were detected. Then, the index depth profiles were determined by reproducing the observed WG modes. The index changes were observed at three different depth regions; (i) a sharp dip at 13 μm in depth, which is attributed to the nuclear stopping S_n peak, (ii) a plateau near the surface between 0 and 3 μm in depth, which can be ascribed to the electronic stopping S_e, since S_e has a very broad peak at the surface, and (iii) a broad peak at 6 μm in depth. Since the last peak is ascribed to neither of S_e nor S_n peak, it could be attributed to the synergy effect of S_e and S_n.

Keywords: swift heavy ion; YAG ($Y_3Al_5O_{12}$); refractive index profiling; synergy effect; optical waveguide

---

1. Introduction

Optical waveguides (WGs) are elements to confine light wave inside and to guide the waves along them, which are considered as important parts for future optical integrate circuits [1]. Some special WGs possess additional functions such as lasing [2,3], the second harmonic generation [3,4], and the photorefractive effect [3,5]. While the most known WGs are optical fibers, here we discuss WGs of the slab type, which consist of thin film layer(s) on a substrate (See Figure 1). The optical WGs of the slab type are easily formed, when a transparent material B (guiding layer), having the highest index (n_B), is sandwiched with materials A (cladding layer: n_A) and C (substrate layer: n_C), both having lower indices than the material B (the guiding layer), i.e., n_A, $n_C < n_B$. At the boundaries A–B and B–C, the total reflections are repeated with the reflection angles higher than certain values. Light could be confined in the material B due to the total reflections at both the boundaries A–B and B–C. The material A (cladding layer) can be replaced by air or a vacuum, since either of them has the index of ~1, i.e., lower than most of the guiding layer material B. Consequently, the simplest WGs consist of two layers: (i) a higher index layer deposited on a lower index substrate can act as a slab-type WG. (ii) another strategy is to decrease the refractive index of a certain depth region of a transparent material without decreasing the index of the shallower layer.

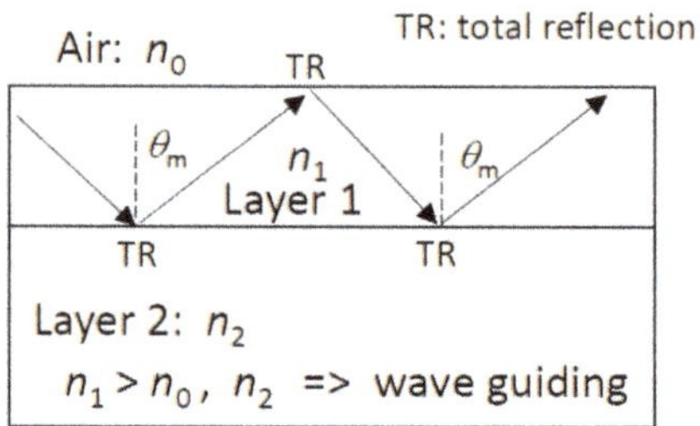

Figure 1. Schematically depicted cross-section of a slab waveguide consisting two layers: An upper layer with higher refractive index and lower one with lower index. Air (or vacuum) on the upper layer acts as a cladding layer. At light-propagating angles at discrete values, the total reflections are induced at the boundaries between air and the layer 1 and between layers 1 and 2. Consequently light in the layer 1 is confined inside of the layer 1 and guided along the inside of the layer.

Ion irradiation can realize the latter structures (ii). According to the Lorentz–Lorenz's (LL) formula (Equation (1)), the relative change of the refractive index $\Delta n/n$ is described as the following relation [3],

$$\frac{\Delta n}{n} = \frac{\left(n^2 - 1\right)\left(n^2 + 2\right)}{6n}\left[-\frac{\Delta V}{V} + \frac{\Delta \alpha}{\alpha} + F\right] \tag{1}$$

or approximately

$$\frac{\Delta n}{n} \propto -\frac{\Delta V}{V} \tag{2}$$

where $\Delta V/V$, $\Delta \alpha/\alpha$, and F denote relative changes of volume, of polarizability, and of other factors such as phase transition, respectively. From the first term in the right side of Equation (1), it is expected that the ensemble of Frenkel pairs, i.e., pairs of vacancies and interstitial atoms, could induce local volume expansion ($\Delta V > 0$), which results in the index reduction ($\Delta n < 0$) in some transparent solids. Moreover, lattice expansion and contraction ($\Delta V > 0$ and < 0) results in the index reduction and enhancement ($\Delta n > 0$ and < 0), respectively.

Of course, ion irradiation could exchange atomic arrangements so that newly formed chemical bonds may change the (bond) polarizability α, i.e., the second term of Equation (1). Furthermore, some phase transitions, e.g., amorphization, could suddenly change the relative index (the third term of Equation (1)).

In this paper, however, the first term in the right side of Equation (1) is only considered as an approximation, and the second and the third terms are, at the moment, neglected. Therefore, instead of the Equation (1), the Equation (2) is used in this paper. Our main concern is the detection of the defect formation and/or the stress generation via the index changes. We will not discuss the volume changes quantitatively.

The original idea to produce optical WGs in transparent crystals by ion beams, was to utilize the nuclear stopping process of light ions, e.g., ~1–2 MeV He ions [6]. As schematically shown in Figure 2, the nuclear energy loss (S_n) reaches the maximum at several micrometers beneath the crystal's surface, i.e., the Bragg peak, and leaves serious damage. Much shallower region than the Bragg peak is preserved negligible damage. From Equation (2), the refractive index decreases ($\Delta n < 0$) around the Bragg peak depth due to serious damage, i.e., $\Delta V > 0$, while the index almost preserves in the layer shallower than the Bragg peak. This is the WG structures of the type (ii).

Since the first optical WG was successfully formed by ion irradiations [6], various studies have been developed: This methodology has been applied in various materials [3,7]. In some glasses, it was confirmed that ion irradiation induced the increase of the density, i.e., $\Delta V < 0$, i.e., $\Delta n > 0$ [8].

Olivares et al. clarified that not only the nuclear energy loss (S_n) but also the electronic energy loss (S_e) of swift heavy ions reduces the refractive index of crystals [9]. Rodriguez et al. studied ion tracks in Nd-doped yttrium aluminum garnet (Nd:YAG) formed by 2.2 GeV Au ion irradiation, using transmission electron microscopy (TEM) and small angle X-ray scattering (SAXS) [10]. They concluded that the ion tracks were in an amorphous phase with a hard-cylinder density distribution, other than the core/shell types. We have confirmed the refractive index changes of Nd:YAG induced by 15 MeV Au^{5+} irradiation to 8×10^{14} ions/cm^2, and found that the amorphous phase showed a lower index [11].

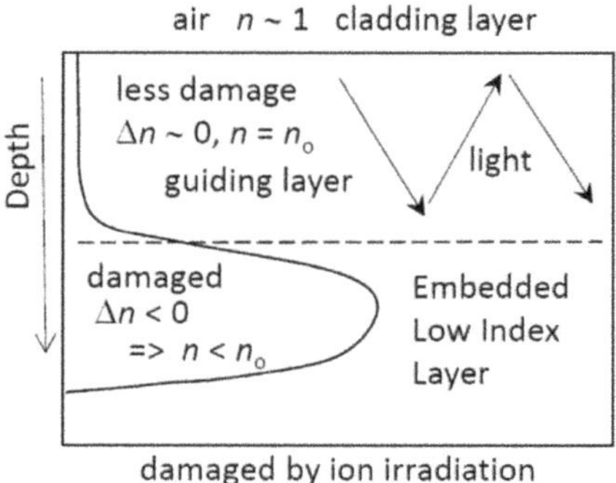

Figure 2. Schematically depicted depth profiles of damage induced by a few MeV light ions. The refractive index is reduced around the depth of the Bragg peak due to strong damage, while the index in shallower region seldom changes. Consequently, a slab-type waveguide is formed only by the irradiation of a few MeV light ions.

As shown in Equation (2), the studies of the refractive index change provide information on the density changes, which are induced by ion irradiation, via damage or stress change. According to a naive image of WG shown in Figure 1, the waveguiding is possible for any angles higher than a certain value. However, this is not correct. Because of the interference of light, the guiding is possible only for discrete values of the angles, each of which corresponds to the WG mode. Furthermore, when the light confinement is surely maintained between the cladding and the substrate layer, the guiding is possible for inhomogeneous index profile in the guiding layer, with different distribution of the modes. Contrary, with measuring the mode angles using, e.g., the prism coupling method, the depth profile of the refractive index can be reconstructed. This paper reports the fluence dependence of the refractive index profiles of yttrium-aluminum-garnet (YAG) crystals irradiated with swift heavy ions (SHIs) of 200 MeV Xe^{14+} ions, at various fluences ranging from 1×10^{11} to 5×10^{13} ions/cm^2.

2. Materials and Methods

2.1. Material

Single crystals of undoped yttrium-aluminum-garnet ($Y_3Al_5O_{12}$, YAG) with the dimensions of 10 mm by 10 mm by 1 mm were purchased from ATOM Optics Co. Ltd., Shanghai, China. The double sides of them were mirror-polished. Here the "undoped YAG" means that Nd ions or other rare-earth impurities have not been intentionally doped. While the samples were single crystals in unirradiated state, they seemed to transform to polycrystalline and finally to amorphous, as shown later. The crystalline structure of YAG ($Y_3Al_5O_{12}$) is the garnet type, i.e., in the cubic symmetry but including many atoms in a unit cell.

2.2. Ion Irradiation

The {0 0 1} plane of the single crystals of YAG was irradiated at room temperature with 200 MeV $^{136}Xe^{14+}$ ions from the tandem accelerator in the Japan Atomic Energy Agency (JAEA), Tokai Research and Development Center. The fluence ranged from 1×10^{11} to 5×10^{13} ions/cm^2, with maintaining the beam current density at ~60 nA/cm^2, except the fluences of 1×10^{11} and 3×10^{11} ions/cm^2 at ~5 nA/cm^2 to avoid inaccurate fluences due to too short irradiation time. Noted that the particle current density is 1/14 of the above-described values. The beam was rasterized at ~40 Hz (horizontal) and ~2 Hz (vertical). An important point was the frequencies were approximated values, which were not commensurate. The area of 10 mm square was irradiated through a square slit.

The stopping powers and the projected range of the 200 MeV Xe ions in YAG was calculated using SRIM 2013 code [12] with the mass density of 4.56 g/cm^3, and shown in Table 1. The displacement energy, the bulk binding energy, and the surface binding energy of each element used for the calculations were summarized in Table 2. The 200 MeV Xe ion provides the electronic stopping power S_e of 24.3 keV/nm, which was much higher than the track formation threshold of 7.5 keV/nm [13].

Table 1. Electronic and nuclear stopping powers (at the surface), and the projected range of 200 MeV ^{136}Xe^{14+}. ions in yttrium-aluminum-garnet (YAG) crystal, calculated by SRIM 2013. The track formation threshold and the X-ray penetration depth of the Cr Kα line used for fixed incident angle X-ray diffractometry (FIA-XRD) are also shown.

200 MeV ^{136}Xe^{14+} => Undoped YAG	
Electronic stopping power (keV/nm)	24.3
Track formation threshold (keV/nm)	7.5
Nuclear stopping power (keV/nm)	0.091
Projected Range (μm)	13.2
X-ray penetration used for FIA-XRD measurements at 15 incidence (μm)	2.76

Table 2. The displacement energies, the bulk binding energies, and the surface binding energies of Y, Al, and O atoms in YAG crystal, used for the SRIM 2013 calculations.

Element	Y	Al	O
Displacement energy (eV)	25	25	28
Bulk binding energy (eV)	3	3	3
Surface binding energy (eV)	4.24	3.36	3

2.3. FIA-XRD

The fixed incident angle X-ray diffractometry (FIA-XRD) was carried out by RINT 2500 MDG, (Rigaku Co., Ltd., Tokyo, Japan), with a fixed incident angle of 15° from the sample surface using the Cr Kα line ($\lambda = 0.22896$ nm) from an X-ray source. Different from the conventional θ-2θ method, the penetration depth of the incoming X-ray is constant for FIA-XRD, irrespective of the scattering angle 2θ. Since the ion irradiation effects often depend on the sample depth, the FIA-XRD method could be advantageous. The incident angle of 15° corresponds to the X-ray penetration depth of 2.76 μm normal from the sample surface. Here, the penetration depth defines where the incident X-ray intensity decreases to $1/e$, where e is the Napier's constant. The density of 4.56 g/cm^3 was used for the penetration calculation.

2.4. Index Measurements

Depth profiles of the refractive index was determined by two methods, i.e., the prism-coupling and the end-face coupling. In former, all the WG mode angles were consistently fitted. Instead of the mode angles θ_m, the effective refractive index n_m is usually used in this community,

$$n_m = n_1 \sin\theta_m \tag{3}$$

where n_1 denotes the index in the guiding layer.

After measuring the angles or the effective refractive indices where WG modes are excited, a depth profile of the index is determined in order to reproduce all the mode angles using the reflectivity calculation method (RCM). See Reference [3] for the RCM. The mode angles where WG modes are excited were measured by the prism-coupling method. Figure 3a schematically depicts a setup of the prism-coupling apparatus. A laser line of 632.8 nm from a He-Ne laser was used as a light source. A polarizer was inserted to select a polarization plan of the WG mode from the transverse electric (TE) or the transverse magnetic (TM). In this paper, most of the data were detected in the TM configuration. A prism was attached on a part of the WG sample in order to couple with the light in air (a vacuum) and the light in the WG via evanescent wave in a gap between the prism and the WG. The prism, the sample, and a charge-coupled device (CCD) detector facing at the end-surface of the sample, were attached on a rotating table, in order to change an incident angle of the laser light to the WG layer without changing the configurations of the prism, the sample, and the CCD detector. The light intensity reflected at the boundary between the prism and the WG was monitored with inserting a beam splitter between the polarizer and the prism. The reflected light was transferred via the beam splitter and detected by "a dark-mode detector". When a mode is excited with adjusting the modal angle θ_m, the light intensity detected at the CCD detector facing the end-surface increases because a part of light is guided to the end-surface. Simultaneously, when the mode is matched, the incident light is efficiently introduced into the WG. Consequently, the reflection intensity at the prism-WG boundary decreases, i.e., the signal at the dark-mode detector decreases.

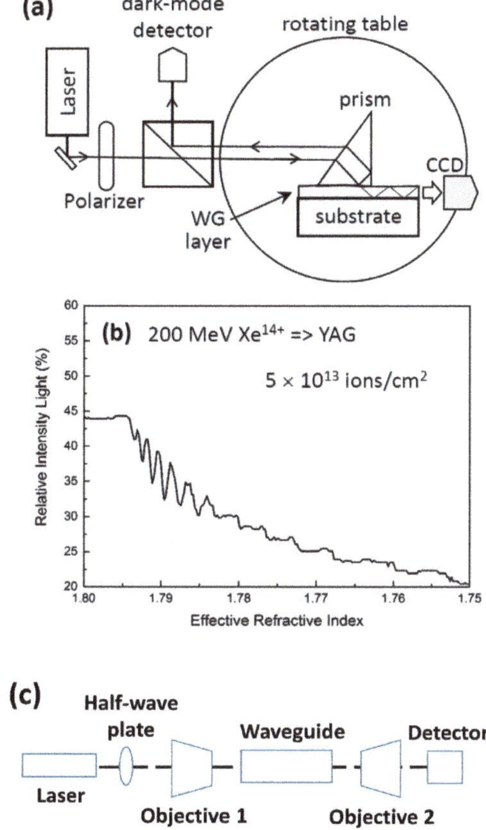

Figure 3. (**a**) a setup for detection of waveguide modes using prism coupling and the dark-mode detection. See text for details. (**b**) a typical data of the dark-mode detection from YAG sample irradiated with 200 MeV Xe ions in the TE mode. Each dip corresponds to a waveguides (WG) modes. (**c**) a setup of the end-face coupling. The polarization of the laser is controlled by a half-wave plate.

The WG modes were searched with scanning the WG angle or equivalently the effective index. Figure 3b shows an example of the effective index (i.e., the WG angle) dependence of the light intensity detected by the dark-mode detector. In this case, dips in the signal correspond to the WG modes. In principle, the same modes should be detected also by the "bright-mode detection" using the CCD detector facing at the end-surface. However, as a rule of thumb, the dark-mode detection provides sharper dips and more sensitive. We applied this method.

As schematically shown in Figure 3c, the end-face coupling method was also applied. With controlling the polarization of laser by a half-wave plate, the laser light was introduced by the objective lens 1. The guided light pattern was detected through the objective lens 2. The spatial intensity patterns at the end-surface were another important information, which were experimentally detected by a CCD camera and calculated by the beam propagation method (BPM) using BeamPROP code (RSoft, Co. Ltd., Tokyo, Japan).

3. Experimental Results

3.1. X-ray Diffraction

Figure 4 shows FIA-XRD patterns from undoped YAG crystals irradiated with 200 MeV Xe^{14+} ions to various fluences ranging from 0 to 5×10^{13} ions/cm^2. Before the irradiation, relatively strong four peaks were observed, all of which were assigned to diffractions from the garnet structure: 55.8° for (4 2 2), 124.6° for (9 2 1), 135.7° for (9 3 2), and 150.3° for (7 7 2) [14]. Because the garnet structure includes many atoms in a unit cell, many peaks are reported in the powder diffraction patterns. However, our samples showed only limited peaks due to high crystallinity. After the irradiation to 1×10^{11} ions/cm^2, the same peaks were observed, while the intensity ratios and the diffraction angles slightly changed. At the fluence of 1×10^{12} ions/cm^2, new four peaks were added, which were also assigned to the garnet structure: 65.1° to (4 4 0), 72.0° to (5 3 2), 88.6° to (5 5 2), and 105.2° to (6 5 3) [14]. Because of the partial amorphization, the single crystal was partly broken down to smaller grains facing various directions, i.e., poly-crystallization. Consequently, the new peaks were allowed to be observed. At the fluence of 1×10^{13} ions/cm^2 and higher, all the peaks disappeared indicating the full amorphization of the YAG crystal. Amorphization of Nd-doped YAG crystals was already reported by Rutherford backscattering spectrometry (RBS)-channeling [15] and TEM [10].

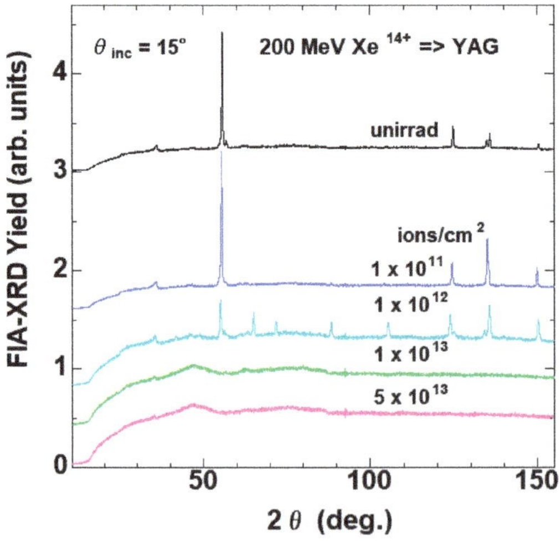

Figure 4. FIA-XRD patterns of undoped YAG samples in unirradiated state and irradiated with 200 MeV Xe^{14+} ions to various fluences ranging from 1×10^{11} to 5×10^{13} ions/cm^2. The patterns are shifted for vertical direction for clarity.

3.2. Refractive Index Profiling

Figure 5 exhibits depth profiles of the refractive index in undoped YAG crystals at various ion fluences. The data in the TM configuration are shown, while those in the TE mode are mentioned later. A chained horizontal line at 1.8295 indicates the index in the unirradiated state at the wavelength of 632.8 nm from literature [16]. After the irradiation to 1×10^{11} ions/cm^2, a very weak and broad enhancement peak was observed at 6 μm in depth with the full width at half maximum (FWHM) of ~4 μm. Since the deviation from the unirradiated value was so small for the data at 1×10^{11} and 3×10^{11} ions/cm^2 that the deviation from the unirradiated value was plotted with five times magnification around the unirradiated value.

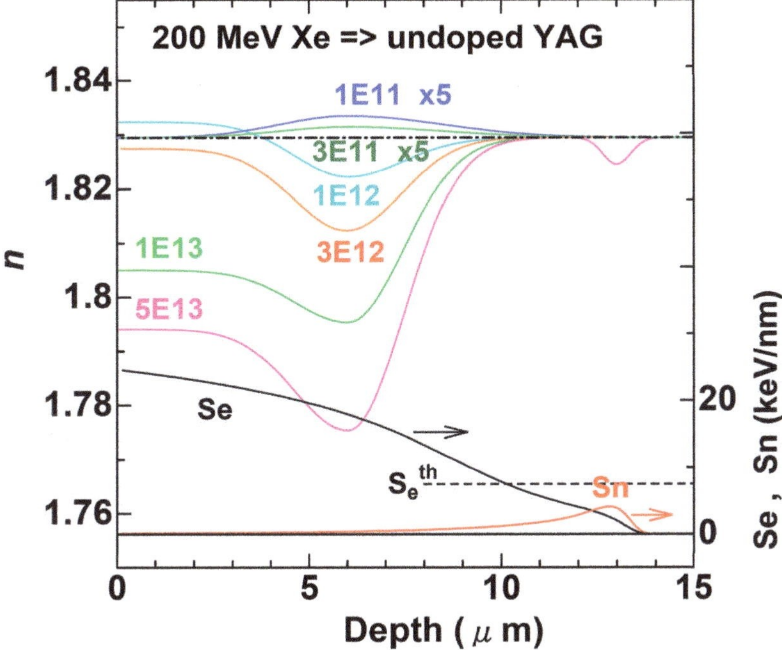

Figure 5. Depth profiles of refractive indices of YAG irradiated with 200 MeV Xe^{14+} ions to various fluences ranging from 1×10^{11} ions/cm^2 to 5×10^{13} ions/cm^2, determined by the prism coupling and the end-face coupling method. All the polarizations are in the transverse magnetic (TM) configuration. A chained horizontal line indicates the index in the unirradiated state of 1.8295 at the wavelength of 632.8 nm from literature. Since the deviation from the unirradiated value was so small for the data of 1×10^{11} and 3×10^{11} ions/cm^2, the deviations were 5 times multiplied and added to the unirradiated value. For references, the depth profiles of electronic and nuclear stopping powers S_e and S_n are plotted with the right axis in the units of keV/nm. A horizontal broken line denotes the threshold S_e value for the track formation in YAG.

Figure 6 shows four different images (i)–(iv) at four different fluences (1×10^{11}, 1×10^{12}, 1×10^{13}, and 5×10^{13} ions/cm^2). From left to right, (i) the experimental index depth profiles used for the end-surface intensity calculations, which are the same as shown in Figure 5, (ii) the optical microscopy images of the end-surface without the guided light, (iii) experimental, and (iv) calculated results of the spatial distributions of the guided light intensity at the end-surfaces. In the optical images of the end-surfaces without the guided light (ii), the near surface regions exhibit different color compared with the bulk parts, which are ascribed to the different indices induced by the ion irradiation. Color changes are stronger for higher ion fluences.

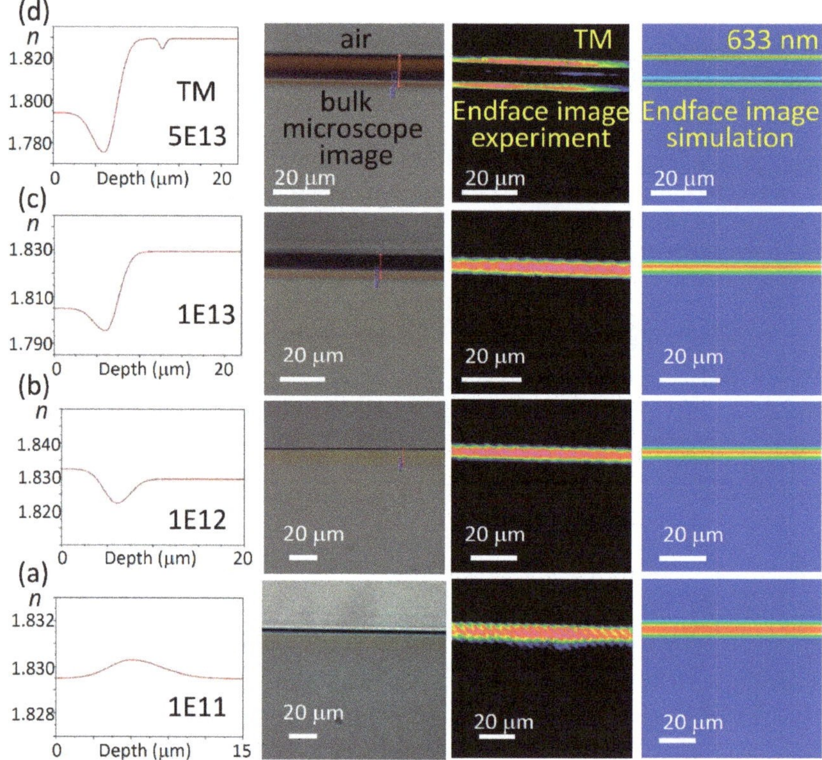

Figure 6. (Left to right) Index profiles used for the calculations of the intensity profile at the end-surfaces; optical microscopy images at the end-surface without wave guiding laser; experimental spatial intensity profiles of the wave guiding laser at the end-surface; and corresponding calculation results, for four different ion fluences (**a**) 1×10^{11}, (**b**) 1×10^{12}, (**c**) 1×10^{13}, and (**d**) 5×10^{13} ions/cm^2. The spatial intensity profiles were calculated by the beam propagation method (BPM) using BeamPROP code.

As shown in Figure 6a, the wave guiding is induced even at the lowest fluence of 1×10^{11} ions/cm^2, since the guided light is buried beneath the surface. The irradiation made index enhancement of 0.001 around the depth of 6 μm as shown in Figure 5. Even though the very small change in the index, the wave guiding was confirmed from the experimental emission image at the end-surface. Furthermore, the calculation shows that the observed index profile predicted the same emission image from the small change of the index profile.

The peak at 6 μm reduced the height at the fluence of 3×10^{11} ions/cm^2. At 1×10^{12} ions/cm^2, the peak further decreased and turned to a dip. Simultaneously, the index at the surface region (0–3 μm in depth) became slightly higher than the unirradiated value. However, the index at the surface region turned to decrease and became lower than the unirradiated value at 3×10^{12} ions/cm^2. Likewise, the bottom value at the dip at 6 μm further decreased. Since the index at the surface region steeply decreased at 1×10^{13} ions/cm^2, the low index region near the surface and the 6 μm dip were almost merged. The WG function of SHI-irradiated YAG crystals is also confirmed at 1×10^{12} and 1×10^{13} ions/cm^2, as shown in Figure 6b,c. The spatial images of the guided light at the end-surface by experiments are well matched with those calculated from the index profiles. Similarly, the microscopy images show further color changes due to the further index changes.

At the fluence of 5×10^{13} ions/cm^2, the index further decreased. Moreover, a new dip appeared at 13 μm in depth, which could be ascribed to the collisional damage, because the depth of the dip well matches with the peak of the nuclear energy loss S_n. As shown in Figure 6d, the index profile of the sample irradiated with 5×10^{13} ions/cm^2 exhibits interesting waveguiding different from other fluences. Two WG modes are propagated in shallow and deep layers, which were, respectively, formed by the electronic and the nuclear damages. The two-branch guiding at the end-surface was also confirmed by the calculations. The calculated emission image is very similar with the experimental one.

Figure 7 shows the incident polarization dependence of the output power at the end-surface of the WGs. These data were collected with rotating a half-wave plate in the end-face coupling schematically shown in Figure 3c. The low fluence samples (1×10^{11} and 1×10^{12} ions/cm^2) show the "8"-shaped (i.e., dipole-shaped) polarization angle dependence. While the strongest power was observed for the TM polarization, little signal (no WG modes) was observed for the TE polarization. Contrary, the high fluence samples (1×10^{13} and 5×10^{13} ions/cm^2) exhibit nearly isotropic angle dependences, i.e., the output power almost does not depend on the polarization angle. At the moment, the mechanism has not been clarified yet. It should be noted that Figure 4 showed that the low fluence samples (1×10^{11} and 1×10^{12} ions/cm^2) are cubic symmetry crystals but the high fluence samples (1×10^{13} and 5×10^{13} ions/cm^2) are amorphous. Amorphization could be an origin of the change of the polarization angle dependence. However, the detailed mechanism is under investigation.

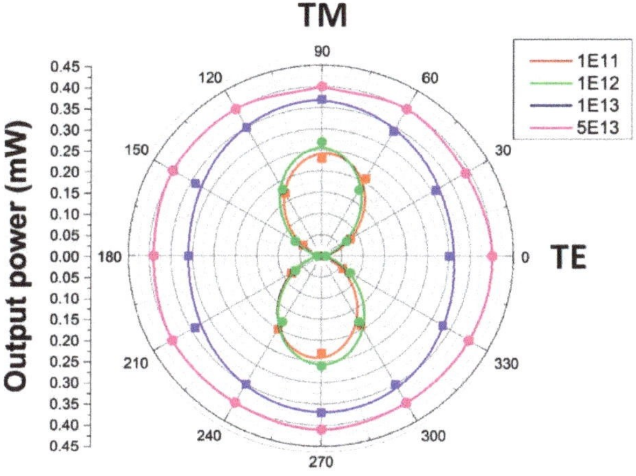

Figure 7. The total output power of the waveguide through the end-surface was plotted with changing the polarization angle of the incident laser. The incident laser power was set to 1.8 mW. The samples were irradiated with 200 MeV Xe^{14+} ions at four different fluences; 1×10^{11}, 1×10^{12}, 1×10^{13}, and 5×10^{13} ions/cm^2.

4. Discussion

As shown in Figure 5, the index dip at 13 µm in depth was only observed at the high fluence of 5×10^{13} ions/cm^2 and higher fluence (not shown in this paper) by end-face coupling, which well matches at the peak of the nuclear energy loss S_n. Consequently, the index reduction at 13 µm in depth is ascribed to the damage induced by the nuclear energy loss. Since S_n is much lower than S_e, the dip appeared only at high fluences. The existence of this mode is clearly evidenced by the deep WG mode shown in Figure 6d.

The index reduction in the surface plateau region between 0 and ~3 µm can be ascribed to the electronic energy loss S_e, which has the highest at the surface. Consequently, both S_e and S_n contribute the index reduction. However, the origin of the dip/peak at 6 µm in depth is not clear. Neither of S_e nor S_n has a peak around 6 µm. Rather, S_e decreases and S_n increases around 6 µm.

It is known that amorphous ion tracks are formed in YAG crystals. The amorphization was confirmed by XRD as shown in Figure 4. The track threshold is reported as 7.5 keV/nm. The S_e of

200 MeV Xe ions in YAG crystal calculated by SRIM 2013 amounts to 24.3 keV/nm at the surface. With increasing the depth, the S_e gradually decreases with making a track. The S_e finally becomes below the threshold value of 7.5 keV/nm [13] and stops forming the tracks around the depth of ~10 μm. The threshold value of 7.5 keV/nm is indicated by a broken line in Figure 5. Deeper than the depth of ~10 μm, tracks are no longer formed. As shown in Figure 5, no index changes are induced in the region deeper than ~10 μm, except the dip at 13 μm. This is another evidence that the dip at 13 μm is not due to S_e but S_n.

Consequently, the index reductions at the surface plateau region and the dip at 13 μm can be described to the electronic and nuclear energy losses, because either of them matches with the maximum depths of S_e and S_n, respectively. However, the origin of the dip/peak at 6 μm is not clear. Neither of S_e nor S_n has a peak around 6 μm. Since the enhancement of the index was firstly observed, the irradiation introduces density increase around 6 μm depth. The density enhancement soon turns to the density reduction probably introduction of defects. Further the index at 6 μm decreases combined with the reduction in the surface plateau. Since the 6 μm peak matches neither of the peaks of S_e and S_n, a possible candidate could be the synergy effects between S_e and S_n. Since the synergy effects are not included in SRIM code, the deviation at 6 μm between the index profiles and SRIM calculations is a matter of course.

The refractive index profiling is a relatively new method to study the irradiation effects of SHIs, which is sensitive to defect formation and/or stress (strain) generation, and detectable the depth profiles at any fluences. Figure 5 clearly shows complex evolutions of the index, i.e., the defect formation and/or the stress generation, along the fluence. At the fluence of 1×10^{11} ions/cm^2, a weak index enhancement, i.e., compressive stress, is generated around 6 μm in depth due to the synergy effect. At 1×10^{12} ions/cm^2, the index turns to decrease, i.e., the compressive stress turns to the defect formation. At the same fluence, the index enhancement, i.e., the compressive stress is induced at 0–3 μm by S_e, but turns to decrease at higher fluences, i.e., the defect formation. At 5×10^{13} ions/cm^2, S_n also contributes the index change via the defect formation. This kind of complex evolutions are only accessible by the index profiling.

5. Conclusions

The evolution of the depth profiles with the fluence of the refractive index in YAG crystals were studied under 200 MeV ^{136}Xe^{14+} ion irradiation by the prism-coupling and the end-face coupling methods, in which various WG modes were detected. The index depth profiles were determined so that the observed WG modes were reproduced. Since the index can be changed with the damage and/or the stress change induced by the irradiation, the index depth profiles provide the depth profiles of damage and/or stress changes induced by ion irradiation and the evolutions with the ion fluence.

At the lowest fluence of 1×10^{11} ions/cm^2, a weak index enhancement was induced at the depth of 6 μm, which does not match either of S_e maximum nor S_n maximum. With increasing the fluence, the peak turns to decrease. At 1×10^{12} ions/cm^2, the index enhancement is induced at the index plateau near the surface. The enhancement soon turned to decrease with the fluence. Then both the surface plateau and the 6 μm dip reduce the index with the fluence. At the highest fluence of 5×10^{13} ions/cm^2, a new dip appeared at the depth of 13 μm, which matched to the peak of S_n.

Index changes were perceived at three different depth regions; (i) a sharp dip at 13 μm in depth, which is ascribed to the nuclear stopping S_n peak, (ii) a plateau near the surface at 0–3 μm, which can be ascribed to the electronic stopping S_e, since S_e has a very broad peak at the surface, and (iii) a broad peak at 6 μm in depth. Since the last peak is ascribed to neither of S_e nor S_n peak, it could be attributed to the synergy effect of S_e and S_n.

Author Contributions: Conceptualization, XRD measurements, Manuscript writing, H.A.; Sample preparation, F.C.; SHI irradiation, N.O. and N.I.; Refractive index measurements, R.L. and F.C. All authors have read and agreed to the published version of the manuscript.

Funding: This study was supported by JSPS-KAKENHI Grant Number 18K04898.

Quantum Beam Sci. **2020**, *4*, 39

Acknowledgments: A part of the study was supported by the Inter-organizational Atomic Energy Research Program through an academic collaborative agreement among JAEA, QST, and the Univ. of Tokyo. Also the other part was carried out under the Common-Use Facility Program of JAEA. The authors are grateful to the staff of the accelerator facility at JAEA-Tokai for their help. The authors thank China Scholarship Council for supporting RL to stay for one year in NIMS, Japan.

Conflicts of Interest: The authors declare no conflict of interest.

References

1. Nishihara, H.; Haruna, M.; Suhara, T. *Optical Integrated Circuits*; McGraw-Hill: New York, NY, USA, 1989.
2. Field, S.J.; Hanna, D.C.; Large, A.C.; Shepherd, D.P.; Tropper, A.C.; Chandler, P.J.; Townsend, P.D.; Zhang, L. An efficient, diode-pumped, ion-implanted Nd: GGG planar waveguide laser. *Opt. Commun.* **1991**, *86*, 161. [CrossRef]
3. Townsend, P.D.; Chandler, P.J.; Zhang, L. *Optical Effects of Ion Implantation*; Cambridge University Press: Cambridge, UK, 1994.
4. Fluck, D.; Binder, B.; Küpfer, M.; Looser, H.; Buchal, C.; Günter, P. Phase-matched second harmonic blue light generation in ion implanted $KNbO_3$ planar waveguides with 29% conversion efficiency. *Optics Commun.* **1992**, *90*, 304. [CrossRef]
5. Youden, K.E.; James, S.W.J.; Eason, R.W.; Chandler, P.J.; Zhang, L.; Townsend, P.D. Photorefractive planar waveguides in $BaTiO_3$ fabricated by ion-beam implantation. *Opt. Lett.* **1992**, *17*, 1509. [CrossRef] [PubMed]
6. Townsend, P.D. Ion implantation in optical materials. *Inst. Phys. Conf. Ser.* **1976**, *28*, 104.
7. Chen, F. Micro- and submicrometric waveguiding structures in optical crystals produced by ion beams for photonic applications. *Laser Photonics Rev.* **2012**, *6*, 622. [CrossRef]
8. Devine, R.A.B. Ion Implantation-Induced and Radiation-Induced Structural Modifications in Amorphous SiO_2. *J. Non-Cryst. Solids* **1993**, *152*, 50. [CrossRef]
9. Olivares, J.; García-Navarro, A.; Méndez, A.; Agulló-López, F.; García, G.; García-Cabañes, A.; Carrascosa, M. Novel optical waveguides by in-depth controlled electronic damage with swift ions. *Nuclear Instrum. Methods Phys. Res. Sect. B Beam Interact. Mater. Atoms* **2007**, *257*, 765. [CrossRef]
10. Rodriguez, M.D.; Li, W.X.; Chen, F.; Trautmann, C.; Bierschenk, T.; Afra, B.; Schauries, D.; Ewing, R.C.; Mudie, S.T.; Kluth, P. SAXS and TEM investigation of ion tracks in neodymium-doped yttrium aluminium garnet. *Nuclear Instrum. Methods Phys. Res. Sect. B Beam Interact. Mater. Atoms* **2014**, *326*, 150. [CrossRef]
11. Amekura, H.; Akhmadaliev, S.; Zhou, S.; Chen, F. A possible new origin of long absorption tail in Nd-doped yttrium aluminum garnet induced by 15 MeV gold-ion irradiation and heat treatment. *J. Appl. Phys.* **2016**, *119*, 173104. [CrossRef]
12. Ziegler, J.F.; Biersack, J.P.; Ziegler, M.D. *SRIM—The Stopping and Range of Ions in Matter*; SRIM Co.: Chester, MD, USA, 2008.
13. Itoh, N.; Duffy, D.M.; Khakshouri, S.; Stoneham, A.M. Making tracks: Electronic excitation roles in forming swift heavy ion tracks. *J. Phys. Condens. Matter* **2009**, *21*, 474205. [CrossRef] [PubMed]
14. Ovanesyan, K.L.; Petrosyan, A.G.; Shirinyan, G.O.; Avetisyan, A.A. Optical dispersion and thermal expansion of garnets $Lu_3Al_5O_{12}$, $Er_3Al_5O_{12}$, and $Y_3Al_5O_{12}$. *Inorg. Mater.* **1981**, *17*, 308.
15. Meftah, A.; Djebara, M.; Khalfaoui, N.; Stoquert, J.P.; Studer, F.; Toulemonde, M. Thermal Spike Description of the Damage Creation in $Y_3Al_5O_{12}$ Induced by Swift Heavy Ions. *Mater. Sci. Forum (Proc. of ICDS-18)* **1997**, *53*, 248–249.
16. Zelmon, D.E.; Small, D.L.; Page, R. Refractive-index measurements of undoped yttrium aluminum garnet from 0.4 to 5.0 µm. *Appl. Opt.* **1998**, *37*, 4933. [CrossRef] [PubMed]

Publisher's Note: MDPI stays neutral with regard to jurisdictional claims in published maps and institutional affiliations.

Article

Effects of Beam Conditions in Ground Irradiation Tests on Degradation of Photovoltaic Characteristics of Space Solar Cells

Mitsuru Imaizumi [1,*], Takeshi Ohshima [2], Yosuke Yuri [2], Kohtaku Suzuki [3] and Yoshifumi Ito [3]

[1] Japan Aerospace Exploration Agency (JAXA), Tsukuba 305-8505, Japan
[2] National Institutes for Quantum and Radiological Science and Technology (QST), Takasaki 370-1292, Japan; ohshima.takeshi@qst.go.jp (T.O.); yuri.yosuke@qst.go.jp (Y.Y.)
[3] The Wakasa Wan Energy Research Center (WERC), Tsuruga 914-0192, Japan; ksuzuki@werc.or.jp (K.S.); yito@werc.or.jp (Y.I.)
* Correspondence: imaizumi.mitsuru@jaxa.jp; Tel.: +81-70-3117-7500

Abstract: We investigated the effects of irradiation beam conditions on the performance degradation of silicon and triple-junction solar cells for use in space. The fluence rates of electron and proton beams were varied. Degradation did not depend on the fluence rate of protons for both cells. A higher fluence rate of electrons caused greater degradation of the Si cell, but the dependence was due to the temperature increase during irradiation. Two beam-area expansion methods, defocusing and scanning, were examined for proton irradiation of various energies (50 keV–10 MeV). In comparing the output degradation from irradiation with defocused and scanned proton beams, no significant difference in degradation was found for any proton energy. We plan to reflect these findings into ISO standard of irradiation test method of space solar cells.

Keywords: solar cell; space application; irradiation test; beam condition; degradation; standardization; ISO

Citation: Imaizumi, M.; Ohshima, T.; Yuri, Y.; Suzuki, K.; Ito, Y. Effects of Beam Conditions in Ground Irradiation Tests on Degradation of Photovoltaic Characteristics of Space Solar Cells. *Quantum Beam Sci.* **2021**, 5, 15. https://doi.org/10.3390/qubs5020015

Academic Editor: Akihiro Iwase

Received: 30 March 2021
Accepted: 18 May 2021
Published: 20 May 2021

Publisher's Note: MDPI stays neutral with regard to jurisdictional claims in published maps and institutional affiliations.

1. Introduction

Most spacecraft are powered by electricity generated by photovoltaic cells mounted on solar panels. In space, solar cells are exposed to radiation environment, and electrons and protons contained in the environment degrade them and reduce their power output. Electrons and protons are relatively light particles that easily penetrate a solar cell, which is generally made of single-crystal semiconductor material. These electrons and protons create crystal defects in the material that form minority-carrier recombination centers or majority-carrier traps. The effect of these two types of defect states is to reduce the output power of a solar cell. Thus, radiation resistance is of the utmost importance to space solar cells.

To determine a material's radiation resistance (i.e., the radiation degradation characteristics) irradiation tests need to be performed on the ground with electron and proton using accelerators. A ground test is mandatory because the obtained degradation characteristics are necessary to predict the degradation of generated power during a mission in space; therefore, it is essential for a test to reflect the actual degradation in space accurately.

Space solar cell irradiation tests are executed in various facilities by many organizations worldwide using their own test protocols. However, to ensure accurate degradation predictions, the test results must be identical regardless of the test procedure and facility. Additionally, investigating the radiation resistance of a new solar cell requires that test results are perfectly reproducible, so standardizing solar cell irradiation tests is crucial.

To this end, the Japan Aerospace Exploration Agency (JAXA), European Space Agency (ESA), National Aeronautics and Space Administration of the United States (NASA), and US Naval Research Laboratory (NRL) have been collaborating to standardize the irradiation test procedures for space solar cells [1]. As a result, an international standard (ISO) of irradiation test methods was published in 2005 and revised in 2015 [2]. However, no quantitative definition of irradiation test conditions is contained in the standard because the effects of conditions of electron/proton beam irradiation on testing had not clarified when the standard was published. Currently, each facility/organization performs solar cell irradiation tests under its own test conditions.

Our paper aims to provide information for standardizing radiation specifications for testing space solar cells. This study clarifies the effects of two typical irradiation beam conditions, fluence rate, and beam-area expansion technique, on a solar cell's degradation for use in space.

2. Experimental

Two typical irradiation beam conditions were selected in this study, dose rate and beam-area expansion methods, for the following reasons: First, in the ground irradiation test, the dose rate is generally selected to complete the test within an acceptable duration. However, the rate of actual radiation exposure in space is several orders of magnitude lower than the rate for ground irradiation tests. A lower dose might be gentler for degradation because recovery of radiation damage can be expected [3]. Therefore, the dose rate employed for a ground test is likely to over-estimate degradation. Second, to secure a beam area for an objective solar cell, scanning with a focused spot beam is the usual technique in a ground irradiation test. On the other hand, radiation particles fall into a solar cell uniformly in space. In the case of scanning, intense particles in the focused beam spot are irradiated momentarily, which may induce greater degradation than a uniform defocus beam. Therefore, clarifying the potential difference in output degradation due to the difference of beam condition is the primary purpose of this study.

An InGaP/GaAs/Ge triple-junction (3J) space solar cell and a high-efficiency silicon (Si) space solar cell were used in this study, both made by SHARP Corporation. Figure 1 shows the schematic cross-section of the solar cells. The size of both cells was 2 cm × 2 cm. The typical efficiency of a 3J cell is 27%; for a Si cell, it is 17%.

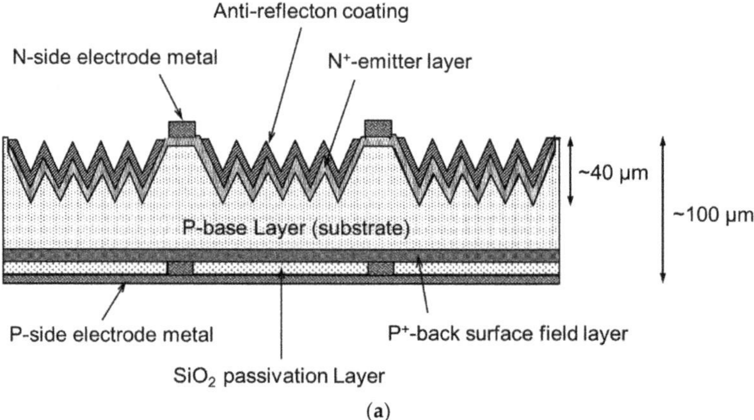

Figure 1. *Cont.*

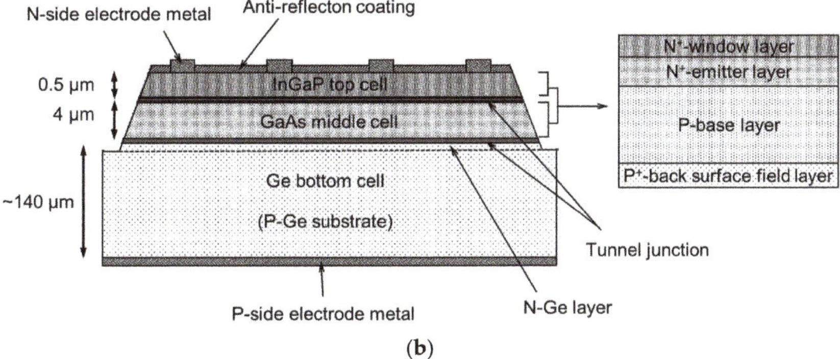

Figure 1. Cross-section of (**a**) a high-efficiency silicon solar cell and (**b**) an InGaP/GaAs/Ge triple-junction solar cell. Note that the figures are not to scale.

The irradiation experiments were carried out at the National Institute for Quantum and Radiological Science and Technology (QST) [4], Takasaki Lab, and the Wakasa Wan Energy Research Center (WERC) [5]. High-energy (10 MeV) proton irradiation was executed using the cyclotron accelerator at QST Takasaki. Low-energy (50–150 keV) proton irradiation was performed using the ion implanters at QST Takasaki and WERC. Electron irradiation was carried out using the Cockcroft–Walton accelerator at QST Takasaki. Current–voltage output characteristics of the solar cells under light illumination (LIV) before and after irradiation were measured at the Japan Aerospace Exploration Agency (JAXA), Tsukuba Space Center. The LIV measurement was made at 25 °C using a dual-source (xenon and halogen lamps) solar simulator with an Air Mass 0 (AM0) spectrum and light intensity equivalent to the Sun in space around the globe. Figure 2 explains the current-voltage characteristics of a solar cell, which is fundamentally a p–n junction diode and has the rectifying characteristic shown as "Dark I-V" in Figure 2. Once the solar cell is illuminated, a photocurrent is generated and the I-V curve shifts downward ("Light I-V") since the direction of the photocurrent is the opposite to that of the injected current. Three typical output parameters, the short-circuit current (Isc), open-circuit voltage (Voc), and maximum power (Pmax), are indicated in the figure.

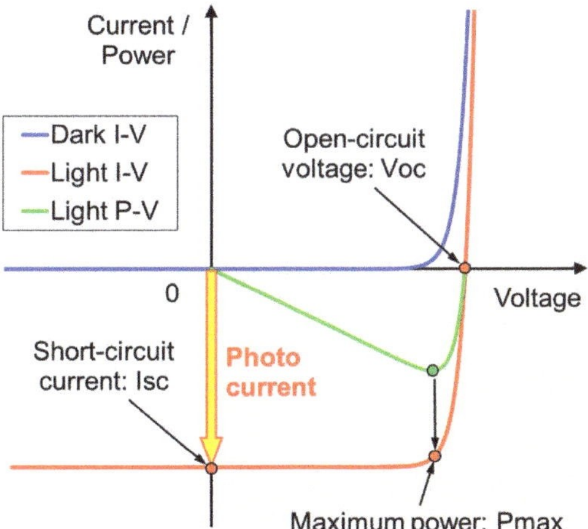

Figure 2. A typical solar cell's current-voltage characteristics in the dark ("Dark I-V") and under illumination ("Light I-V"). Three typical output parameters are indicated: the short-circuit current (Isc), open-circuit voltage (Voc), and maximum power (Pmax).

We carried out two types of radiation experiments. First, we irradiated solar cells with electrons and protons to a specific fluence at different fluence rates. Second, we irradiated solar cells using defocused or scanned proton beams with various proton energies. The sample's output performance was characterized before and after the irradiation tests, and degradation of the output was compared. Details of the experimental conditions are described in the following sections.

2.1. Fluence Rate

This investigation aims to clarify the effects of the dose rate by fluence (i.e., the fluence rate) on solar cell degradation. The adopted radiation particles are 10 MeV protons or 1 MeV electrons; both are regarded as the standard radiation particles for radiation resistance evaluation of a space solar cell. Since the particles' energies are sufficiently high to pass through each of the solar cells, irradiation of the particles forms uniform damage throughout the entire device structures. Table 1 lists the beam conditions. The fluence rate was varied over two orders of magnitude in both the 10 MeV proton and 1 MeV electron irradiation experiments: in five levels for the proton beam and in four levels for the electrons. The samples were placed in air for electron irradiation and in a vacuum for proton one.

Table 1. Beam conditions of high-energy proton and electron experiments for degradation dependence on fluence rate.

Particle	Proton	Electron
Acceleration energy	10 MeV	1.0 MeV
Beam-area expansion method	Scan	Scan
Fluence	5.0×10^{12} cm^{-2}	1.0×10^{15} cm^{-2}
Fluence rate	6.6×10^8, 1.0×10^9, 1.0×10^{10}, 6.6×10^{10}, 1.0×10^{11} cm^{-2} s^{-1}	1.7×10^{11}, 1.0×10^{12}, 4.0×10^{12}, 1.7×10^{13} cm^{-2} s^{-1}
Institute	QST Takasaki	

In addition, the effects of fluence rate with low-energy protons (50–150 keV) were examined for the 3J solar cell. A 3J cell has a stacked structure with an InGaP top cell, a GaAs middle cell, and a Ge bottom cell as shown in Figure 1b. Low energy protons stop in the cell, so radiation damage is localized around the Bragg peak of the protons. According to the TRIM simulation [3], 50, 100, and 150 keV protons stop in the InGaP top cell, around the interface between the InGaP top cell and GaAs middle cell, and in the GaAs middle-cell, respectively. Therefore, the effects of the damage may differ. Table 2 shows the beam conditions.

Table 2. Beam conditions of low-energy proton experiment for degradation dependence on fluence rate.

Particle	Proton		
Acceleration Energy	**50 keV**	**100 keV**	**150 keV**
Beam-area expansion method		Defocus	
Fluence		1.0×10^{12} cm^{-2}	
Fluence rate		1.4×10^{10}, 3.5×10^{10}, 1.4×10^{11}, 4.2×10^{11} cm^{-2} s^{-1}	
Institute		WERC	

2.2. Beam-Area Expansion Method

This examination determines the difference in solar cell degradation between the two beam-area expansion techniques, scanning and defocusing. Only proton was use in this examination because no electron accelerator capable of providing a defocused beam was available. High (10 MeV) and low (50–100 keV) energy levels were selected for the proton beams. Tables 3 and 4 summarize these test conditions. In the case of the defocus beam of 10 MeV proton, a primary beam with a Gaussian distribution over a diameter of ~80 mm was uniformized using multipole magnets to form a rectangular exposure area up to 80 mm × 100 mm. This unique technique was developed at QST Takasaki [6–8].

Table 3. Beam conditions of high-energy proton experiment for difference in degradation between scanned and defocused beams.

Particle	Proton	
Acceleration Energy	**10 MeV**	
Fluence (Φ)	$1.0 \times 10^{11}, 3.0 \times 10^{11}, 6.0 \times 10^{11}, 1.0 \times 10^{12}, 2.0 \times 10^{12}, 3.0 \times 10^{12}, 5.0 \times 10^{12}, 7.0 \times 10^{12}, 1.0 \times 10^{13}, 2.0 \times 10^{13}, 3.0 \times 10^{13}, 5.0 \times 10^{13}$ cm^{-2}	
Beam-area expansion method	Scan	Defocus
Beam area	100 mm × 100 mm	80 mm × 100 mm
Scan frequency	Horizontal: 50 Hz, Vertical: 0.5 Hz	–
Beam spot size	~10 mm ϕ	–
Fluence rate	1.0×10^9 cm^{-2} s^{-1} ($\Phi \leq 1.0 \times 10^{12}$ cm^{-2}), 5.0×10^9 cm^{-2} s^{-1} ($1.0 \times 10^{12} < \Phi < 2.0 \times 10^{13}$ cm^{-2}), 1.0×10^{10} cm^{-2} s^{-1} ($\Phi \geq 2.0 \times 10^{13}$ cm^{-2})	8.8×10^{10} cm^{-2} s^{-1}
Institute	QST Takasaki	

Table 4. Beam conditions of low-energy proton experiment for difference in degradation between scanned and defocused beams.

Particle	Proton	
Acceleration Energy	**50, 100 keV**	
Fluence	1.0×10^{12} cm^{-2}	
Beam area	30 mm × 30 mm	
Beam-area expansion method	Scan	Defocus
Scan frequency	Horizontal: 89 Hz, Vertical: 502 Hz	–
Fluence rate	3.5×10^9 cm^{-2} s^{-1}	1.0×10^{11} cm^{-2} s^{-1}
Beam spot size	~2 mm ϕ	–
Institute	QST Takasaki	WERC

For the 10 MeV proton experiment, proton irradiation and LIV measurements were alternately implemented in an irradiation chamber. Irradiation was interrupted for each fluence step, and LIV measurements were made using a single-source (xenon lamp) AM0 solar simulator. Schematic top-view of the chamber is posted in Figure 3. This unique setup enables us to collect solar cell's degradation characteristics from a single sample if the obtained data are properly corrected for sample temperature using the temperature coefficients of its output parameters.

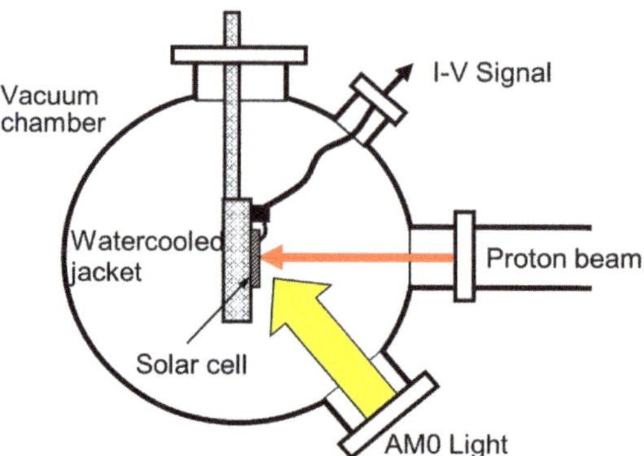

Figure 3. Schematic top view of the 10 MeV-proton irradiation vacuum chamber. A test solar cell can be illuminated by simulated solar light (AM0) to measure output characteristics instantly.

For both the experiments, beam profiles were observed by using an alumina fluorescent plate right before the irradiations. We confirmed that the adopted scan frequencies and the beam spot sizes provided uniform proton exposure without voids on the entire beam area.

3. Results and Discussion

3.1. Fluence Rate

Figure 4 exhibits the LIV characteristics of the (a) Si and (b) 3J solar cells before and after 10 MeV proton irradiation with a fluence of 5.0×10^{12} cm^{-2}. The fluence rates for the blue and red curves are 6.6×10^8 and 6.6×10^{10} cm^{-2} s^{-1}. The Si cell shows noticeable degradation, but there is no difference in LIV after irradiation. On the other hand, the 3J cell does not degrade as greatly as the Si, but there is a slight difference in Isc after irradiation. Figure 5 indicates the remaining factors of Isc, Voc, and Pmax for the (a) Si and (b) 3J solar cells as a function of 10 MeV proton fluence rate [9]. Two samples were irradiated for each beam condition. From the results, no significant dependence in degradation on fluence rate can be observed within the range of the experiment for either the Si or 3J solar cell. This fact confirms that the difference in Isc observed in Figure 4b is not significant and comes from a fluctuation of the sample characteristics.

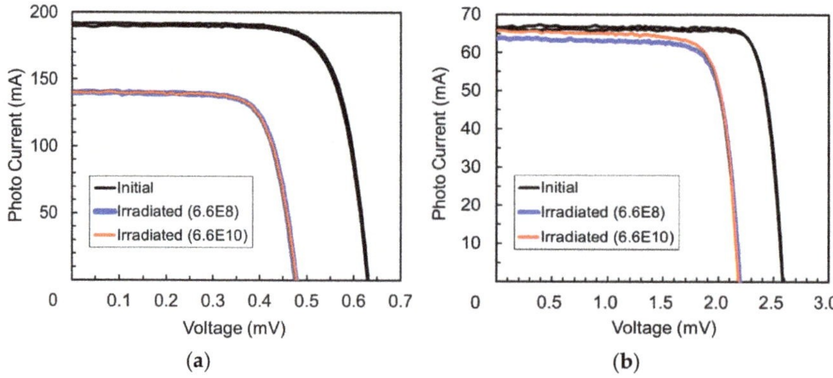

Figure 4. Light current–voltage characteristics of (**a**) high-efficiency silicon and (**b**) InGaP/GaAs/Ge triple-junction solar cells before and after 10 MeV proton irradiation with a fluence of 5.0×10^{12} cm^{-2}. The adopted fluence rates for blue and red curves are 6.6×10^8 (6.6E8) and 6.6×10^{10} (6.6E10) cm^{-2} s^{-1}.

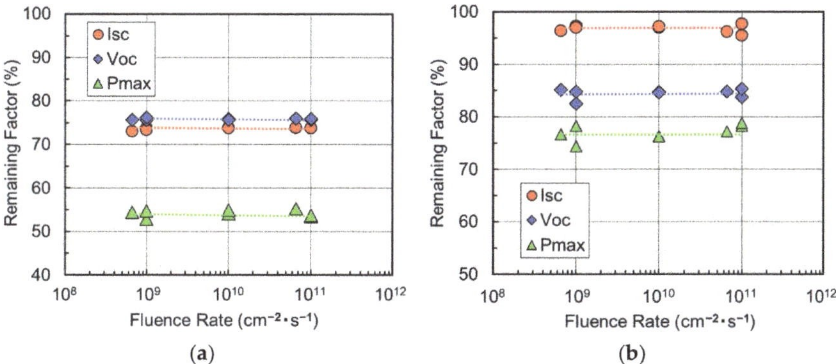

Figure 5. Remaining factors of short-circuit current (Isc), open-circuit voltage (Voc), and maximum power (Pmax) for (**a**) high-efficiency silicon and (**b**) InGaP/GaAs/Ge triple-junction solar cells as a function of 10 MeV proton fluence rate. The fluence is 5.0×10^{12} cm^{-2}.

Figure 6 presents the LIV characteristics of the (a) Si and (b) 3J solar cells before and after 1 MeV electron irradiation with a fluence of 1.0×10^{15} cm^{-2}. The fluence rates for blue and red curves are 1.7×10^{11} and 1.7×10^{13} cm^{-2} s^{-1}. Again, the Si cell shows considerable degradation. Furthermore, there is a clear difference between the blue and red curves. The 3J cell exhibits far less degradation and has no significant difference between its blue and red curves. Figure 7 depicts the remaining factors of Isc, Voc, and Pmax of the (a) Si and (b) 3J solar cells as a function of 1 MeV electron fluence rate [9]. Two samples were used for each beam condition as well. No dependence in degradation on fluence rate is conformed for the 3J cell. However, the degradation in the Si cell becomes greater as the fluence rate increases. We attributed this to the sample temperature rising during the electron irradiation; therefore, this tendency is not pertinent to this study. The sample temperatures were 30 and 90 °C at the fluence rates of 1.7×10^{11} and 1.7×10^{13} cm^{-2} s^{-1}. Details of this temperature effect have been described elsewhere [9]. However, this fact demonstrates that controlling the sample temperature is important especially for electron irradiation tests on Si solar cells.

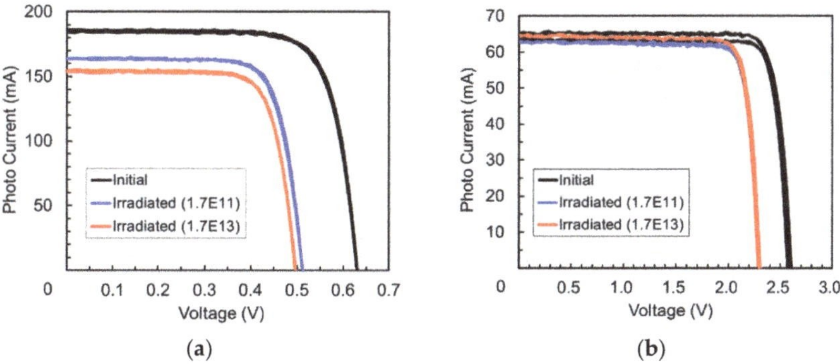

Figure 6. Light current-voltage characteristics of (**a**) high efficiency silicon and (**b**) InGaP/GaAs/Ge triple-junction solar cells before and after 1 MeV electron irradiation with a fluence of 1.0×10^{15} cm^{-2}. The adopted fluence rates for blue and red curves are 1.7×10^{11} (1.7E11) and 1.7×10^{13} (1.7E13) cm^{-2} s^{-1}.

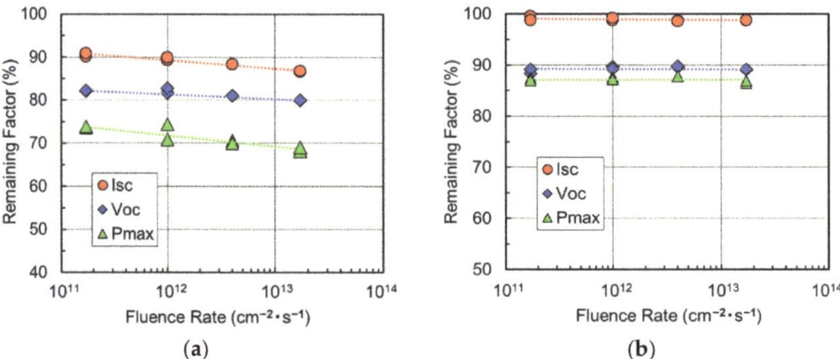

Figure 7. The remaining factors of short-circuit current (Isc), open-circuit voltage (Voc), and maximum power (Pmax) for (**a**) high-efficiency silicon and (**b**) InGaP/GaAs/Ge triple-junction solar cells as a function of 1 MeV electron fluence rate. The fluence is 1.0×10^{15} cm^{-2}.

The remaining factors of Isc, Voc, and Pmax of the 3J solar cells as a function of the fluence rate of protons with the energies of (a) 50, (b) 100, and (c) 150 keV are exhibited in Figure 8. Although there is some scattering in the remaining factor of Pmax of the (a) 50 and (b) 100 keV results, the degradation is likely to be independent of the fluence rate in the examined range.

The adopted fluence rates in this investigation are reasonable values for radiation tests of solar cells. The results confirmed that degradation of Si and 3J solar cells due to electrons and protons is insensitive to the fluence rate that is generally used in actual radiation tests. Furthermore, this independence of degradation on fluence rate is established for both uniform and localized damage.

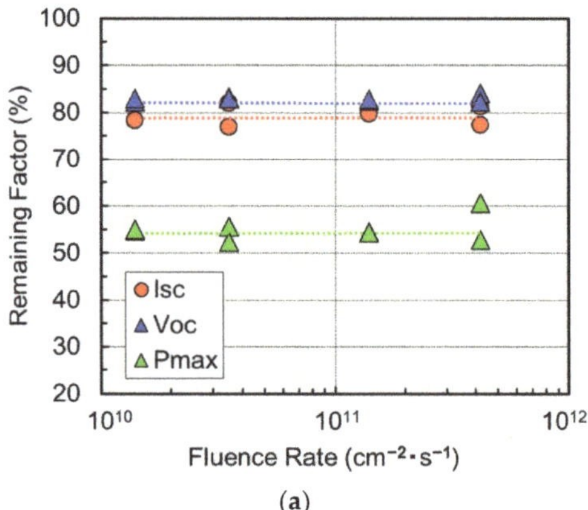

Figure 8. *Cont.*

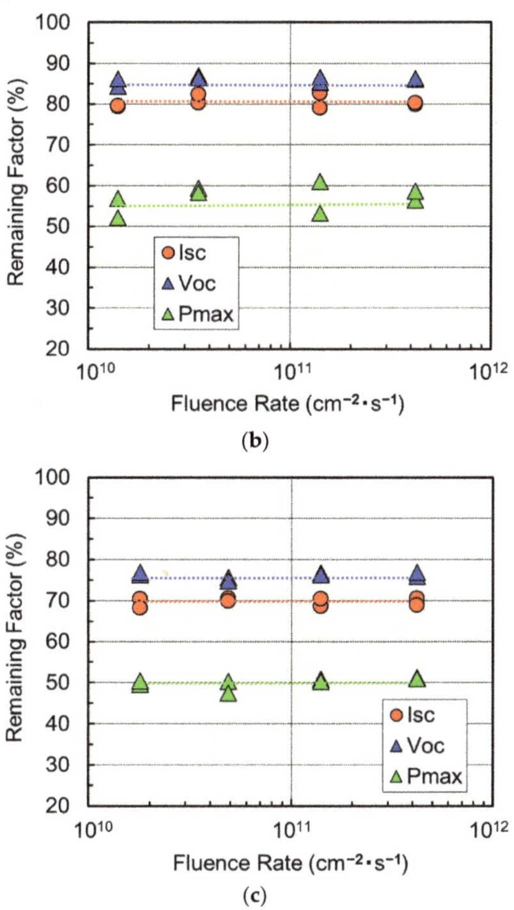

Figure 8. The remaining factors of short-circuit current (Isc), open-circuit voltage (Voc), and maximum power (Pmax) of InGaP/GaAs/Ge triple-junction solar cell as a function of fluence rate of protons with energies of (**a**) 50, (**b**) 100 and (**c**) 150 keV.

3.2. Beam-Area Expansion Method

Figure 9 shows a typical set of LIV data collected using the 10 MeV proton irradiation chamber equipped with a solar simulator of the previous section. This is the case of the Si solar cell and the scanned beam, but temperature correction is not applied. Fluence was varied from 0 (initial) to 5.0×10^{13} cm^{-2} with twelve steps. This figure displays representative deterioration in output as LIV characteristics of a solar cell due to radiation damage.

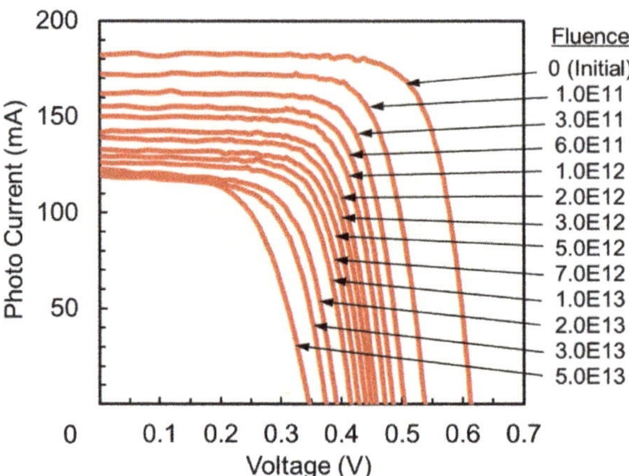

Figure 9. Typical set of LIV data for high-efficiency silicon solar cell collected using the 10 MeV proton irradiation chamber equipped with a solar simulator. No temperature correction was applied.

Figure 10 compares the radiation degradation trends of Isc, Voc, and Pmax with scanned (solid symbols) and the defocused (open symbols) beams on the (a) Si and (b) 3J solar cells due to 10 MeV protons. All the degradations in the remaining factors are nearly the same except for Isc of the 3J solar cell. However, the degradation of Pmax, the product of current and voltage, of the 3J cell does not differ between the scan and defocus results. We conclude that the discrepancy in the Isc degradation is insignificant. The Isc of Si solar cell shows an anomalous increase at the last fluence point of 5.0×10^{13} cm^{-2}. This phenomenon can be seen right before a catastrophic degradation of a solar cell [10–12]. Thus, it is an indication of the life of the Si solar cell in terms of radiation degradation.

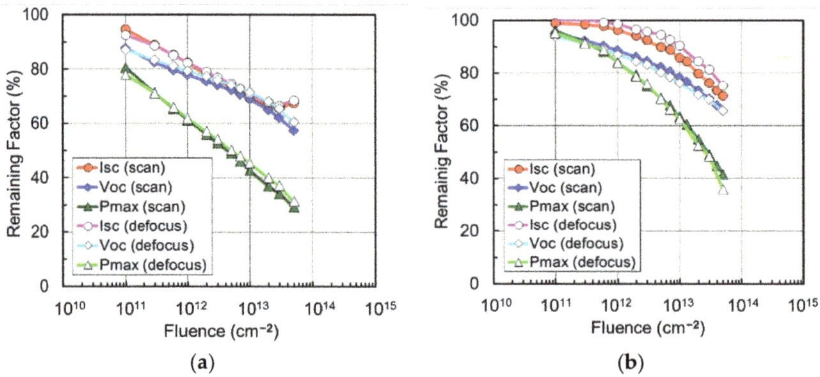

Figure 10. Comparison of degradation characteristics of short-circuit current (Isc), open-circuit voltage (Voc), and maximum power (Pmax) of (**a**) high-efficiency silicon and (**b**) InGaP/GaAs/Ge triple-junction solar cells by irradiations with scanned and defocused 10 MeV proton beams.

Figures 11 and 12 present the degradation of LIV characteristics due to (a) 50 and (b) 100 keV protons with scanned and defocused beams for the Si and 3J solar cells, respectively. The fluence was 1.0×10^{12} cm^{-2} for both energies. Two samples were used for each condition. In Figure 10a and a,b, one result indicates different degradations, but the other three results show almost the same degradation. Therefore, the different degradations of the three solar cells are not thought to be caused by the difference in the beam-area expansion technique. Figure 10 shows that 50 keV protons inflicted greater damage on the Si solar cell than 100 keV protons because the 50 keV protons stop at shallower position in the Si solar cell and create dense defects there. The p–n junction where most photocarriers are generated is located below ~0.5 µm from the surface. Therefore, 50 keV protons induce greater degradation especially on the photocurrent.

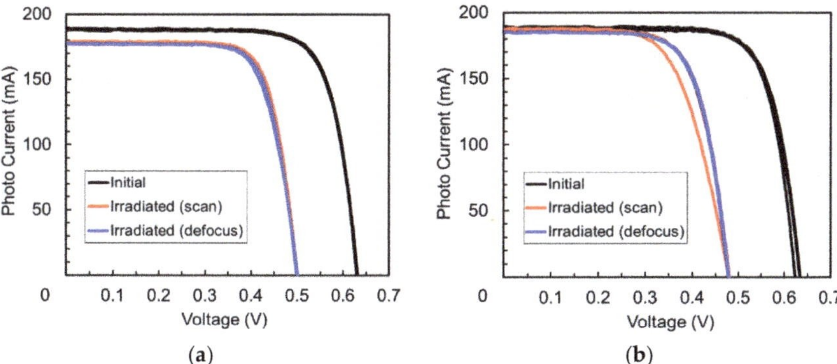

Figure 11. Light current–voltage characteristics of high-efficiency silicon solar cell before and after (**a**) 50 and (**b**) 100 keV proton irradiation with the fluence of 1.0×10^{12} cm^{-2}. The beam-area expansion techniques for red and blue curves are scanning and defocusing, respectively.

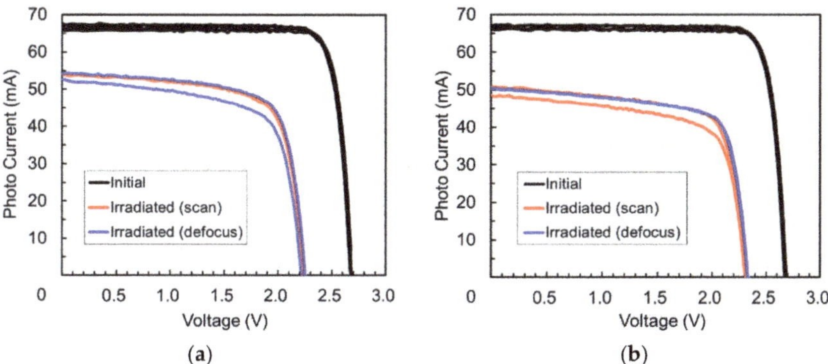

Figure 12. Light current–voltage characteristics of InGaP/GaAs/Ge triple-junction solar cell before and after (**a**) 50 and (**b**) 100 keV proton irradiation with the fluence of 1.0×10^{12} cm^{-2}. The beam-area expansion techniques for red and blue curves are scanning and defocusing, respectively.

The results led us to conclude that there is little difference in solar cell performance degradation between the two beam-area expansion techniques of scanning and defocusing, suggesting that the intense spot beam in the scanning method has no significant influence. We can use either technique for a radiation test of solar cells.

4. Summary

This study investigated the effects of two typical beam conditions in a space solar cell radiation test, the fluence rate, and beam-area expansion techniques (scanning and defocusing) on radiation degradation. Both high-efficiency silicon and InGaP/GaAs/Ge triple-junction solar cells were employed in this study. The 10 MeV protons and 1 MeV electrons were applied with different fluence rates. In addition, low-energy (50–150 keV) protons were irradiated to the triple-junction solar cell to understand the effects of localized damage on its stacked structure. All the results confirmed that the fluence rate did not affect the degradation of output performance of the two kinds of solar cell. However, cell temperature during the irradiation influenced the degradation of the silicon solar cell, a greater degradation arising from higher temperatures. Thus, sample temperature is an essential factor to be controlled in irradiation tests. High- (10 MeV) and low-energy (50 and 100 keV) protons were used with the two beam-area expansion techniques. The two sets of results proved that the radiation degradation is same regardless of the beam-area expansion technique. The obtained information will be reflected in the international standard on radiation test methods for space solar cells.

Author Contributions: Conceptualization, M.I., T.O. and Y.I.; methodology, M.I., Y.Y., K.S. and Y.I.; validation, M.I. and K.S.; formal analysis, M.I.; investigation, M.I. and T.O.; resources, M.I.; data curation, M.I.; writing—original draft preparation, M.I.; writing—review and editing, T.O., Y.Y. and K.S.; visualization, M.I. and Y.Y.; supervision, T.O. and K.S.; project administration, M.I.; funding acquisition, M.I. All authors have read and agreed to the published version of the manuscript.

Funding: This research received no external funding.

Acknowledgments: We would like to express our sincere appreciation to M. Saito of the Advanced Engineering Services Co., Ltd. for his tremendous contribution to the irradiation experiments. We would also like to thank the members of the space photovoltaics community for many fruitful discussions.

Conflicts of Interest: The authors declare no conflict of interest.

References

1. Jenkins, P.; Scheiman, D.; Goodbody, C.; Baur, C.; Sharps, P.; Imaizumi, M.; Yoo, H.; Sahlstrom, T.; Walters, R.; Lorentzen, J.; et al. Results from an International Measurement Round Robin of III-V Triple-Junction Solar Cells under Air Mass Zero. In Proceedings of the 2006 IEEE 4th World Conference on Photovoltaic Energy Conference, Waikoloa, HI, USA, 7–12 May 2006; pp. 1968–1970. [CrossRef]
2. *International Standard: IS23038, "Space Systems—Space Solar Cells—Electron and Proton Irradiation Test Methods"*; The International Organization for Standardization (ISO): Geneva, Switzerland, 2015.
3. Ziegler, J.F. SRIM—The Stopping and Range of Ions in Matter. Available online: http://srim.org/ (accessed on 12 April 2020).
4. Kurashima, S.; Satoh, T.; Saitoh, Y.; Yokota, W. Irradiation Facilities of the Takasaki Advanced Radiation Research Institute. *Quantum Beam Sci.* **2017**, *1*, 2. [CrossRef]
5. Hatori, S.; Ishigami, R.; Kume, K.; Suzuki, K. Ion Accelerator facility of the Wakasa Wan Energy Research Center for the Study of Irradiation Effects on Space Electronics. *Quantum Beam Sci.* **2021**, *5*, 14. [CrossRef]
6. Yuri, Y.; Miyawaki, N.; Kamiya, T.; Yokota, W.; Arakawa, K. Uniformization of the transverse beam profile by means of nonlinear focusing method. *Phys. Rev. ST Accel. Beams* **2007**, *10*, 104001. [CrossRef]
7. Yuri, Y.; Yuyama, T.; Ishizaka, T.; Ishibori, I.; Okumura, S. Transformation of the Beam Intensity Distribution and Formation of a Uniform Ion Beam by Means of Nonlinear Focusing. *Plasma Fusion Res.* **2014**, *9*, 4406106. [CrossRef]
8. Yuri, Y.; Fukuda, M.; Yuyama, T. Formation of hollow ion beams of various shapes using multipole magnets. *Prog. Theor. Exp. Phys.* **2019**, *2019*, 053G01. [CrossRef]
9. Saito, M.; Imaizumi, M.; Ohshima, T.; Takeda, Y. Effects of irradiation beam conditions on radiation degradation of solar cells. In Proceedings of the 2010 35th IEEE Photovoltaic Specialists Conference, Honolulu, HI, USA, 20–25 June 2010; pp. 2616–2619. [CrossRef]
10. Yamaguchi, M.; Taylor, S.J.; Yang, M.; Matsuda, S.; Kawasaki, O.; Hisamatsu, T. High-energy and high-fluence proton irradiation effects in silicon solar cells. *J. Appl. Phys.* **1996**, *80*, 4916. [CrossRef]
11. Taylor, S.J.; Yamaguchi, M.; Yamaguchi, T.; Watanabe, S.; Ando, K.; Matsuda, S.; Hisamatsu, T.; Kim, S.I. Comparison of the effects of electron and proton irradiation on n+–p–p+ silicon diodes. *J. Appl. Phys.* **1998**, *83*, 4620. [CrossRef]
12. Imaizumi, M.; Taylor, S.J.; Yamaguchi, M.; Ito, T.; Hisamatsu, T.; Matsuda, S. Analysis of structure change of Si solar cells irradiated with high fluence electrons. *J. Appl. Phys.* **1999**, *85*, 1916. [CrossRef]

quantum beam science

MDPI

Review

Modification of Critical Current Density Anisotropy in High-T_c Superconductors by Using Heavy-Ion Irradiations

Tetsuro Sueyoshi 

Department of Electrical Engineering, Kyushu Sangyo University, 2-3-1 Matsukadai Higashi-ku, Fukuoka 813-8503, Japan; s.teturo@ip.kyusan-u.ac.jp; Tel.: +81-92-673-5632; Fax: +81-92-673-5091

Abstract: The critical current density J_c, which is a maximum value of zero-resistivity current density, is required to exhibit not only larger value but also lower anisotropy in a magnetic field B for applications of high-T_c superconductors. Heavy-ion irradiation introduces nanometer-scale irradiation tracks, i.e., columnar defects (CDs) into high-T_c superconducting materials, which can modify both the absolute value and the anisotropy of J_c in a controlled manner: the unique structures of CDs, which significantly affect the J_c properties, are engineered by adjusting the irradiation conditions such as the irradiation energy and the incident direction. This paper reviews the modifications of the J_c anisotropy in high-T_c superconductors using CDs installed by heavy-ion irradiations. The direction-dispersion of CDs, which is tuned by the combination of the plural irradiation directions, can provide a variety of the magnetic field angular variations of J_c in high-T_c superconductors: CDs crossing at $\pm\theta_i$ relative to the c-axis of YBa$_2$Cu$_3$O$_y$ films induce a broad peak of J_c centered at $B \mid\mid c$ for $\theta_i < \pm45°$, whereas the crossing angle of $\theta_i \geq \pm45°$ cause not a J_c peak centered at $B \mid\mid c$ but two peaks of J_c at the irradiation angles. The anisotropy of J_c can also modified by tuning the continuity of CDs: short segmented CDs formed by heavy-ion irradiation with relatively low energy are more effective to improve J_c in a wide magnetic field angular region. The modifications of the J_c anisotropy are discussed on the basis of both structures of CDs and flux line structures depending on the magnetic field directions.

Keywords: high-T_c superconductors; critical current density; flux pinning; heavy-ion irradiation; columnar defects; anisotropy

Citation: Sueyoshi, T. Modification of Critical Current Density Anisotropy in High-T_c Superconductors by Using Heavy-Ion Irradiations. *Quantum Beam Sci.* **2021**, 5, 16. https://doi.org/10.3390/qubs5020016

Academic Editor: Akihiro Iwase

Received: 21 March 2021
Accepted: 18 May 2021
Published: 21 May 2021

1. Introduction

High-T_c superconductors have attracted considerable research activity, especially for electric power applications at high magnetic fields and temperatures, because the zero-resistive current and the high superconducting transition temperature T_c enable us to operate zero-resistance devices at liquid-nitrogen temperature. Nowadays, coated conductors based on biaxially textured REBa$_2$Cu$_3$O$_y$ (REBCO, RE: rare earth elements) thin films have been significantly developed as second generation high-T_c superconducting tapes and have become commercially available now [1,2].

The critical current density J_c in magnetic field (in-field J_c), which is a maximum current density with zero-resistivity, is the most important parameter in REBCO-coated conductors for the practical applications. The absolute values of J_c for REBCO-coated conductors, however, have still remained below the practical level for high magnetic field applications [3]. In addition, the electronic mass anisotropy in the layered structure of CuO$_2$ planes for high-T_c superconductors induces a large anisotropy of J_c against a magnetic field orientation [4], which gives rise to obstacles to the superconducting magnet applications: a minimum in the magnetic field angular variation of J_c, which is usually located at the magnetic field B parallel to the c-axis, limits the operation current [5,6].

The in-field J_c can be controlled by immobilization of nano-sized quantized-magnetic-flux-lines (flux lines) penetrating into superconductors in a magnetic field. The motion of flux lines is suppressed by crystalline defects and impurities in the specimen, which are called pinning centers (PCs). Thus, artificially embedding crystalline defects as effective PCs is just a key strategy to improve the in-field performance of superconductors [1,3,7]. For the last fifteen years or so, doping of non-superconducting secondary phases such as $BaMO_3$ (M = Zr, Sn, Hf, etc.) and RE_2O_3 has been attempted to form those into effective PCs in REBCO thin films [8–12].

The flux pinning effect depends on the shape (dimensionality), orientation, size, and distribution of PCs. In particular, the dimensionality of PCs significantly affects the feature of flux pinning, as shown in Figure 1. For example, one-dimensional PCs such as columnar defects (CDs) exhibit a preferential direction for the flux pinning: the strong flux pinning occurs in the magnetic field direction along their long axis. Three-dimensional PCs such as nano-particles, on the other hand, have the morphology with no correlated orientation for flux pinning, resulting in the isotropic pinning force against any direction of magnetic field. These features of PCs play an important role in the modification of the J_c properties in REBCO films: those parameters of PCs such as their shape and size, should be designed to meet the requirements for each application.

Swift-heavy-ion irradiation to high-T_c superconductors produces amorphous CDs of damaged material parallel to the projectile direction through the electron excitation process rather than the nuclear collision process. The CDs produced by the irradiation effectively work as one-dimensional PCs [13–15]. The orientation of one-dimensional PCs determines the preferential direction of flux pinning [13,16]. Therefore, heavy-ion irradiation can be expected to modify the anisotropy of J_c in high-T_c superconductors by tuning the irradiation direction. In addition, the size and shape of CDs strongly depends on the electronic stopping power S_e, which is defined as energy loss of the incident ion per unit length via electronic excitation in the target material [17]: continuous CDs with thick diameter are formed at higher S_e than a certain value and discontinuous CDs with thin diameter are located at intervals along the ion path at lower S_e [18–20]. In particular, discontinuous CDs may provide more effective flux pinning in a wide magnetic field angular range, because the ends of discontinuous CDs can act as PCs even in magnetic field directions tilted from their long axis [21,22]. Thus, the discontinuity of CDs is also one of the important factors for the modification of the J_c anisotropy in high-T_c superconductors, as well as the direction-dispersion of CDs.

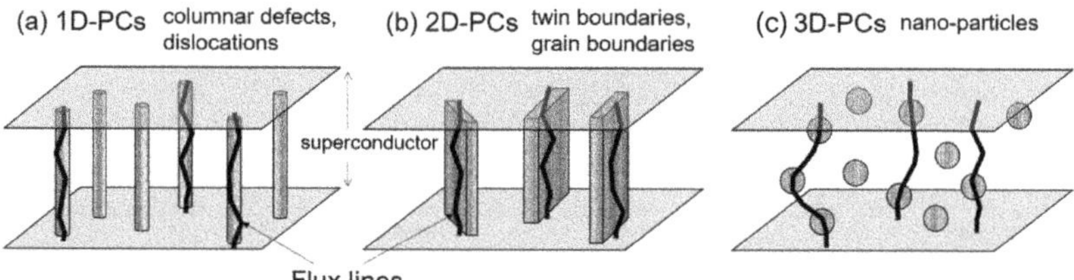

Figure 1. Sketch of the different dimensional categories for PCs: (**a**) 1D columnar (linear) defects, (**b**) 2D planar defects such as twin boundaries, and (**c**) 3D nano-particles.

A major advantage of using heavy-ion irradiation for the formation of CDs is that any CD configuration can be prepared by tuning the irradiation energy and the incident direction [23,24], independently from a fabrication process of samples (see Figure 2): the pinning structure can be efficiently designed to meet the requirements for different applications, which would be valuable for the development of high-performance coated conductors. In addition, unique pinning structures architected by the irradiations may enable us to find new physics of flux line dynamics. Therefore, heavy-ion irradiation to high-T_c superconductors can provide the design criteria for the supreme pinning landscape making the most of the potential for flux pinning, which leads to J_c close to the theoretical limit of critical current density, i.e., the pair-breaking critical current density.

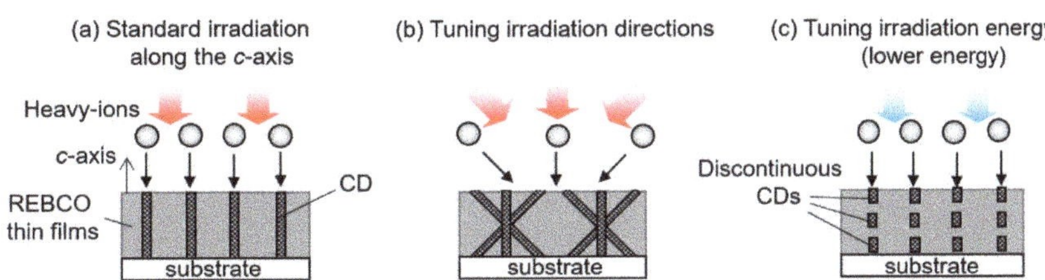

Figure 2. Schematic illustration of various configurations of CDs designed by heavy-ion irradiation: (**a**) continuous CDs parallel to the c-axis produced by standard irradiation, (**b**) direction-dispersed CDs installed by tuning the irradiation directions, (**c**) discontinuous CDs formed by adjusting irradiation energy to lower value.

In this paper, we describe the results of the modification of the J_c properties in REBCO thin films and coated conductors, which were obtained by our studies through heavy-ion irradiation under various irradiation conditions. Most of previous works of other researchers using heavy-ion irradiation have focused on the improvement of J_c at $B \parallel c$ where J_c usually shows the minimum [13–15,18,19]. On the other hand, heavy-ion irradiation effects over a wide magnetic field angular range have not been well studied so far. By contrast, we focus especially on modification of the J_c anisotropy in high-T_c superconductors by using heavy-ion irradiation: our aim in this review is to improve J_c in all magnetic field angular range from $B \parallel c$ to $B \parallel ab$ by using CDs and to explore breakthroughs for strong and isotropic pinning landscape in REBCO coated conductors. To meet the aim in this paper, we selected Xe ions as the irradiation ion species: the Xe-ion irradiation to REBCO thin films can provide large increase of J_c without heavily damaging crystallinity even at a large amount of doses, 5.0×10^{11} ions/cm^2 [24] and easily enables us to tune the morphology of CDs through the adjustment of the irradiation energy at a tandem accelerator of Japan Atomic Energy Agency (JAEA) used in our works. Firstly, we present the reduction of the J_c anisotropy by using the direction-dispersed CDs, which are introduced by controlling the irradiation direction. Secondly, we report the influence of CDs tilted at small angle(s) relative to the *ab*-plane on the J_c properties near $B \parallel ab$, which is one of key factors to improve J_c in all magnetic field directions. In particular, we show the influence of CDs along the *ab*-plane on J_c at $B \parallel ab$ by preparing an in-plane aligned *a*-axis-oriented YBCO film. Finally, we clarify the potential of discontinuous CDs for flux pinning in comparison with continuous CDs, where the morphology of CDs is controlled by the irradiation energy.

2. Experimental

The samples used in our works were mostly *c*-axis oriented YBCO thin films and GdBCO coated conductors. The *c*-axis oriented YBCO thin films were fabricated by a pulsed laser deposition (PLD) technique on (100) surface of SrTiO$_3$ single crystal substrates. The thickness of the films was about 300 nm. The GdBCO coated conductor, on the other hand, was fabricated on an ion-beam-assisted deposition (IBAD) substrate by a PLD method (Fujikura Ltd., Tokyo, Japan). The thickness of GdBCO layer is 2.2 µm and the self-field critical current I_c of this tape with 5 mm width is about 280 A. The samples were cut from the tape of the GdBCO coated conductor. The Ag stabilizer layer on the superconducting layer was removed by a chemical process. The YBCO thin films and the samples cut from the GdBCO coated conductor were patterned into a shape of about 40 µm wide and 1 mm long micro-bridge before the irradiation.

The heavy-ion irradiations with Xe ions were performed using the tandem accelerator of JAEA in Tokai, Japan. Tuning of the discontinuity of CDs along the *c*-axis can be controlled by the irradiation energy. The values of S_e for the Xe-ion irradiation energies above 200 MeV are above 2.9 keV/Å, which is above the threshold value of $S_e = 20$ keV/nm to create continuous CDs along the *c*-axis over the whole sample thickness for YBCO [17].

Thus, the irradiation with 200 MeV Xe ions was performed to install continuous CDs into YBCO thin films. In addition, the Xe-ion irradiation with 270 MeV was applied in order to create continuous CDs for GdBCO coated conductors, where the projectile length was longer than the thickness of 2.2 µm. Discontinuous CDs, on the other hand, were formed into YBCO thin films and GdBCO coated conductors by the irradiation with 80 MeV Xe ions, where the value of S_e is below 20 keV/nm: the radius of CDs strongly fluctuates along the ion path and CDs are shortly segmented at intervals in their longitudinal direction when the S_e is lower than the threshold value, as shown in Figure 3 [19,20,25]. All of the irradiation energies used in our works are enough for the projectile ranges to exceed the thickness of the samples: the incident ions pass through the superconducting layer completely.

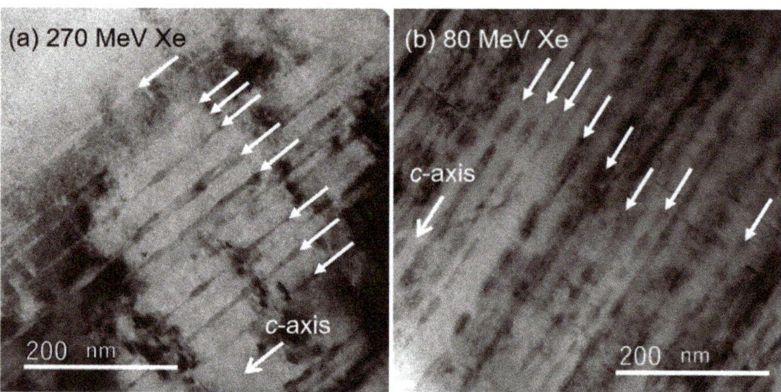

Figure 3. Cross-sectional TEM images of GdBCO coated conductors irradiated with (**a**) 270 MeV and (**b**) 80 MeV Xe ions, respectively. The irradiation dose is 1.94×10^{11} ions/cm^2. The arrows indicate several ion tracks. Reprinted with permission from [25], copyright 2015 by IEEE.

The direction of CDs was adjusted by controlling the incident ion beam direction tilted off the *c*-axis by θ_i, which was always directed perpendicular to the bridge direction of the sample (see Figure 4). When the irradiation directions are dispersed, the fluence in each irradiation direction is calculated by dividing the total fluence by the number of the irradiation directions. The fluence of the irradiation is often represented as a matching field B_φ: B_φ is the magnetic field where the density of flux lines is equal to that of CDs, e.g., the fluence of 4.84×10^{10} ions/cm^2 corresponds to $B_\varphi = 1$ T.

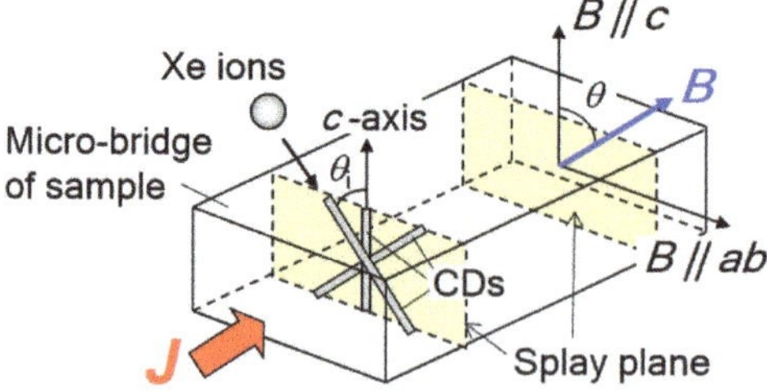

Figure 4. Sketch of the experimental arrangement in this work.

It should be noted that the introduction of irradiation defects causes a lattice distortion of the host matrix, which affects the superconducting properties such as critical temperature (T_c). The strain induces the oxygen vacancies [26], resulting in the reduction of T_c: the value of T_c decreases when the fluence of the irradiation increases [24]. The strain also affects the J_c properties through the influence on T_c: J_c decreases largely, when the influence of the strain increases excessively. Therefore, the irradiation fluences were adjusted to avoid heavy damage to the crystallinity in our works.

The cross sections of the irradiated samples were observed by conventional transmission electron microscopy (TEM) with a JEM-2000 EX instrument (JEOL, Tokyo, Japan) operating at 200 kV. The thin TEM specimens were prepared by a focused ion beam method using an Quanta 3D system (FEI, Hillsboro, Oregon, USA). The J_c properties were measured through the transport properties by using a four-probe method. The J_c was defined by a criterion of electric field, 1 μV/cm. The transport current was always perpendicular to the magnetic field and the c-axis (maximum Lorentz force configuration). The magnetic field angular dependences of J_c were evaluated as a function of the angle θ between the magnetic field and the c-axis of the samples (see Figure 4).

3. Results and Discussion

3.1. Modification of J_c Around B ‖ c by Controlling Heavy-Ion Irradiation Angles

Heavy-ion irradiation can introduce CDs in any direction in a controlled manner, so we can install CDs at the magnetic field angles where the J_c shows a minimum, one by one: the material processing with heavy-ions is one of effective ways to modify the J_c anisotropy in high-T_c superconductors, which enables us to push up overall J_c, as shown in Figure 5.

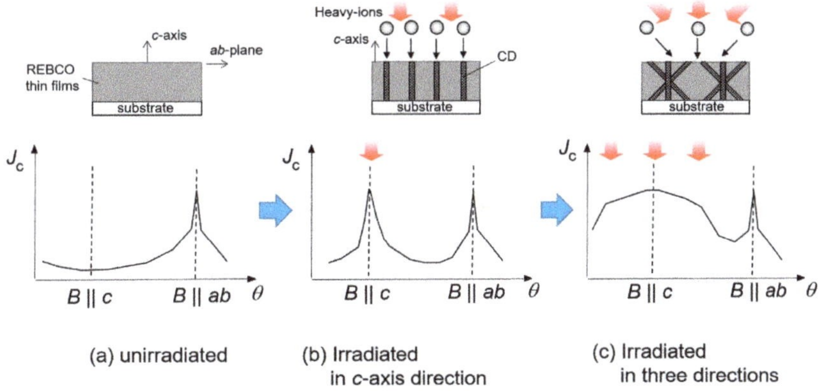

Figure 5. Schematic image of modification of the J_c anisotropy by controlling the irradiation directions ((**a**) typical J_c anisotropy of unirradiated high-T_c superconductors, (**b**) modified J_c anisotropy with CDs along the c-axis, (**c**) modified J_c anisotropy with direction-dispersed CDs).

We first examined the influence of bimodal angular distribution of CDs consisting of CDs crossing at $\pm\theta_i$ relative to the c-axis on the J_c properties in a wide magnetic field angular range [27,28]. Figure 6 shows the magnetic-field angular dependence of J_c normalized by the self-magnetic-field critical current density J_{c0} for YBCO thin films with the crossed CDs, which were installed by 200 MeV Xe ion irradiation with $B_\varphi = 2$ T (c10-2: $\theta_i = \pm10°$, c25-2: $\theta_i = \pm25°$, c45-2: $\theta_i = \pm45°$, p06-2: parallel CD configuration of $\theta_i = 6°$, and Pure: unirradiated samples). The magnetic field was rotated in the splay plane where the two parallel CD families are crossing each other, as shown in Figure 4. All the irradiated samples show an additional peak of the normalized J_c around $B ‖ c$ ($\theta = 0°$) for lower magnetic fields: the values of the normalized J_c are enhanced around $B ‖ c$ compared to the unirradiated one. This indicates that CDs with any crossing angle work as effective PCs, pushing up the J_c around $B ‖ c$. The influence of the crossing angle of CDs is evident in the shape of the additional peak around $B ‖ c$: the width of the normalized J_c peak becomes broader when the crossing angle is larger. Therefore, the bimodal angular distribution of CDs can expand the magnetic field angular range where the normalized J_c increases, by controlling the crossing angle.

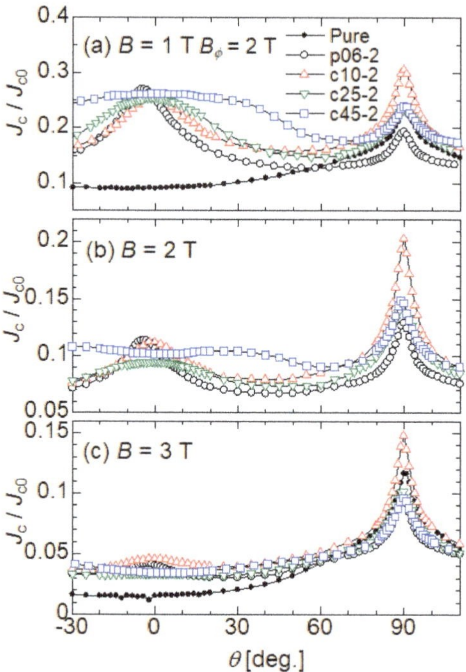

Figure 6. Magnetic field angular dependence of J_c normalized by the self-magnetic-field critical current density J_{c0} for YBCO thin films with the crossed CDs (c10-2: $\theta_i = \pm10°$, c25-2: $\theta_i = \pm25°$, c45-2: $\theta_i = \pm45°$, p06-2: parallel CD configuration of $\theta_i = 6°$, and Pure: unirradiated samples). Reprinted with permission from [28], copyright 2016 by IOP.

It is noteworthy that the crossover phenomenon from the broad-plateau-like behavior to the double peak emerges on the normalized J_c around $B \parallel c$ for c45-2 when the magnetic field increases across the matching field of $B_\varphi = 2$ T: the normalized J_c more rapidly reduces at $B \parallel c$ with increasing magnetic field, which results in a dip structure at $B \parallel c$ for c45-2 at 2 T, as shown in Figure 6. In general, the J_c peak in the magnetic field angular dependence of J_c is a sign of long-axis correlated flux pinning of CDs. Their-long-axis correlated flux pinning is maintained up to higher magnetic fields [29,30]. For the crossing angle of $\theta_i = \pm45°$, by contrast, the influence of the long-axis correlated flux pinning is weakened at $B \parallel c$, since the directions of CDs are far from the c-axis direction. Thus, the dip behavior at $B \parallel c$ is a sign of disappearance of their-long axis correlated flux pinning at $B \parallel c$.

The effective magnetic field angular region for flux pinning of CDs is described by a trapping angle φ_t, at which flux lines begin to be partially trapped by CDs [4]. The general formula of φ_t is expressed as:

$$\varphi_t = \sqrt{2\,\varepsilon_p/\varepsilon_l} \tag{1}$$

where ε_p is the pinning energy of CDs and ε_l is the line tension of flux lines. The line tension energy of flux lines in anisotropic superconductors is given by the following equation:

$$\varepsilon_l(\Theta) \propto \varepsilon_0/\gamma^2\varepsilon(\Theta)^3 \tag{2}$$

where Θ is the angle between the magnetic field and the ab-plane, ε_0 is a basic energy scale, γ is the mass anisotropy, and $\varepsilon(\Theta) = (\sin^2\Theta + \gamma^{-2}\cos^2\Theta)^{1/2}$ [4]. The trapping angle φ_t is experimentally estimated as the difference in the angle between the peak value and the minimum one on the magnetic field angular dependence of J_c [31]. For p06-2, the value of φ_t is ~55° at $B < B_\varphi$, which is estimated from Figure 6a. Using this value of φ_t as the trapping angle of CDs parallel to the *c*-axis approximately and $\gamma = 5$ together with equations (1) and (2), the value of φ_t for CDs tilted at $\theta_i = 45°$ is about 37°. Therefore, CDs tilted at $\theta_i = 45°$ hardly contribute to trapping flux lines at $B \mid\mid c$: CDs tilted at $\theta_i = 45°$ does not work as their-long-axis correlated PCs for $B \mid\mid c$.

The bimodal angular distribution of CDs for $\theta_i \pm 45°$ gives rise to the drop in J_c at the mid-direction of the crossing angle. Secondly, we investigated the flux pinning properties for a trimodal angular distribution of CDs consisting of CDs crossing at $\theta_i = 0°$ and $\pm 45°$ (referred to as the "standard" trimodal-configuration), in order to obtain high J_c with no drop over a wide magnetic field angular region [32]. In addition, another geometry for the trimodal configuration was prepared, where a splay plane defined by the three irradiation angles is parallel to the transport current direction (referred to as "another" trimodal-configuration), as shown in Figure 7: the two trimodal configurations enable us to elucidate the influence of the splay plane direction on the J_c properties directly. Figure 8 shows the magnetic field angular dependence of normalized J_c by J_{c0} ($= j_c$) at several magnetic fields from 1 T up to 5 T for YBCO thin films with the trimodal angular configurations of CDs. A large enhancement of j_c centered at $B \mid\mid c$ can be seen for all the irradiated samples. In particular, both the trimodal angular configurations show a much broader peak with larger j_c than that of the parallel CD configuration. It should be noted that there is no drop of j_c at $B \mid\mid c$ for both the trimodal configurations. This result indicates that the three parallel CD families tilted at $\theta_i = 0°$ and $\pm 45°$ effectively work as strong PCs in each irradiation direction: flux pinning at $B \mid\mid c$ where CDs tilted at $\theta_i = \pm 45°$ slightly contribute to trapping flux lines, is reinforced by CDs along the *c*-axis.

"Another" trimodal-configuration

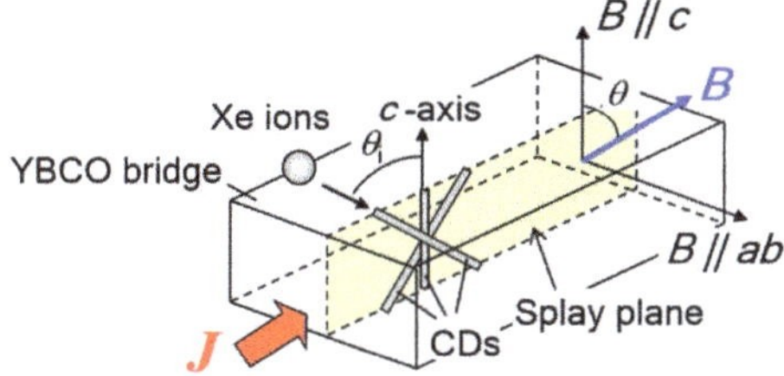

Figure 7. Sketch of CDs dispersed in geometry of "another" trimodal-configuration, where a splay plane defined by the three irradiation angles is parallel to the transport current direction. Reprinted with permission from [32], copyright 2016 by IOP.

Interestingly, the behaviour of j_c around $B \mid\mid c$ strongly depends on the direction of the splay plane for the trimodal configuration of CDs: the j_c of another trimodal-configuration shows a peak at $B \mid\mid c$, whereas standard one exhibits not so much a peak as a plateau-shaped curve. In addition, the height of j_c peak for another tirmodal-configuration is higher than the value of j_c at $B \mid\mid c$ for standard one. For standard trimodal-configuration, sliding motion of flux lines occurs along the tilted CDs at $B \mid\mid c$ because of the splay plane parallel to the Lorentz force, resulting in the reduction of the pinning efficiency [33]. The crossed CDs for another trimodal-configuration, by contrast, suppress the motion of flux lines efficiently, since flux lines move across the crossed CDs by the Lorentz force. Thus, the splay plane parallel to the transport current direction provides stronger flux pinning at $B \mid\mid c$, like planar PCs. Furthermore, the j_c of another trimodal-configuration is the highest even when the magnetic field is tilted from the c-axis. This is probably due to the entanglement of flux lines induced in a mesh of the splay plane tilted from the magnetic field, where the motion of flux lines is suppressed [34]. These results suggest that the direction of the splay plane is one of key factors for flux pinning of direction-dispersed CDs, as well as the degree of the direction-dispersion [32,35].

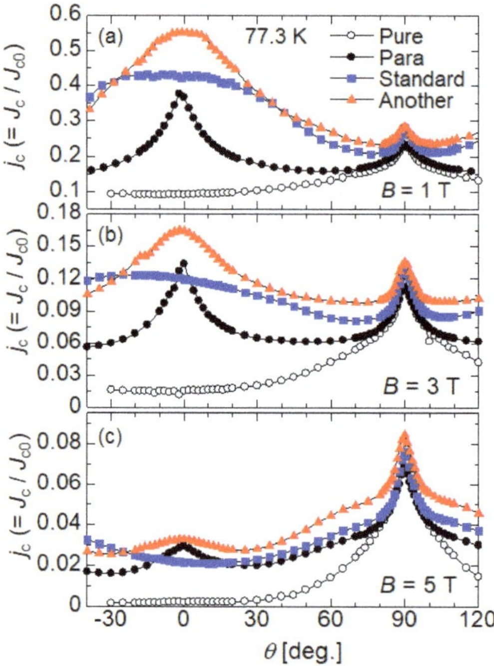

Figure 8. Magnetic-field angular dependence of J_c normalized by the self-magnetic-field critical current density J_{c0} for YBCO thin films with various CD configurations (Pure: unirradiated samples, Para: parallel CD configuration of $\theta_i = 0°$, Standard: standard trimodal-configuration, and Another: another trimodal-configuration). Reprinted with permission from [32], copyright 2016 by IOP.

3.2. Modification of J_c Anisotropy by Controlling Number of Heavy-Ion Irradiation Directions

We further increased the number of the directions of CDs by controlling the irradiation directions (see Figure 9), in order to spread the strong pinning effect of CDs over a wider magnetic field angular range. Figure 10 shows the magnetic field angular dependence of J_c and n-values for YBCO thin films with direction-dispersed CDs, where the number of directions of CDs was applied from one to five every 30 degrees [36]. The n-value is estimated from a linear fit to empirical formula of electric field (E) versus current density (J), $E \sim J^n$ in the range of 1 to 10 μV/cm. The n-value is equivalent to $U_0 / k_B T$ (U_0: pinning potential energy) [37,38], representing thermal activation for flux motion. When the number of CD directions is increased, the angular region with high J_c is more expanded. Note that the height of the J_c peak at $B \parallel c$ declines, since the density of CDs decreases with increasing number of irradiation directions in this work. Thus, the large direction-dispersion of CDs is effective for the enhancement of J_c over a wider magnetic field angular region centered at $B \parallel c$.

The J_c around $B \parallel ab$, on the other hand, does not seems to be affected by flux pinning of the direction-dispersed CDs: both J_c and n-value at $\theta = 90°$ rather tend to decrease with increasing the degree of the direction-dispersion of CDs. One of the reasons for the reduction of J_c at $B \parallel ab$ by the introduction of CDs is the damage on the superconductivity and/or the ab-plane-correlated PCs [16]. It should be noted that the sample of Quintmodal contains CDs crossing at $\pm30°$ relative to the ab-plane (i.e., $\theta_i = \pm60°$); nevertheless, the crossed CDs do not seem to contribute to the pinning interaction around $B \parallel ab$. Figure 11 represents the magnetic field angular dependence of J_c for YBCO thin films including bimodal angular configurations of CDs with $\theta_i = \pm30°$ and $\pm60°$ relative to the c-axis, respectively [39]. The crossing angle of $\pm30°$ relative to the c-axis induces the enhancement of J_c over a wide angular region centered at $B \parallel c$. The crossing of CDs at $\pm30°$ relative to the *ab* plane, i.e., $\theta_i = \pm60°$, by contrast, is ineffective in pushing up the J_c at the mid-direction of the crossing angle, i.e., at $B \parallel ab$, whereas the peak of J_c emerges at $\theta = \pm60°$. These results indicate that the flux pinning around $B \parallel ab$ is hardly affected even by CDs tilted toward the ab-plane, which significantly differs from the flux pinning of CDs at $B \parallel c$. Thus, the flux pinning of CDs around $B \parallel ab$ is a new issue for the complete reduction of the J_c anisotropy.

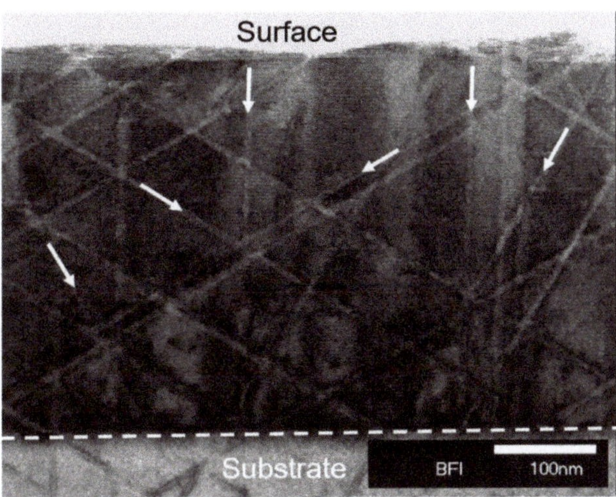

Figure 9. Bright field TEM image showing CDs tilted at $\theta_i = 0°$, $\pm30°$ and $\pm60°$ relative to the *c*-axis of the YBCO film, which are installed by 200 MeV Xe ion irradiation. The arrows indicate several ion tracks. Reprinted with permission from [36], copyright 2018 by IOP.

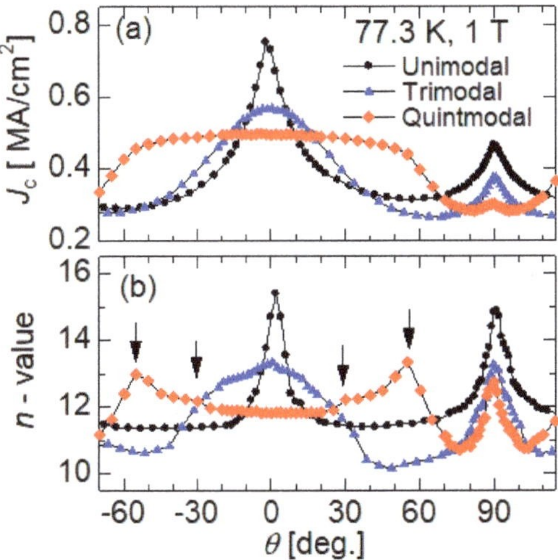

Figure 10. Magnetic-field angular dependence of J_c (upper, (**a**)) and *n*-value (lower, (**b**)) for YBCO thin films with various CD configurations (Unimodal: parallel CD configuration with $\theta_i = 0°$, Trimodal: trimodal-configuration with $\theta_i = 0°$ and $\pm30°$, and Quintmodal: quintmodal-configuration with $\theta_i = 0°$, $\pm30°$ and $\pm60°$). The arrows indicate the peaks or the shoulder on the $n(\theta)$ curve for Quintmodal. Reprinted with permission from [37], copyright 2013 by IEEE.

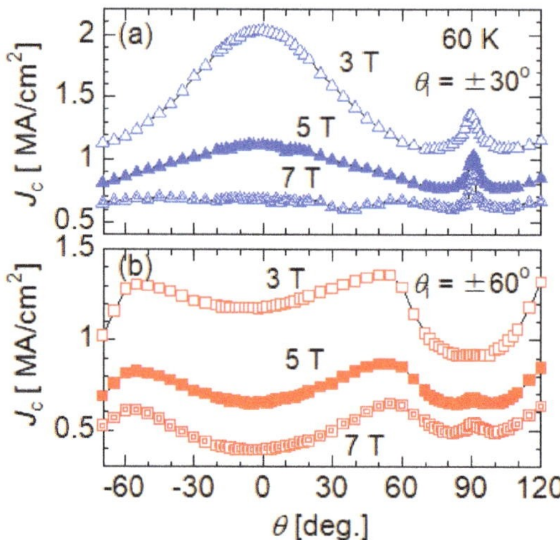

Figure 11. Magnetic-field angular dependence of J_c at temperature of 60 K and magnetic field of 3 T to 7 T for YBCO thin films with CDs crossing at (**a**) $\theta_i = \pm 30°$ and (**b**) $\pm 60°$ relative to the c-axis, respectively. Reprinted with permission from [36], copyright 2018 by IOP.

3.3. Modification of J_c Around $B \parallel ab$ by Controlling Heavy-Ion Irradiation Directions

A significant enhancement of J_c at $B \parallel c$ has been caused by the introduction of artificially PCs, which is much higher than J_c at $B \parallel ab$ now [40]. Thus, the improvement of J_c at $B \parallel ab$ has been required at the next step, in order to increase overall J_c. The influence of CDs on the flux pinning at $B \parallel ab$, however, has not been well studied so far, because the J_c at $B \parallel ab$ is the highest innately due to the electronic mass anisotropy in high-T_c superconductors [4] and the introduction of CDs is generally difficult in the direction close to the ab-plane. In contrast, heavy-ion irradiation can be an effective tool even for exploring the flux pinning effect of CDs at $B \parallel ab$, because CDs can be installed in any direction by adjusting the irradiation direction.

GdBCO-coated conductors were irradiated with 270 MeV Xe-ions, where the irradiation angle Θ_i relative to the ab-plane was controlled in the range from $\pm 5°$ to $\pm 15°$ relative to the ab-plane in order to install crossed CDs around the ab-plane [41]. The cross-sectional TEM image of the GdBCO-coated conductor irradiated at $\Theta_i = \pm 10°$, Figure 12, shows the formation of continuous CDs along the irradiation directions. At the bottom part of the GdBCO layer, by contrast, some CDs become thinner and indicate angular dispersion in the irradiation directions. This is due to smaller value of S_e than the threshold value of 20 keV/nm for the formation of continuous CDs [17,42], because the S_e changes from 29.1 to 7.40 keV/nm through the GdBCO layer for the oblique irradiation at $\Theta_i = 10°$.

Figure 13 shows the magnetic field angular dependence of J_c for the irradiated samples with $\Theta_i = \pm 5°$, $\pm 10°$, and $\pm 15°$, respectively. The CD crossing-angles of $\Theta_i \leq \pm 15°$ significantly affect the magnetic field angular variation of J_c around $B \parallel ab$. The introduction of crossed CD at $\Theta_i = \pm 15°$ provides a triple peak of J_c centered at $B \parallel ab$, where a large J_c peak exists at $B \parallel ab$ and the other two J_c peaks emerge around $\theta = 75°$ and $105°$, independently each other. This behavior is in contrast to the case of CDs crossing at $\theta_i \leq \pm 30°$ relative to the c-axis, which shows a single peak of J_c centered at $B \parallel c$, as represented in Figures 6 and 11. As the crossing-angle of Θ_i decreases, the two divided peaks of J_c at $\pm \Theta_i$ overlap with the central J_c peak at $B \parallel ab$: a single peak centered at $\theta = 90°$ occurs for the crossing angles of $\Theta_i \leq \pm 10°$. In particular, the crossing angle of $\Theta_i = \pm 5°$ provides the large and sharp J_c peak at $B \parallel ab$, showing the highest value of all the samples at $B \parallel ab$. To our knowledge, it is the first confirmation that CDs contribute to the improvement of J_c at $B \parallel ab$.

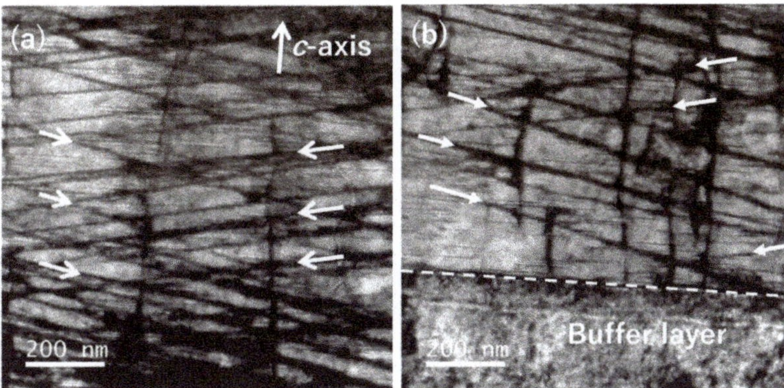

Figure 12. Cross-sectional TEM images of a GdBCO coated conductor irradiated with 270 MeV Xe ions at $\Theta_i = \pm10°$ relative to the *ab*-plane (**a**) near the surface and (**b**) at the bottom part of GdBCO layer, respectively. The arrows show several ion tracks. Reprinted with permission from [41], copyright 2017 by IEEE.

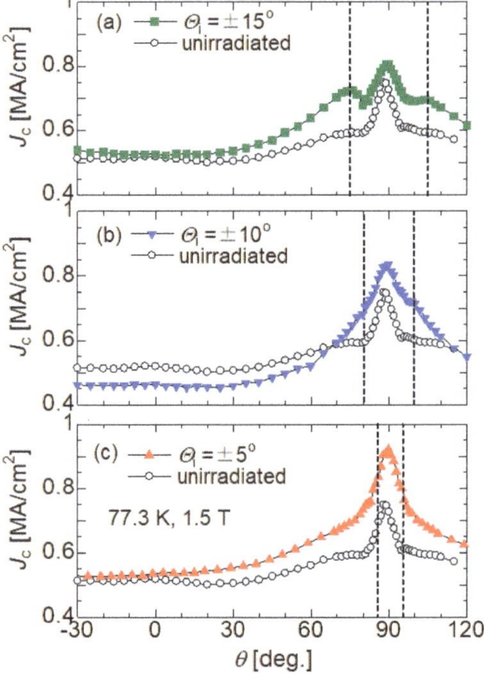

Figure 13. Magnetic-field angular dependence of J_c at 77.3 K, 1.5 T in GdBCO coated conductors with crossed CDs at $\pm\Theta_i$ relative to the *ab*-plane ((**a**) $\Theta_i = \pm15°$, (**b**) $\Theta_i = \pm10°$, and (**c**) $\Theta_i = \pm5°$). The broken lines are drawn at the positions of the irradiation angles. Reprinted with permission from [41], copyright 2017 by IEEE.

These behaviors are closely associated with the elastic properties of flux lines around $B \parallel ab$. The line tension energy of flux lines becomes very strong at $B \parallel ab$, where the core of flux lines shows the elliptical nature in anisotropic superconductors. The strong line tension of flux lines significantly affects the trapping angle φ_t of CDs tilted toward the *ab*-plane. In general, the value of φ_t for CDs along the *c*-axis (i.e., at $\Theta_i = 90°$) is about 65° in GdBCO coated conductors [25]. The φ_t for CDs tilted at small angle of Θ_i, on the other hand, can be evaluated by substituting the value of $\varphi_t \sim 65°$ for $\Theta_i = 90°$ and $\gamma = 5$ together with equations (1) and (2): $\varphi_t \sim 6.6°$ for $\Theta_i = 5°$, $\varphi_t \sim 8.7°$ for $\Theta_i = 10°$, and $\varphi_t \sim 11.9°$ for $\Theta_i = 15°$. Thus, the trapping angles of CDs tilted toward the *ab*-plane becomes very small: flux lines are hardly trapped along CDs when the magnetic field direction is displaced from the direction of CDs even slightly. In particular, the trapping angle for CDs tilted at $\Theta_i \geq 10°$ is smaller than the CD tilt-angle Θ_i, suggesting that the tilted CDs hardly affect the flux pinning at $B \parallel ab$. Therefore, CDs tilted at $\Theta_i \geq 10°$ and the *ab*-plane correlated PCs provide flux pinning independently. The CDs tilted at $\Theta_i = 5°$, on the other hand, can fully contribute to the improvement of J_c at $B \parallel ab$, because the trapping angle exceeds the value of Θ_i.

3.4. Modification of J_c at $B \parallel ab$ by Heavy-Ion Irradiation along the a-Axis

An in-plane aligned *a*-axis-oriented YBCO film offers an excellent opportunity for further exploration into the influence of CDs on the flux pinning at $B \parallel ab$, since we can easily install CDs along the *ab*-plane with the ion-beam normal to the film [43]. We prepared the in-plane aligned *a*-axis-oriented YBCO film by a PLD technique with an ArF excimer laser, where a (100) $SrLaGaO_4$ substrate with Gd_2CuO_4 buffer layer was used to promote the in-plane orientation of YBCO thin film [44]. The film was patterned into the shape of a microbridge so as to make the bridge direction parallel to the *b*-axis, where transport current can be applied along the *ab*-plane (see Figure 14). Both the in-plane-aligned texture of the film and the experimental arrangement enable us to remove the extra effect such as the interlayer Josephson current and the channel flow of flux lines along the CuO_2 plane, providing deeper insights on the nature of flux pinning of CDs along the *ab*-plane.

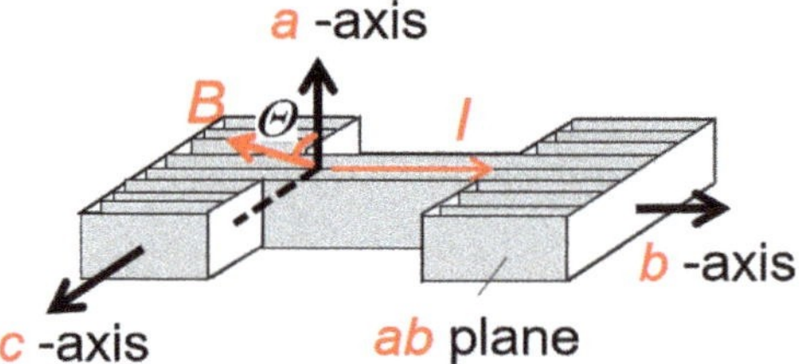

Figure 14. Sketch of the experimental arrangement using the in-plane aligned *a*-axis-oriented YBCO film. Reprinted with permission from [43], copyright 2019 by IEEE.

The in-plane aligned *a*-axis-oriented YBCO thin film showed good *a*-axis orientations without other orientations for the X-ray θ-2θ diffraction pattern, as shown in Figure 15. In addition, X-ray diffraction φ scanning using the (102) plane of the YBCO film before the irradiation indicated two-fold symmetry, since strong peaks stood out at around 90° and 270° in the inset of Figure 15. Therefore, the in-plane aligned *a*-axis-orientated microstructure can be confirmed on the film used in this work.

A cross-sectional TEM image of the in-plane aligned *a*-axis oriented YBCO film after the irradiation with 200 MeV Xe ions is shown in Figure 16a. The straight CDs along the *a*-axis are elongated through the thickness of the YBCO film. Figure 16b shows the plan-view TEM image of the *a*-axis oriented YBCO thin film after the irradiation. The CDs formed by the ion beam along the a-axis are roughly elliptical in shape, whereas CDs parallel to the *c*-axis are usually circular [17,45]. In general, the shape of CDs depends on the direction of the incident ions relative to the crystallographic axes in high-T_c superconductors, because the anisotropy of thermal diffusivity causes more severe irradiation damage for the creation of CDs along the *a*- and/or the *b*-axis [17].

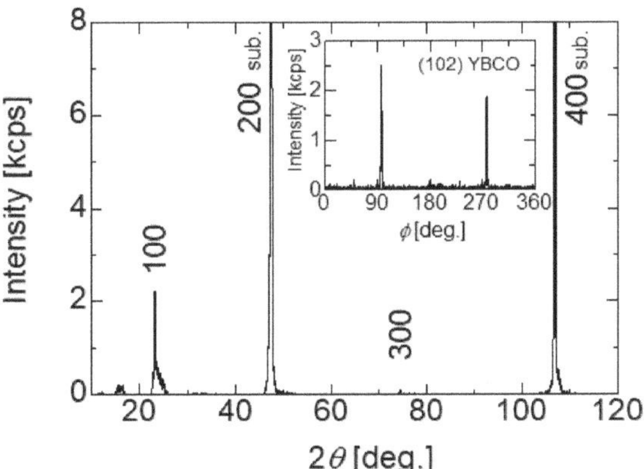

Figure 15. X-ray diffraction θ-2θscan of the in-plane aligned a-axis oriented YBCO thin film before the irradiation. Inset: X-ray φ scan using (102) plane of the YBCO thin film before the irradiation. Reprinted with permission from [43], copyright 2019 by IEEE.

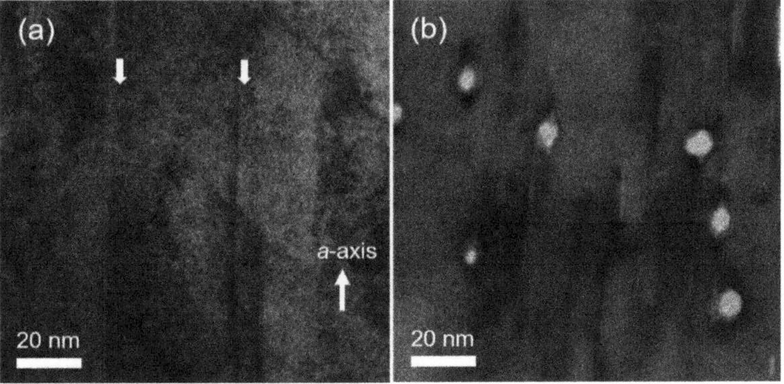

Figure 16. (**a**) Cross-sectional and (**b**) plan-view TEM images of the a-axis oriented YBCO thin film irradiated with 200 MeV Xe ions along the a-axis. The arrows indicate several ion tracks. Reprinted with permission from [43], copyright 2019 by IEEE.

Figure 17 represents the magnetic field dependence of J_c at 72 K for the a-axis oriented YBCO film before and after the irradiation. The J_c at $B \mid \mid c$ is reduced by the introduction of CDs along the a-axis, especially for high magnetic fields. The CDs along the a-axis hardly interact with flux lines at $B \mid \mid c$, since the CDs are perpendicular to the magnetic field direction. Moreover, CDs perpendicular to magnetic field direction create easy channel for flux lines to creep along the length of the CDs [46]. In addition to these deterioration effects, the irradiation damage to the host matrix causes the pronounced reduction of J_c at $B \mid \mid c$.

The introduction of CDs along the a-axis, on the other hand, hardly reduces the absolute value of J_c at $B \mid \mid a$, even though the J_c is affected by the local irradiation damage to the CuO_2 planes as well as the J_c at $B \mid \mid c$. It should be noted that the normalized J_c by J_{c0} increases after the irradiation, especially for high magnetic fields (see the inset of Figure 17). This behavior suggests that CDs contribute to the flux pinning at $B \mid \mid ab$. For low magnetic fields, by contrast, the pinning effect of CDs along the a-axis is hardly visible even on the normalized J_c. This is attributed to the presence of the naturally growth defects such as stacking faults in the film: Such pre-existing defects act as ab-plane correlated PCs both before and after the irradiation, which obscures the pinning effect of CDs, especially for low magnetic fields.

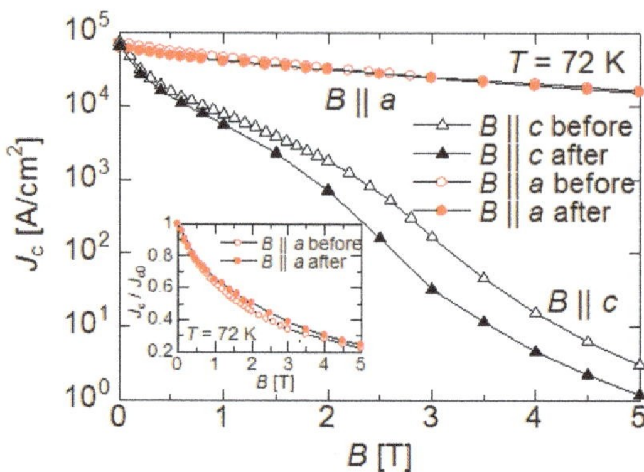

Figure 17. Magnetic field dependence of J_c at $B \parallel c$ and at $B \parallel a$ in the a-axis oriented YBCO thin film before and after the irradiation. Inset: J_c normalized by self-field critical current density J_{c0} as a function of magnetic field along the a-axis. Reprinted with permission from [43], copyright 2019 by IEEE.

3.5. Modification of the J_c Anisotropy by Controlling the Heavy-Ion Irradiation Energy

The modification of the J_c anisotropy in high-T_c superconductors is sensitive to direction-dispersions of CDs, as mentioned in the previous sections. Another way to modify the J_c properties by CDs is to tune the morphologies of CDs. Especially for the morphology of short segmented (i.e., discontinuous) CDs, the ends of the discontinuous CDs can provide a variety of additional pinning effects: the ends of the segmented CDs can trap flux lines in magnetic field tilted from their long axis [21,22] and the existence of gaps in the segmented CDs can suppress thermal motion of flux lines, as shown in Figure 18. Furthermore, the volume fraction of CDs relative to the superconducting area can be minimized for discontinuous CDs, since CDs are shortly segmented: the reduction of the volume fraction of the crystalline defects suppresses the degradation of the superconductivity associated with the introduction of PCs, leading to the improvement of the absolute value of J_c in a whole magnetic field angular region [19,20]. For iron-based superconductors, the morphology of CDs formed by heavy-ion irradiation tends to be discontinuous, which induces the remarkable improvement of J_c [47–49]. The morphology of CDs in high-T_c superconductors can be tuned by adjusting the irradiation energy for heavy-ion irradiation. In addition, the pinning effect of discontinuous CDs can be compared directly with that of continuous ones under same irradiation conditions except for the irradiation energy: the heavy-ion irradiations with different irradiation energies enable us to clarify the superiority of discontinuous CDs in the flux pinning effect over continuous CDs.

We first compared the flux pinning properties of discontinuous CDs with those of continuous ones when their long axis is parallel to the c-axis: GdBCO-coated conductors were irradiated with 80 MeV and 270 MeV Xe-ions along the c-axis, respectively [25]. For the irradiated sample with 270 MeV Xe ions, the straight and continuous CDs with the diameter of 4-11 nm penetrate the superconducting layer along the c-axis, as shown in Figure 3a. The value of S_e calculated using SRIM code varies from 3.0 to 2.8 keV/Å through the superconducting layer with the thickness of 2.2 μm for the 270 MeV Xe-ion irradiation, so that the continuous CDs are formed over the whole sample. The 80 MeV Xe-ion irradiation, by contrast, produces short segmented CDs in their longitudinal direction along the c-axis, as shown in Figure 3b: the length of the segmented CDs with the diameter of 5-10 nm varies from 15 to 50 nm along their length, while the gaps between the segmented CDs is also variable, ranging between 15 and35 nm. The formation of discontinuous CDs is attributed to the value of S_e changing from 2.0 to 1.4 keV/Å for the 80 MeV Xe ions into REBCO thin films [17,19,20].

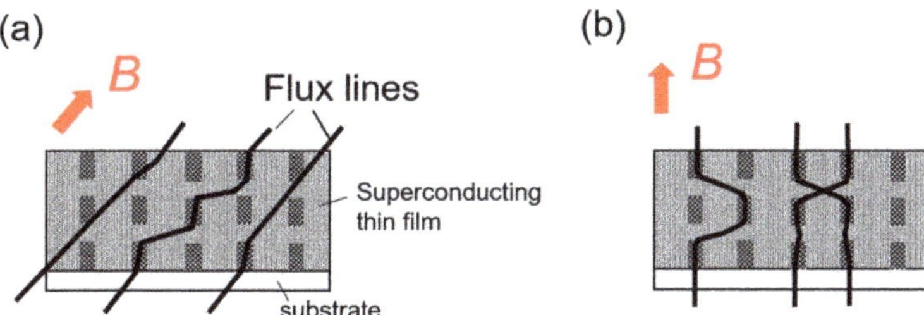

Figure 18. Schematic images of flux pinning peculiar to discontinuous CDs. (**a**) Ends of the discontinuous CDs work as PCs in magnetic field tilted off their long axis, which provide effective flux pinning over a wide magnetic field angular range. (**b**) Existence of gaps in the discontinuous CDs can suppress thermal motion of kinks of flux lines, which further improve J_c in comparison with continuous CDs.

Figure 19 shows the magnetic field angular dependences of J_c at 70 K and 84 K in GdBCO-coated conductors irradiated with 80 MeV and 270 MeV Xe ions, respectively. The 80 MeV irradiation causes higher J_c in all magnetic field directions compared to the 270 MeV irradiation, which becomes more pronounced at lower temperature of 70 K. The high J_c at $B \mid \mid c$ for the 80 MeV irradiation is attributed to the existence of gaps in the segmented CDs, which induce the suppression of thermal motion of flux lines (see Figure 18b). In addition, the ends of discontinuous CDs can trap flux lines in magnetic field tilted from their long axis, as shown in Figure 18a. These flux pinning effects of discontinuous CDs become more remarkable at lower temperature where a core size of flux line approaches the thin diameter of the discontinuous CDs. Moreover, discontinuous CDs more minimize the degradation of the superconductivity associated with the introduction of PCs compared with continuous CDs. Thus, the discontinuity of CDs can contribute to further enhancement of J_c.

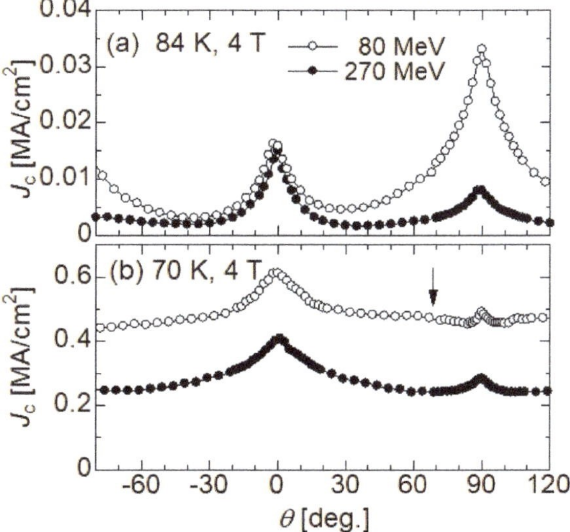

Figure 19. Magnetic field angular dependence of J_c at 4 T for the irradiated samples with 80 MeV and 270 MeV Xe ions ((**a**) 84 K and (**b**) 70K). Reprinted with permission from [25], copyright 2015 by IEEE.

The superior flux pinning effect of discontinuous CDs can be further modified by tuning the direction-dispersion. We irradiated GdBCO coated conductors with 80 MeV Xe ions, where the incident ion beams were tilted from the c-axis by θ_i to introduce various kinds of direction-dispersed CDs: a parallel configuration composed of CDs parallel to the c-axis, bimodal angular configuration composed of CDs tilted at $\theta_i = \pm45°$ relative to the c-axis, and trimodal angular configuration composed of CDs tilted at $\theta_i = 0°$ and $\pm45°$ [50].

Figure 20a shows a cross-sectional TEM image of the GdBaCuO-coated conductor irradiated with 80 MeV Xe-ions at $\theta_i = 0°$ and $\pm45°$. The morphologies of CDs are schematically emphasized in Figure 20b. Interestingly, the 80 Me V Xe-ion beams create CDs with different morphologies depending on the irradiation angles of θ_i: thick and elongated CDs are formed along the ion path at $\theta_i = 45°$, whereas the 80 MeV ions at $\theta_i = 0°$ creates short segmented CDs along their length. In general, the morphology of CDs is determined by the value of S_e, which is the energy transferred from the incident ions for the electronic excitation. A thermal spike model [51,52], which is one of models to interpret the formation of irradiation defects through the electronic excitation, can describes the direction-dependent morphologies of CDs in high-T_c superconductors by considering the anisotropy of thermal diffusivity [17,50]. According to the thermal spike model, the energy of the electronic excitation is converted into the thermal energy of lattice, which is the source for the formation of irradiation defects. In high-T_c superconductors, the thermal diffusivity along the c-axis is smaller than that along other crystallographic axes, which results in the suppression of a temperature spread in the planes containing the c-axis. Thus, the incident ion beam tilted from the c-axis causes more severe structural damage, resulting in the formation of elongated CDs with a thicker diameter.

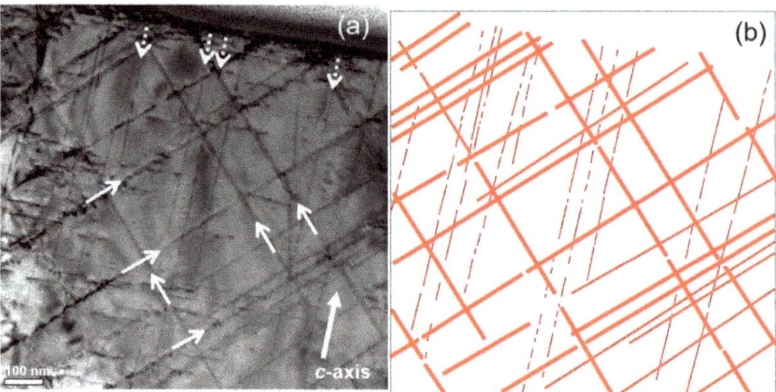

Figure 20. (a) Cross-sectional TEM image of the GdBCO coated conductor irradiated with 80 MeV Xe ions at $\theta_i = 0°$ and $\pm45°$. Several continuous CDs are indicated by the solid arrows and discontinuous CDs along the c-axis are indicated by the dotted arrows. (b) Schematic image emphasizing the morphologies of CDs in the cross-sectional TEM image. Reprinted with permission from [50], copyright 2020 by the Japan Society of Applied Physics.

Figure 21 shows the magnetic field angular dependence of J_c for GdBCO coated conductors irradiated with 80 MeV and 270 MeV Xe ions, where the irradiation angles are $\theta_i = 0°$ for the parallel CD configurations and $\theta_i = 0°$, $\pm45°$ for the trimodal angular configuration, respectively. The trimodal angular distribution shows higher J_c values than the parallel CD configuration at 70 K under the same irradiation energy. This suggests that the direction-dispersion of CDs is more effective to enhance the flux pinning over a wide magnetic field angular region, as mentioned in Section 3.2. It is noteworthy that the trimodal angular configuration produced by 80 MeV Xe ions shows the highest J_c in all the CD configurations over the whole magnetic field angular region at 70 K. The 80 MeV trimodal configuration consists of short segmented CDs along the c-axis and elongated CDs crossing at $\theta_i = \pm45°$, as shown in Figure 20a. For $B \parallel c$, the motion of double kinks of flux lines peculiar to one-dimensional PCs is suppressed by the gaps between the segmented CDs, as shown in Figure 18b. Furthermore, continuous CDs crossing at $\theta_i = \pm45°$ assist in trapping the unpinned segments of flux lines, as shown in Figure 22a. The pinning of kinks of flux lines is effective for further improvement of J_c [53,54]. Therefore, the combination of discontinuous CDs and continuous ones crossing at $\theta_i = \pm45°$ provides the enhancement of J_c at $B \parallel c$.

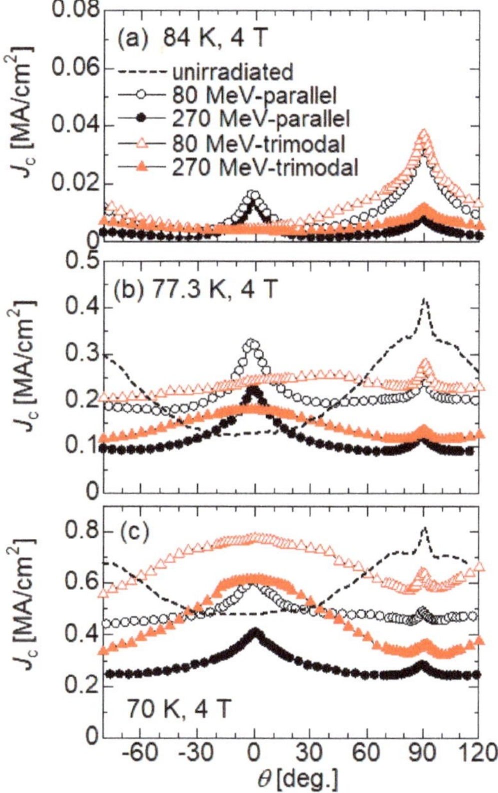

Figure 21. Magnetic field angular dependence of J_c at magnetic field of 4 T and temperatures of (**a**) 84 K, (**b**) 77.3 K, and (**c**) 70 K for GdBCO coated conductors irradiated with 80 MeV and 270 MeV Xe ions, where the irradiation angles are $\theta_i = 0°$ for parallel CD configurations and $\theta_i = 0°$, $\pm45°$ for trimodal angular configurations, respectively. The broken lines for (**b**) 77.3 K and (**c**) 70 K show the J_c properties of the unirradiated sample as reference data. Reprinted with permission from [50], copyright 2020 by the Japan Society of Applied Physics.

At the intermediate angles between $B \parallel c$ and $B \parallel ab$, on the other hand, the continuous CDs crossing at $\theta_i = \pm45°$ predominantly trap flux lines. In addition, the flux pinning of CDs crossing at $\theta_i = \pm45°$ is further enhanced through the suppression of the motion of kinks of flux lines by the gaps in discontinuous CDs, as shown in Figure 22b. Thus, the hybrid flux pinning by the two different kinds of PCs causes the large enhancement of J_c even at the intermediate magnetic field angles.

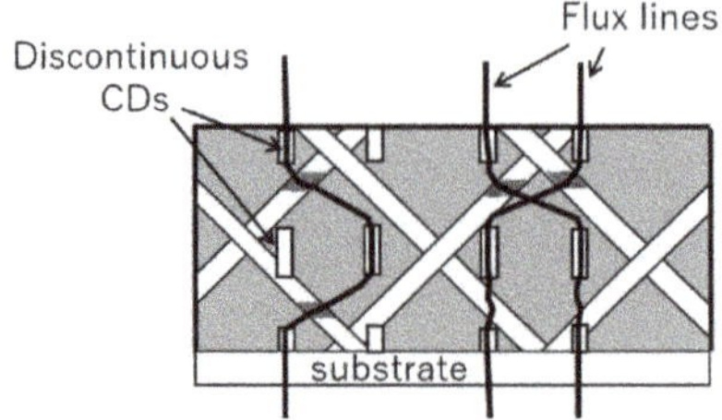
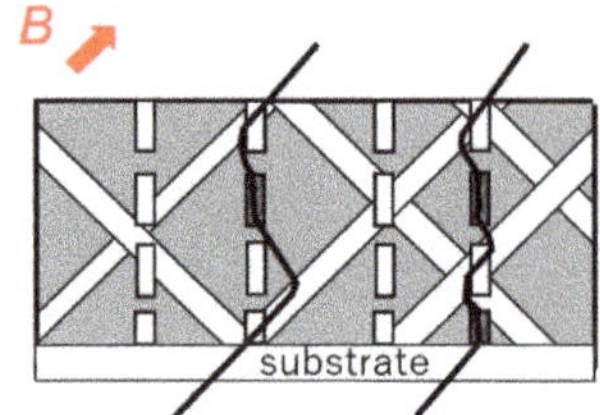

(a) Magnetic field parallel to the *c*-axis   (b) Magnetic field tilted off the *c*-axis

Figure 22. Schematic images of flux pinning for the 80 MeV trimodal angular configuration in a magnetic field B (**a**) along the *c*-axis and (**b**) in the intermediate angular region between $B \parallel c$ and $B \parallel ab$. The hatched regions in the inclined CDs and the discontinuous ones represent the interaction areas with kinks of flux lines. Reprinted with permission from [50], copyright 2020 by the Japan Society of Applied Physics.

There is a possibility that direction-dispersed CDs with "complete discontinuity" further provide a high and isotropic J_c in high-T_c superconductors. In fact, $BaHfO_3$ nanorods tend to grow discontinuously and to be widely dispersed in the directions, causing a significant improvement of J_c in a wide magnetic field angular range for REBCO thin films [55,56]. The irradiation using lighter ions with lower energy, which provides lower S_e for high-T_c superconductors (e.g., Kr-ion irradiation with 80 MeV, where $S_e = 16.0$ keV/nm), may produce discontinuous CDs even in directions tilted from the *c*-axis. However, there is a trade-off between the discontinuity of CDs and the thickness of CDs for the formation of CDs by heavy-ion irradiations: discontinuous CDs tend to be thin diameter [18,25], where the elementary pinning force of one segmented column with thin diameter becomes weak. Thus, the discontinuity of CDs does not always provide the strong pinning landscape for the heavy-ion irradiation process. The introduction of direction-dispersed, discontinuous, and thick CDs by the ion irradiation process can be the key to further making high J_c fairly isotropic in high-T_c superconductors.

4. Conclusions

We have systematically examined the modification of the anisotropy of J_c in REBCO thin films by using heavy-ion irradiations: the morphology and the configuration of the irradiation defects were controlled by the irradiation conditions such as the irradiation energy and the incident direction. The direction-dispersed CDs were designed in REBCO thin films to push up the J_c in the magnetic field angular region from $B \parallel c$ to $B \parallel ab$, by controlling the irradiation directions. When the directions of CDs were extensively dispersed around the *c*-axis, the J_c was enhanced over a wider magnetic field angular region centered at $B \parallel c$. The J_c at $B \parallel ab$, on the other hand, was hardly affected even by CDs tilted toward the *ab*-plane, which is attributed to the strong line tension energies of flux lines around $B \parallel ab$ in the anisotropic superconductors. We demonstrated the improvement of J_c at $B \parallel ab$ by the introduction of CDs, where the angle of CDs relative to the *ab*-plane were controlled down to $\Theta_i = 5°$. These results suggest that direction-dispersed CDs can provide the isotropic enhancement of J_c over all magnetic field angular region when the angles of CDs are matched with the anisotropic line tension energy of flux lines.

Another promising morphology of CDs, i.e., discontinuous CDs, which can be introduced by heavy-ion irradiation with relatively low energy, showed large potential for the enhancement of J_c over a wide magnetic field angular region. In particular, the combination of the discontinuity and the direction-dispersion lead to further enhancement of J_c: the gaps in discontinuous CDs provide the suppression of the motion of flux lines, while the direction-dispersion of CDs produces the strong flux pinning over a wide magnetic field angular region.

The heavy-ion irradiation to high-T_c superconductors can provide the flux pinning structure with higher and more isotropic J_c, by further tuning the irradiation process: the systematic studies using the ion irradiation process may lead to the approach to the theoretical limit of J_c, i.e., pair-breaking critical current density.

Author Contributions: Conceptualization, T.S. All authors have read and agreed to the published version of the manuscript.

Funding: This work was supported by KAKENHI (25420292, 16K06269 and 19K04474) from the Japan Society for the Promotion of Society.

Data Availability Statement: The data presented in this study are available on request from the author.

Acknowledgments: The author thanks N. Ishikawa, M. Mukaida, A. Ichinose, K. Yasuda, T. Fujiyoshi, S. Semboshi, T. Ozaki, and H. Sakane for their constant support during the researches. Experimental works were partly performed under the Common-Use Facility Program of JAEA.

Conflicts of Interest: The authors declare no conflict of interest.

References

1. Obradors, X.; Puig, T. Coated conductors for power applications: Materials challenges. *Supercond. Sci. Technol.* **2014**, *27*, 044003. [CrossRef]
2. Shiohara, Y.; Yoshizumi, M.; Takagi, Y.; Izumi, T. Future prospects of high T_c superconductors-coated conductors and their applications. *Physica C* **2013**, *484*, 1–5. [CrossRef]
3. Matsumoto, K.; Mele, P. Artificial pinning center technology to enhance vortex pinning in YBCO coated conductors. *Supercond. Sci. Technol.* **2010**, *23*, 014001. [CrossRef]
4. Blatter, G.; Feigelman, M.V.; Geshkenbein, V.B.; Larkin, A.I.; Vinokur, V.M. Vortices in high-temperature superconductors. *Rev. Mod. Phys.* **1994**, *66*, 1125–1388. [CrossRef]
5. Awaji, S.; Namba, M.; Watanabe, K.; Kai, H.; Mukaida, M.; Okayasu, S. Flux pinning properties of correlated pinning at low temperatures in ErBCO films with inclined columnar defects. *J. Appl. Phys.* **2012**, *111*, 013914. [CrossRef]
6. Foltyn, S.R.; Civale, L.; MacManus-Driscoll, J.L.; Jia, Q.X.; Maiorov, B.; Wang, H.; Maley, M. Materials science challenges for high-temperature superconducting wire. *Nat. Mater.* **2007**, *6*, 631–642. [CrossRef] [PubMed]
7. Feighan, J.P.F.; Kursumovic, A.; MacManus-Driscoll, J.L. Materials design for artificial pinning centers in superconductor PLD coated conductors. *Supercond. Sci. Technol.* **2017**, *30*, 123001. [CrossRef]
8. MacManus-Driscoll, J.L.; Foltyn, S.R.; Jia, Q.X.; Wang, H.; Serquis, A.; Civale, L.; Maiorov, B.; Hawley, M.E.; Maley, M.P.; Peterson, D.E. Strongly enhanced current densities in superconducting coated conductors of $YBa_2Cu_3O_{7-x}+BaZrO_3$. *Nat. Mater.* **2004**, *3*, 439–443. [CrossRef]
9. Gapud, A.A.; Kumar, D.; Viswanathan, S.K.; Cantoni, C.; Varela, M.; Abiade, J.; Pennycook, S.J.; Christen, D.K. Enhancement of flux pinning in $YBa_2Cu_3O_7$ thin films embedded with epitaxially grown Y_2O_3 nanostructures using a multi-layering process. *Supercond. Sci. Technol.* **2005**, *18*, 1502–1505. [CrossRef]
10. Mele, P.; Matsumoto, K.; Horide, T.; Ichinose, A.; Mukaida, M.; Yoshida, Y.; Horii, S.; Kita, R. Ultra-high flux pinning properties of $BaMO_3$-doped $YBa_2Cu_3O_{7-x}$ thin films (M = Zr, Sn). *Supercond. Sci. Technol.* **2008**, *21*, 032002. [CrossRef]
11. Miura, S.; Tsuchiya, Y.; Yoshida, Y.; Ichino, Y.; Awaji, S.; Matsumoto, K.; Ibi, A.; Izumi, T. Strong flux pinning at 4.2 K in $SmBa_2Cu_3O_y$ coated conductors with $BaHfO_3$ nanorods controlled by low growth temperature. *Supercond. Sci. Technol.* **2017**, *30*, 084009. [CrossRef]
12. Mele, P.; Guzman, R.; Gazquez, J.; Puig, T.; Obradors, X.; Saini, S.; Yoshida, Y.; Mukaida, M.; Ichinose, A.; Matsumoto, K.; et al. High pinning performance of $YBa_2Cu_3O_{7-x}$ films added with Y_2O_3 nanoparticulate defects. *Supercond. Sci. Technol.* **2015**, *28*, 024002. [CrossRef]
13. Civale, L.; Marwick, A.D.; Worthington, T.K.; Kirk, M.A.; Thompson, J.R.; Krusin-Elbaum, L.; Sun, Y.; Clem, J.R.; Holtzberg, F. Vortex confinement by columnar defects in $YBa_2Cu_3O_7$ crystals: Enhanced pinning at high fields and temperatures. *Phys. Rev. Lett.* **1991**, *67*, 648–651. [CrossRef] [PubMed]
14. Budhani, R.C.; Suenaga, M.; Liou, S.H. Giant suppression of flux-flow resistivity in heavy-ion irradiated $Tl_2Ba_2Ca_2Cu_3O_{10}$ films: Influence of linear defects on vortex transport. *Phys. Rev. Lett.* **1992**, *69*, 3816–3819. [CrossRef] [PubMed]
15. Matsushita, T.; Isobe, G.; Kimura, K.; Kiuchi, M.; Okayasu, S.; Prusseit, W. The effect of heavy ion irradiation on the critical current density in DyBCO coated conductors. *Supercond. Sci. Technol.* **2008**, *21*, 054014. [CrossRef]
16. Holzapfel, B.; Kreiselmeyer, G.; Kraus, M.; Saemann-Ischenko, G.; Bouffard, S.; Klaumunzer, S.; Schultz, L. Angle-resolved critical transport-current density of $YBa_2Cu_3O_{7-\delta}$ thin films and $YBa_2Cu_3O_{7-\delta}/PrBa_2Cu_3O_{7-\delta}$ superlattices containing columnar defects of various orientations. *Phys. Rev. B* **1993**, *48*, 600–603. [CrossRef]
17. Zhu, Y.; Cai, Z.X.; Budhani, R.C.; Suenaga, M.; Welch, D.O. Structures and effects of radiation damage in cuprate superconductors irradiated with several-hundred-MeV heavy ions. *Phys. Rev. B* **1993**, *48*, 6436–6450. [CrossRef]

18. Kuroda, N.; Ishikawa, N.; Chimi, Y.; Iwase, A.; Ikeda, H.; Yoshizaki, R.; Kambara, T. Effects of defect morphology on the properties of the vortex system in $Bi_2Sr_2CaCu_2O_{8+\delta}$ irradiated with heavy ions. *Phys. Rev. B* **2001**, *63*, 224502. [CrossRef]

19. Fuchs, G.; Nenkov, K.; Krabbes, G.; Weinstein, R.; Gandini, A.; Sawh, R.; Mayes, B.; Parks, D. Strongly enhanced irreversibility fields and Bose-glass behaviour in bulk YBCO with discontinuous columnar irradiation defects. *Supercond. Sci. Technol.* **2007**, *20*, S197–S204. [CrossRef]

20. Strickland, N.M.; Talantsev, E.F.; Long, N.J.; Xia, J.A.; Searle, S.D.; Kennedy, J.; Markwitz, A.; Rupich, M.W.; Li, X.; Sathyamurthy, S. Flux pinning by discontinuous columnar defects in 74 MeV Ag-irradiated $YBa_2Cu_3O_7$ coated conductors. *Physica C* **2009**, *469*, 2060–2067. [CrossRef]

21. Matsumoto, K.; Tanaka, I.; Horide, T.; Mele, P.; Yoshida, Y.; Awaji, S. Irreversibility fields and critical current densities in strongly pinned $YBa_2Cu_3O_{7-x}$ films with $BaSnO_3$ nanorods: The influence of segmented $BaSnO_3$ nanorods. *J. Appl. Phys.* **2014**, *116*, 163903. [CrossRef]

22. Horide, T.; Sakamoto, N.; Ichinose, A.; Otsubo, K.; Kitamura, T.; Matsumoto, K. Hybrid artificial pinning centers of elongated-nanorods and segmented-nanorods in $YBa_2Cu_3O_7$ films. *Supercond. Sci. Technol.* **2016**, *29*, 105010. [CrossRef]

23. Chikumoto, N.; Konczykowski, M.; Terai, T.; Murakami, M. Modification of flux pinning properties of RE-Ba-Cu-O superconductors by irradiation. *Supercond. Sci. Technol.* **2000**, *13*, 749–752. [CrossRef]

24. Nakashima, N.; Chikumoto, N.; Ibi, A.; Miyata, S.; Yamada, Y.; Kubo, T.; Suzuki, A.; Terai, T. Effect of ion-irradiation and annealing on superconductive property of PLD prepared YBCO tapes. *Physica C* **2007**, *463–465*, 665–668. [CrossRef]

25. Sueyoshi, T.; Kotaki, T.; Furuki, Y.; Uraguchi, Y.; Kai, T.; Fujiyoshi, T.; Shimada, Y.; Yasuda, K.; Ishikawa, N. Influence of discontinuous columnar defects on flux pinning properties in GdBCO coated conductors. *IEEE Trans. Appl. Supercond.* **2015**, *25*, 6603004. [CrossRef]

26. Horide, T.; Kametani, F.; Yoshioka, S.; Kitamura, T.; Matsumoto, K. Structural evolution induced by interfacial lattice mismatch in self-organized YBa2Cu3O7-δ nanocomposite film. *ACS Nano* **2017**, *11*, 1780–1788. [CrossRef]

27. Sueyoshi, T.; Furuki, Y.; Kai, T.; Fujiyoshi, T.; Ishikawa, N. Flux Pinning properties of splayed columnar defects ranging from B $\parallel$ c-axis to B $\parallel$ ab-plane in GdBCO coated conductors. *Physica C* **2014**, *504*, 53–56. [CrossRef]

28. Sueyoshi, T.; Sogo, T.; Nishimura, T.; Fujiyoshi, T.; Mitsugi, T.; Ikegami, T.; Awaji, S.; Watanabe, K.; Ichinose, A.; Ishikawa, N. Angular behaviour of critical current density in $YBa_2Cu_3O_y$ thin films with crossed columnar defects. *Supercond. Sci. Technol.* **2016**, *29*, 065023. [CrossRef]

29. Awaji, S.; Namba, M.; Watanabe, K.; Miura, M.; Yoshida, Y.; Ichino, Y.; Takai, Y.; Matsumoto, K. c-axis correlated pinning behavior near the irreversibility fields. *Appl. Phys. Lett.* **2007**, *90*, 122501. [CrossRef]

30. Awaji, S.; Namba, M.; Watanabe, K.; Nojima, T.; Okayasu, S.; Horide, T.; Mele, P.; Matsumoto, K.; Miura, M.; Ichino, Y. Vortex pinning phase diagram for various kinds of c-axis correlated disorders in RE123 films. *J. Phys. Conf. Ser.* **2008**, *97*, 012328. [CrossRef]

31. Civale, L.; Maiorov, B.; Serquis, A.; Willis, J.O.; Coulter, J.Y.; Wang, H.; Jia, Q.X.; Arendt, P.N.; MacManus-Driscoll, J.L.; Maley, M.P.; et al. Angular-dependent vortex pinning mechanisms in $YBa_2Cu_3O_7$ coated conductors and thin films. *Appl. Phys. Lett.* **2004**, *84*, 2121–2123. [CrossRef]

32. Sueyoshi, T.; Nishimura, T.; Fujiyoshi, T.; Mitsugi, F.; Ikegami, T.; Ishikawa, N. High flux pinning efficiency by columnar defects dispersed in three directions in YBCO thin films. *Supercond. Sci. Technol.* **2016**, *29*, 105006. [CrossRef]

33. Schuster, T.; Indenbom, M.V.; Kuhn, H.; Kronmuller, H.; Leghissa, M.; Kreiselmeyer, G. Observation of in-plane anisotropy of vortex pinning by inclined columnar defects. *Phys. Rev. B* **1994**, *50*, 9499–9502. [CrossRef] [PubMed]

34. Hwa, T.; Doussal, P.L.; Nelson, D.R.; Vinokur, V.M. Flux pinning and forced vortex entanglement by splayed columnar defects. *Phys. Rev. Lett.* **1993**, *71*, 3545–3548. [CrossRef] [PubMed]

35. Takahashi, A.; Pyon, S.; Kambara, T.; Yoshida, A.; Tamegai, T. Effects of Asymmetric splayed columnar defects on the anomalous peak effect in $Ba_{0.6}K_{0.4}Fe_2As_2$. *J. Phys. Soc. Jpn.* **2020**, *89*, 094705. [CrossRef]

36. Sueyoshi, T.; Furuki, Y.; Tanaka, E.; Fujiyoshi, T.; Mitsugi, F.; Ikegami, T.; Ishikawa, N. Angular dependence of critical current density in YBCO films with columnar defects crossing at widespread angles. *IEEE Trans. Appl. Supercond.* **2013**, *23*, 8002404. [CrossRef]

37. Zeldov, E.; Amer, N.M.; Koren, G.; Gupta, A.; McElfresh, M.W.; Gambino, R.J. Flux creep characteristics in high-temperature superconductors. *Appl. Phys. Lett.* **1990**, *56*, 680–682. [CrossRef]

38. Civale, L.; Maiorov, B.; MacManus-Driscoll, J.L.; Wang, H.; Holesinger, T.G.; Foltyn, S.R.; Serquis, A.; Arendt, P.N. Identification of intrinsic ab-plane pinning in $YBa_2Cu_3O_7$ thin films and coated conductors. *IEEE Trans. Appl. Supercond.* **2005**, *15*, 2808–2811. [CrossRef]

39. Sueyoshi, T.; Furuki, Y.; Fujiyoshi, T.; Mitsugi, F.; Ikegami, T.; Ichinose, A.; Ishikawa, N. Angular behavior of flux dynamics in YBCO films with crossed columnar defects around the ab-plane. *Supercond. Sci. Technol.* **2018**, *31*, 125002. [CrossRef]

40. Tsuruta, A.; Yoshida, Y.; Ichino, Y.; Ichinose, A.; Matsumoto, K.; Awaji, S. The influence of the geometric characteristics of nanorods on the flux pinning in high-performance $BaMO_3$-doped $SmBa_2Cu_3O_y$ films (M = Hf, Sn). *Supercond. Sci. Technol.* **2014**, *27*, 065001. [CrossRef]

41. Sueyoshi, T.; Iwanaga, Y.; Fujiyoshi, T.; Takai, Y.; Mukaida, M.; Kudo, M.; Yasuda, K.; Ishikawa, N. Angular behavior of J_c in GdBCO-coated conductors with crossed columnar defects around ab plane. *IEEE Trans. Appl. Supercond.* **2017**, *27*, 8001305. [CrossRef]

42. Civale, L.; Krusin-Elbaum, L.; Thompson, J.R.; Wheeler, R.; Marwick, A.D.; Kirk, M.A.; Sun, Y.R.; Holtzberg, F.; Feild, C. Reducing vortex motion in $YBa_2Cu_3O_7$ crystals with splay in columnar defects. *Phys. Rev. B* **1994**, *50*, 4102–4105. [CrossRef]
43. Sueyoshi, T.; Iwanaga, Y.; Fujiyoshi, T.; Takai, Y.; Muta, M.; Mukaida, M.; Ichinose, A.; Ishikawa, N. Flux pinning by columnar defects along *a*-axis in *a*-axis oriented YBCO thin films. *IEEE Trans. Appl. Supercond.* **2019**, *29*, 8003204. [CrossRef]
44. Mukaida, M. Growth of *a*-axis-oriented $YBa_2Cu_3O_x$ films on Gd_2CuO_4 buffer layers. *Jpn. J. Appl. Phys.* **1997**, *36*, L767–L770. [CrossRef]
45. Budhani, R.C.; Zhu, Y.; Suenaga, M. The effects of heavy-ion irradiation on superconductivity in $YBa_2Cu_3O_7$ epitaxial films. *IEEE Trans. Appl. Supercond.* **1993**, *3*, 1675–1678. [CrossRef]
46. Petrean, A.; Paulius, L.; Tobos, V.; Cronk, H.; Kwok, W.K. Vortex dynamics in $YBa_2Cu_3O_{7-\delta}$ with in-plane columnar defects introduced by irradiation. *Physica C* **2014**, *505*, 65–69. [CrossRef]
47. Nakajima, Y.; Tsuchiya, Y.; Taen, T.; Tamegai, T.; Okayasu, S.; Sasase, M. Enhancement of critical current density in Co-doped $BaFe_2As_2$ with columnar defects introduced by heavy-ion irradiation. *Phys. Rev. B* **2009**, *80*, 012510. [CrossRef]
48. Tamegai, T.; Taen, T.; Yagyuda, H.; Tsuchiya, Y.; Mohan, S.; Taniguchi, T.; Nakajima, Y.; Okayasu, S.; Sasase, M.; Kitamura, H.; et al. Effects of particle irradiation on vortex states in iron-based superconductors. *Supercond. Sci. Technol.* **2012**, *25*, 084008. [CrossRef]
49. Fang, L.; Jia, Y.; Chaparro, C.; Sheet, G.; Claus, H.; Kirk, M.A.; Koshelev, A.E.; Welp, U.; Crabtree, G.W.; Kwok, W.K.; et al. High, magnetic field independent critical currents in $(Ba, K)Fe_2As_2$ crystals. *Appl. Phys. Lett.* **2012**, *101*, 012601. [CrossRef]
50. Sueyoshi, T.; Kotaki, T.; Furuki, Y.; Fujiyoshi, T.; Semboshi, S.; Ozaki, T.; Sakane, H.; Kudo, M.; Yasuda, K.; Ishikawa, N. Strong flux pinning by columnar defects with directionally dependent morphologies in GdBCO-coated conductors irradiated with 80 MeV Xe ions. *Jpn. J. Appl. Phys.* **2020**, *59*, 023001. [CrossRef]
51. Toulemonde, M. Nanometric phase transformation of oxide materials under GeV energy heavy ion irradiation. *Nucl. Instrum. Methods Phys. Res. Sect. B Beam Interact. Mater. Atoms* **1999**, *156*, 1–11. [CrossRef]
52. Szenes, G. Ion-velocity-dependent track formation in yttrium iron garnet: A thermal-spike analysis. *Phys. Rev. B* **1995**, *52*, 6154–6157. [CrossRef]
53. Maiorov, B.; Baily, S.A.; Zhou, H.; Ugurlu, O.; Kennison, J.A.; Dowden, P.C.; Holesinger, T.G.; Foltyn, S.R.; Civale, L. Synergetic combination of different types of defect to optimize pinning landscape using $BaZrO_3$-doped $YBa_2Cu_3O_7$. *Nat. Mater.* **2009**, *8*, 398–404. [CrossRef] [PubMed]
54. Hua, J.; Welp, U.; Schlueter, J.; Kayani, A.; Xiao, Z.L.; Crabtree, G.W.; Kwok, W.K. Vortex pinning by compound defects in $YBa_2Cu_3O_{7-\delta}$. *Phys. Rev. B* **2010**, *82*, 024505. [CrossRef]
55. Tobita, H.; Notoh, T.; Higashikawa, K.; Inoue, M.; Kiss, T.; Kato, T.; Hirayama, T.; Yoshizumi, M.; Izumi, T.; Shiohara, Y. Fabrication of $BaHfO_3$ doped $Gd_1Ba_2Cu_3O_{7-\delta}$ coated conductors with the high I_c of 85 A/cm-w under 3 T at liquid nitrogen temperature (77 K). *Supercond. Sci. Technol.* **2012**, *25*, 062002. [CrossRef]
56. Kaneko, K.; Furuya, K.; Yamada, K.; Sadayama, S.; Barnard, J.S.; Midgley, P.A.; Kato, T.; Hirayama, T.; Kiuchi, M.; Matsushita, T.; et al. Three-dimensional analysis of $BaZrO_3$ pinning centers gives isotropic superconductivity in $GdBa_2Cu_3O_{7-\delta}$. *J. Appl. Phys.* **2010**, *108*, 063901. [CrossRef]

quantum beam science

Article

Effect of 1.5 MeV Proton Irradiation on Superconductivity in FeSe$_{0.5}$Te$_{0.5}$ Thin Films

Toshinori Ozaki [1,*], Takuya Kashihara [1], Itsuhiro Kakeya [2] and Ryoya Ishigami [3]

[1] Department of Nanotechnology for Sustainable Energy, Kwansei Gakuin University, 2-1 Gakuen, Sanda 669-1337, Japan; ete03144@kwansei.ac.jp
[2] Department of Electronic Science and Engineering, Kyoto University, Nishikyo-ku, Kyoto 615-8510, Japan; kakeya@kuee.kyoto-u.ac.jp
[3] The Wakasa Wan Energy Research Center, Nagatani, Tsuruga 914-0192, Japan; rishigami@werc.or.jp
* Correspondence: tozaki@kwansei.ac.jp

Abstract: Raising the critical current density J_c in magnetic fields is crucial to applications such as rotation machines, generators for wind turbines and magnet use in medical imaging machines. The increase in J_c has been achieved by introducing structural defects including precipitates and vacancies. Recently, a low-energy ion irradiation has been revisited as a practically feasible approach to create nanoscale defects, resulting in an increase in J_c in magnetic fields. In this paper, we report the effect of proton irradiation with 1.5 MeV on superconducting properties of iron–chalcogenide FeSe$_{0.5}$Te$_{0.5}$ films through the transport and magnetization measurements. The 1.5 MeV proton irradiation with 1×10^{16} p/cm^2 yields the highest J_c increase, approximately 30% at 5–10 K and below 1 T without any reduction in T_c. These results indicate that 1.5 MeV proton irradiations could be a practical tool to enhance the performance of iron-based superconducting tapes under magnetic fields.

Keywords: superconductor; irradiation; critical current

Citation: Ozaki, T.; Kashihara, T.; Kakeya, I.; Ishigami, R. Effect of 1.5 MeV Proton Irradiation on Superconductivity in FeSe$_{0.5}$Te$_{0.5}$ Thin Films. *Quantum Beam Sci.* **2021**, *5*, 18. https://doi.org/10.3390/qubs5020018

Academic Editors: Akihiro Iwase and Lorenzo Giuffrida

Received: 31 March 2021
Accepted: 25 May 2021
Published: 4 June 2021

Publisher's Note: MDPI stays neutral with regard to jurisdictional claims in published maps and institutional affiliations.

1. Introduction

Iron-based superconductors have a reasonably high superconducting transition temperature T_c, very high upper critical magnetic fields H_{c2}, quite a small anisotropy γ and larger critical grain boundary angle than cuprate superconductors, which make them promising for high-field applications such as superconducting magnet and generators [1–5]. The use of superconducting materials for high field applications is limited by the critical current density J_c in magnetic fields, which can be sustained by pinning the vortices (flux pinning) at structural defects with nano-meter sizes such as cracks, voids, grain boundaries and secondary phases [6,7]. The ion irradiation is a useful tool to generate the desired defect structure. Depending on the ion species, ion energy and the properties of the target materials, ion irradiation enables the creation of defects with well-controlled morphology and density, such as point [8], cluster [9–12] and columnar [13–15] defects. Early works on the ion irradiation of cuprate (Cu–O based) high-T_c superconductors (HTS) for improving J_c in the magnetic field have mostly focused on the high-energy, over hundreds of MeV, heavy ion irradiation [13–15]. At this energy range, the irradiation of superconducting materials by the swift heavy ion mainly causes electronic excitation and ionization of the target atoms. As a result, continuous amorphous tracks are formed in a process that can be described as the rapid melting and solidification of nm-sized columns in the path of an ion. Even though the heavy ion tracks proved to be very effective pinning defects, this approach has been limited to fundamental studies of the vortex matter.

Recently, ion irradiation of HTS with a low energy has received a renewed interest as a practical method for increasing J_c in magnetic fields, due to the compact accelerator, lower radioactivity and less costly operation [9–12]. Low-energy ion irradiation utilizes a different mechanism for the creation of vortex pinning defects. The electronic excitation and ionization are low enough so the heat can dissipate without damaging the materials. The low-energy ion irradiation leads to the collision of the ion with the target atom nuclei, resulting in cascade, point and cluster defects. Matsui et al. demonstrated that 3 MeV Au^{2+} ion irradiation to 700 nm thick YBCO films yielded an enhancement in the in-field J_c at 77 K of up to a factor of 4 [9]. Equally impressive results in YBCO commercial tape have been reported by Jia et al. using 4 MeV proton [10]. Recently, we reported a route to raise both T_c and J_c in iron-based superconducting $FeSe_{0.5}Te_{0.5}$ (FST) thin films by low-energy (190 keV) proton irradiation [16,17]. The 190 keV proton irradiation yields the increase in T_c due to the nanoscale compressive strain induced by cascade defects. The irradiation also induced a near doubling of J_c at 4.2 K from the self-field to 35 T through strong vortex pinning by the cascade defects and surrounding nanoscale strain.

In this paper, we report the effect of 1.5 MeV proton irradiation on iron–chalcogenide FST superconducting films. We report the performance of irradiated samples at different temperatures in a magnetic field up to 9 T. We show that 1.5 MeV protons clearly enhance J_c in magnetic fields <1 T with no subsequent reduction in T_c. However, we did not observe a reproducible positive effect in the magnetic fields >1 T. The results are discussed in terms of the spatial distribution of defects produced by fast protons.

2. Materials and Methods

All films in this study were deposited by the pulsed laser deposition (PLD) method using a Nd:YAG laser (λ = 266 nm). We first grew a CeO_2 layer with a thickness of about 80–100 nm on $SrTiO_3$ single-crystal substrate at a substrate temperature of 600–650 °C and oxygen partial pressure of ~115 mTorr. Then, 100–130 nm thick FST films were grown on CeO_2 buffer layers. During the deposition of FST films, the substrate temperature and oxygen partial pressure were kept at 300–360 °C and ~1 × 10^{-6} Torr, respectively.

Superconducting transport properties were measured using the conventional four-probe method in a physical property measurement system (PPMS, Quantum Design). $T_{c,10}$ and J_c were determined from the ρT and I–V curves using 0.1 ρ_n and 1 $\mu V/cm$ criteria, respectively. Here, ρ_n means the normal state resistivity above the transition temperature. The current was applied perpendicularly to the magnetic field. The magnetization was measured using a superconducting quantum interference device (SQUID, Quantum Design) magnetometer. Two FST films (sample A and B) were fabricated under the same deposition condition for different irradiation conditions. Each FST film was cut into 3 pieces: one for magnetization measurement before and after irradiation with same film, another for transport measurement before irradiation (pristine) and the other for transport measurement after irradiation (irradiated).

The FST films were irradiated with 1.5 MeV proton doses of 1 × 10^{15} and 1 × 10^{16} p/cm^2 in vacuum at room temperature using the 5 MV tandem accelerator of the Wakasa Wan Energy Research Center (WERC). The samples were mounted on a copper plate with a double-faced carbon tape. The incident angle of ions was set as normal to the film surface. The flux was kept around 3.2 × 10^{12} p/cm^2·s, corresponding to a beam current density of ~500 nA/cm^2. The surface temperature was monitored by a thermocouple. The surface temperature during the irradiation remained below 40 °C.

Prior to the ion irradiation experiment, we ran Stopping and Range of Ions in Matter (SRIM) [18] to estimate ion range and damage profile in our experiment. Based on the simulation results, 1 × 10^{15} and 1 × 10^{16} p/cm^2 are estimated to be ~3.2 × 10^{-5} and ~3.2 × 10^{-4} dpa (displacement per atm), respectively.

3. Results and Discussion

3.1. Magnetic Measurements

Figure 1a,b compare the temperature dependence of magnetic moment M with $H//c$ for two FST films (film-A and film-B) before and after irradiation with 1 × 10^{15} and 1 × 10^{16} p/cm^2 dose, respectively. Both the zero-field-cooled (ZFC) and field-cooled

(FC) magnetizations in 2 Oe magnetic field parallel to the *c*-axis indicate the appearance of superconductivity (obtained by the bifurcation of ZFC and FC) in pristine FST films at 16.8 K for film-A and 16.6 K for film-B. After the irradiation, the superconducting transitions occurred at 16.8 K for film-A and 16.8 K for film-B, indicating that 1.5 MeV proton irradiations with 1×10^{15} and 1×10^{16} p/cm^2 dose have little impact on T_c^{mag}. However, the diamagnetic signal was enhanced with a sharper superconducting transition in the FST film-B irradiated with 1×10^{16} p/cm^2 dose. A degradation of T_c after the ion irradiation is commonly reported in iron-based superconductors [19], although there have been a few reports on an increased T_c in iron-based superconductors irradiated with proton and electron [16,20,21]. In previous work, the Fe(Se,Te) films were covered by Al foil with 80 µm thickness and irradiated with 3.5 MeV protons at doses of 2.68×10^{16} and 5.35×10^{16} p/cm^2, corresponding to 2.30×10^{-3} and 4.59×10^{-3} dpa, respectively [22–24]. The average bombarding energy of the protons on the Fe(Se,Te) film was calculated to be 1.43 ± 0.07 MeV. As a result, the irradiations to doses of 2.68×10^{16} and 5.35×10^{16} p/cm^2 slightly suppressed T_c from 17.7 K for pristine film to 17.3 K and 17.1 K, respectively. Given these results, the primary reason of the almost same T_cs before and after the irradiation in our study would be a lower fluence than that in the previous works.

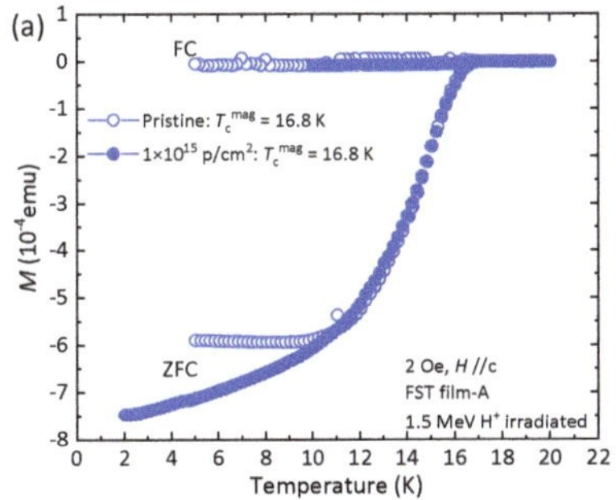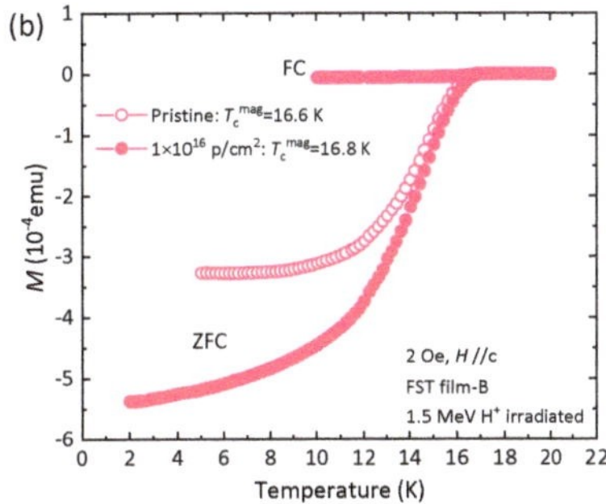

Figure 1. Temperature dependences of magnetic moment *M* for both zero-field-cooled (ZFC) and field-cooled (FC) process at a magnetic field of *H* = 2 Oe applied along the *c*-axis for FST films before and after 1.5 MeV proton irradiation with (**a**) 1×10^{15} and (**b**) 1×10^{16} p/cm^2 dose, respectively.

Figure 2 shows the magnetic field dependence of J_c for the FST film-B at 5, 8, 10 K before and after 1.5 MeV proton irradiation at a dose of 1×10^{16} p/cm^2. The J_c was estimated from the magnetization hysteresis (*M*–*H*) loops using the critical-state Bean model [25,26]. For a rectangular prism-shaped crystal of dimensions $a < b$, we obtained the in-plane critical current density J_c^{ab} in the magnetic field parallel to the *c*-axis as $J_c^{ab} = 20\Delta M/(a(1 - a/3b))$, where ΔM is the difference in magnetization $M(emu/cm^3)$ between the top and bottom branches of the *M-H* loop. In the inset of Figure 2, the *M*–*H* loop in FST film-B at 5 K before and after the irradiation of a dose of 1×10^{16} p/cm^2 is plotted. A large irreversibility is noticeable up to around 4 T at 5 K. We attained a 30% increase in J_c in the magnetic field below 1 T, which indicates that the irradiation defects contribute to vortex pinning. In contrast, we observed almost no change in the in-field J_c above 1 T. Irradiation with MeV protons could produce mostly random point defects and nanocluster [27] due to ion–nucleus collisions. Sylva et al. reported that 3.5 MeV proton irradiation with 6.40×10^{16} p/cm^2 dose (corresponding to 2.27×10^{-3} dpa) yields J_c improvement of about 40% at 4.2 K and 7 T with respect to the pristine film almost without a decrease in T_c [22]. On the contrary, J_c of 3.5 MeV proton irradiated Fe(Se,Te) films covered with 80 µm thick Al foil decreased by up to 80% after irradiation at 4.2 K. The in-field J_c performance in the irradiated FST films in our study could be attributed to the small number of vortex pinning defects created by the irradiation at low fluence.

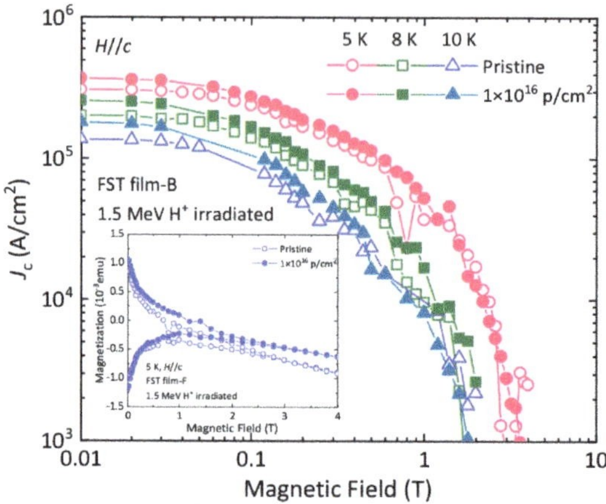

Figure 2. Magnetic field dependence of critical current density $J_c^{ab}(H)$ at 5, 8 and 10 K calculated using the critical-state Bean model for FST film-B pre- and post- 1.5 MeV proton irradiation with 1×10^{16} p/cm^2 dose. The inset shows magnetic hysteresis loop under $H//c$ at 5 K.

3.2. Transport Measurement

In transport measurements, the current is forced to flow through the sample in a particular direction, enabling the direct characterization of superconductivity as a function of temperature, applied magnetic field and field angle. However, we observed an obvious degradation of superconducting properties in the transport measurement of the FST film-B. This could be due to sample degradation, sample handling during mounting and unmounting in a measurement system and possible damage by the laser cutting for patterning the bridge on FST films. In this section, we refer to the FST film-A. Figure 3 presents the temperature dependence of the electrical resistivity before and after irradiation for FST film-A with 1×10^{15} p/cm^2 dose of 1.5 MeV proton. The FST films before and after the irradiation showed metallic behavior below 200 K. Additionally, 1.5 MeV proton irradiation with 1×10^{15} p/cm^2 dose has little effect on normal-state resistivity due to the low dpa. On the contrary, the normal-state resistivity shows nearly upwards parallel-shift upon 6 MeV Au-ion irradiation with a dose of 1×10^{12} Au/cm^2, corresponding to 6.42×10^{-3} dpa [11]. We observed no change in $T_{c,10}$ (=17.5 K) before and after the 1.5 MeV protons irradiation with 1×10^{15} p/cm^2 dose. This could be due to the low fluence, i.e., low dpa.

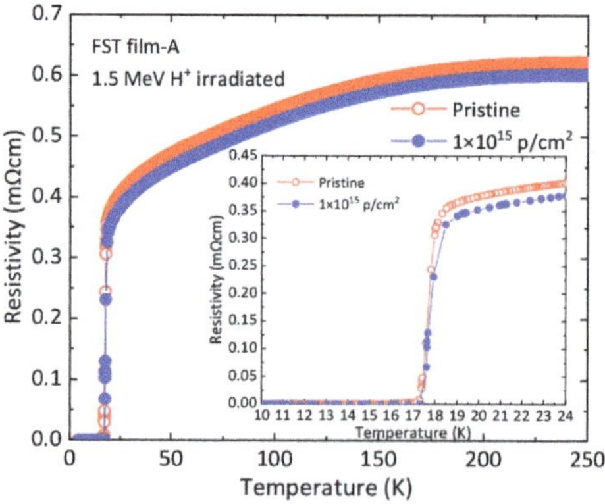

Figure 3. Temperature dependences of electrical resistivity at 0 T for the FST film-A before and after 1.5 MeV proton irradiation with 1×10^{15} p/cm^2 dose. Inset shows a magnified temperature region near T_c.

Figure 4 presents the magnetic field dependence of transport critical current density J_c with $H//c$ for the FST film-A before and after irradiation with 1.5 MeV protons to a dose of 1×10^{15} p/cm^2 at 4.2 K. Comparing J_cs obtained from magnetization and transport measurements, the values of J_c obtained from transport measurement are larger than those of J_c calculated from magnetization measurement. This would come from the difference of criterion to determine the J_c values. The positive effect of the proton irradiation on J_c at 4.2 K is unambiguous in the magnetic field below 1 T. As the magnetic field increased, the difference between pristine and the irradiated FST film became smaller. Similar behavior was observed in $J_c(H)$ (calculated from magnetization measurement in Figure 2) for FST film-B irradiated with 1×10^{16} p/cm^2 dose.

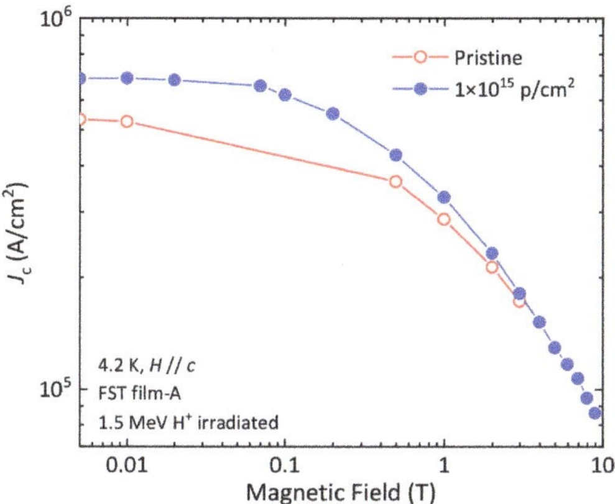

Figure 4. Magnetic field dependence of critical current density J_c obtained from transport measurement at 4.2 K for FST film-A pre- and post-1.5 MeV proton irradiation with 1×10^{15} p/cm^2 dose.

A more detailed representation of the pinning efficiency can be obtained from the angular dependence of J_c. We show $J_c(\theta)$ for the FST film-A irradiated with 1×10^{15} p/cm^2 dose of 1.5 MeV proton beam under 1 and 3 T at 4.2K in Figure 5. The pristine film has a less-anisotropic J_c angular dependence at 1 and 3 T without a prominent J_c peak at $H//c$, which is often observed in YBa$_2$Cu$_3$O$_y$ films [28]. A small J_c-anisotropy, γ_{Jc} ($J_c^{H//ab}/J_c^{H//c}$), of 1.7 is observed at 1 T. This value is smaller than the value of Fe(Se,Te) films grown on Fe-buffered MgO substrates (γ_{Jc} = 2.6) [29] while it is larger than the value of Fe(Se,Te) films grown on CaF$_2$ substrates [30,31]. These differences might arise from the difference of the substrate and buffer layer. Upon irradiation with 1.5 MeV proton, the J_c increases for most of the field orientations, retaining a small γ_{Jc} of 1.7 at 1 T, indicating that the vortex pinning defects would be less anisotropic and randomly distributed. At 3 T, there is a significant decrease in J_c in the angular range $\pm30°$ from $H//ab$. Iron-based and cuprate high-temperature superconductors commonly possess inherent layered structures, consisting of alternating conducting and insulating atomic planes. In general, the strong J_c peak for $H//ab$ could be ascribed to the vortex pinning by the intrinsic pinning and planar defects such as intergrowths and stacking faults, parallel to the ab plane [32–35]. In the iron–chalcogenide Fe(Se,Te) compound, which is composed of only the Fe-Se(Te) layer, $J_c(\theta)$ has a maximum at $H//ab$ due to intrinsic pinning from the Fe–Se(Te) intralayer and Van der Waals interlayer couplings [29,34,35]. Hence, the J_c suppression at around $H//ab$ would occur because of the reduction in the density of intrinsic pinning upon the irradiation.

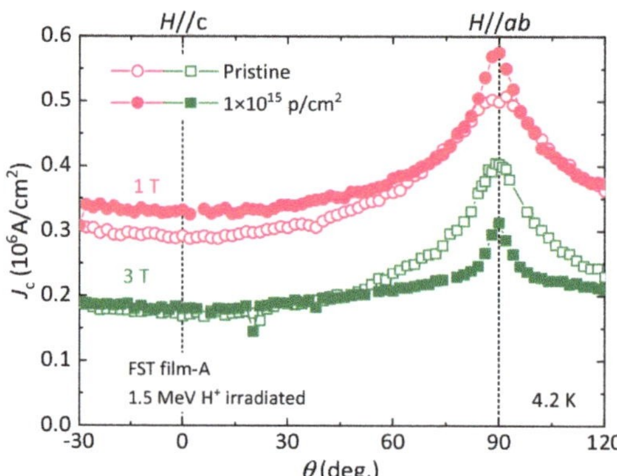

Figure 5. Angular field dependence of the critical current density J_c obtained from transport measurement for FST film-A before and after 1.5 MeV proton irradiation with 1×10^{15} p/cm^2 dose measured at 4.2 K under 1 and 3 T.

4. Conclusions

We conclude a study on the effect of 1.5 MeV proton irradiation on superconducting properties of FST films. Upon the irradiation up to 1×10^{16} p/cm^2 dose, T_c remains virtually unchanged in magnetization as well as in transport measurement. An approximately 30% enhancement of J_c in the magnetic field below 1 T is observed using 1.5 MeV proton irradiation with 1×10^{16} p/cm^2. Transport properties of a pristine film and an irradiated film with a fluence of 1×10^{15} p/cm^2 show a small anisotropy of J_c in the applied magnetic field range at 4.2 K. The enhancement of J_c for almost all the field orientations was accomplished by the irradiation at a dose of 1×10^{15} p/cm^2 at 4.2 K and 1 T. These results indicate that 1.5 MeV proton irradiation is effective in providing less anisotropic pinning defects in the magnetic field below 1 T in iron–chalcogenide superconducting films. Additionally, by fine tuning an irradiation fluence of proton, superconducting properties can be further improved.

Author Contributions: Conceptualization, T.O.; sample preparation, T.K. and T.O.; ion irradiation, R.I., T.K. and T.O.; transport measurement, T.K., I.K. and T.O.; magnetization measurement, T.O. and T.K.; data curation, T.K. and T.O.; writing—original draft preparation, T.O.; writing—review and editing, I.K. and R.I. All authors have read and agreed to the published version of the manuscript.

Funding: This work was partly supported by Foundation of Kinoshita Memorial Enterprise.

Acknowledgments: This research has been performed under the collaboration program between Kwansei Gakuin University, Kyoto University and Wakasa Wan Energy Research Center.

Conflicts of Interest: The authors declare no conflict of interest.

References

1. Putti, M.; Pallecchi, I.; Bellingeri, E.; Cimberle, M.R.; Tropeano, M.; Ferdeghini, C.; Palenzona, A.; Tarantini, C.; Yamamoto, A.; Jiang, J.; et al. New Fe-based superconductors: Properties relevant for applications. *Supercond. Sci. Technol.* **2010**, *23*, 034003. [CrossRef]
2. Gurevich, A. Iron-based superconductors at high magnetic fields. *Rep. Prog. Phys.* **2011**, *74*, 124501. [CrossRef]
3. Katase, T.; Ishimaru, Y.; Tsukamoto, A.; Hiramatsu, H.; Kamiya, T.; Tanabe, K.; Hosono, H. Advantageous grain boundaries in iron pnictide superconductors. *Nat. Commun.* **2011**, *2*, 409. [CrossRef] [PubMed]
4. Si, W.; Zhang, C.; Shi, X.; Ozaki, T.; Jaroszynski, J.; Li, Q. Grain boundary junctions of FeSe$_{0.5}$Te$_{0.5}$ thin films on SrTiO$_3$ bi-crystal substrates. *Appl. Phys. Lett.* **2015**, *106*, 032602. [CrossRef]
5. Iida, K.; Hänisch, J.; Yamamoto, A. Grain boundary characteristics of Fe-based superconductors. *Supercond. Sci. Technol.* **2020**, *33*, 043001. [CrossRef]

6. Larbalestier, D.; Gurevich, A.; Feldmann, D.M.; Polyanskii, A. High-T_c superconducting materials for electric power applications. *Nature* **2001**, *414*, 368. [CrossRef] [PubMed]
7. Foltyn, S.R.; Civale, L.; MacManus-Driscoll, J.L.; Jia, Q.X.; Maiorov, B.; Wang, H.; Maley, M. Materials science challenges for high-temperature superconducting wire. *Nat. Mater.* **2007**, *6*, 631. [CrossRef] [PubMed]
8. Kirk, M.A. Structure and flux pinning properties of irradiation defects in $YBa_2Cu_3O_{7-x}$. *Cryogenics* **1993**, *33*, 235. [CrossRef]
9. Matsui, H.; Ogiso, H.; Yamasaki, H.; Kumagai, T.; Sohma, M.; Yamaguchi, I.; Manabe, T. 4-fold enhancement in the critical current density of $YBa_2Cu_3O_7$ films by practical ion irradiation. *Appl. Phys. Lett.* **2012**, *101*, 232601. [CrossRef]
10. Jia, Y.; LeRoux, M.; Miller, D.J.; Wen, J.G.; Kwok, W.K.; Welp, U.; Rupich, M.W.; Li, X.; Sathyamurthy, S.; Fleshler, S.; et al. Doubling the critical current density of high temperature superconducting coated conductors through proton irradiation. *Appl. Phys. Lett.* **2013**, *103*, 122601. [CrossRef]
11. Ozaki, T.; Wu, L.; Zhang, C.; Si, W.; Jie, Q.; Li, Q. Enhanced critical current in superconducting $FeSe_{0.5}Te_{0.5}$ films at all magnetic field orientations by scalable gold ion irradiation. *Supercond. Sci. Technol.* **2018**, *31*, 024002.
12. Zhang, Y.; Rupich, M.W.; Solovyov, V.; Li, Q.; Goyal, A. Dynamic behavior of reversible oxygen migration in irradiated-annealed high temperature superconducting wires. *Sci. Rep.* **2020**, *10*, 14848. [CrossRef]
13. Sueyoshi, T.; Kotaki, T.; Furuki, Y.; Fujiyoshi, T.; Semboshi, S.; Ozaki, T.; Sakane, H.; Kudo, M.; Yasuda, K.; Ishikawa, N. Strong flux pinning by columnar defects with directionally dependent morphologies in GdBCO-coated conductors irradiated with 80 MeV Xe ions. *Jpn. J. Appl. Phys.* **2020**, *59*, 023001. [CrossRef]
14. Civale, L. Vortex pinning and creep in high-temperature superconductors with columnar defects. *Supercond. Sci. Technol.* **1997**, *10*, A11. [CrossRef]
15. Kirk, M.A.; Yan, Y. Structure and properties of irradiation defects in $YBa_2Cu_3O_{7-x}$. *Micron* **1999**, *30*, 507. [CrossRef]
16. Ozaki, T.; Wu, L.; Zhang, C.; Jaroszynski, J.; Si, W.; Zhou, J.; Zhu, Y.; Li, Q. A route for a strong increase of critical current in nanostrained iron-based superconductors. *Nat. Commun.* **2016**, *7*, 13036. [CrossRef] [PubMed]
17. Ozaki, T.; Wu, L.; Gu, G.; Li, Q. Ion irradiation of iron chalcogenide superconducting films. *Supercond. Sci. Technol.* **2020**, *33*, 094008. [CrossRef]
18. Ziegler, J.F.; Biersack, J.P.; Littmark, U. *The Stopping and Range of Ions in Solids*; Pergamon: Oxford, UK, 1985.
19. Eisterer, M. Radiation effects on iron-based superconductors. *Supercond. Sci. Technol.* **2018**, *31*, 013001. [CrossRef]
20. Teknowijoyo, S.; Cho, K.; Tanatar, M.A.; Gonzales, J.; Böhmer, A.E.; Cavani, O.; Mishra, V.; Hirschfeld, P.J.; Bud'ko, S.L.; Canfield, P.C.; et al. Enhancement of superconducting transition temperature by pointlike disorder and anisotropic energy gap in FeSe single crystals. *Phys. Rev. B* **2016**, *94*, 064521. [CrossRef]
21. Mizukami, Y.; Konczykowski, M.; Matsuura, K.; Watashige, T.; Kasahara, S.; Matsuda, Y.; Shibauchi, T. Impact of Disorder on the Superconducting Phase Diagram in $BaFe_2(As_{1-x}P_x)_2$. *J. Phys. Soc. Jpn.* **2017**, *86*, 083706. [CrossRef]
22. Sylva, G.; Bellingeri, E.; Ferdeghini, C.; Martinelli, A.; Pallecchi, I.; Pellegrino, L.; Putti, M.; Ghigo, G.; Gozzelino, L.; Torsello, D.; et al. Effects of high-energy proton irradiation on the superconducting properties of Fe(Se, Te) thin films. *Supercond. Sci. Technol.* **2018**, *31*, 054001. [CrossRef]
23. Leo, A.; Sylva, G.; Braccini, V.; Bellingeri, E.; Martinelli, A.; Pallecchi, I.; Ferdeghini, C.; Pellegrino, L.; Putti, M.; Ghigo, G.; et al. Anisotropic Effect of Proton Irradiation on Pinning Properties of Fe(Se, Te) Thin Films. *IEEE Trans. Appl. Supercond.* **2019**, *21*, 6601904. [CrossRef]
24. Leo, A.; Grimaldi, G.; Nigro, A.; Ghigo, G.; Gozzelino, L.; Torsello, D.; Braccini, V.; Sylva, G.; Ferdeghini, C.; Putti, M. Critical current anisotropy in Fe(Se, Te) films irradiated by 3.5 MeV protons. *J. Phys. Conf. Ser.* **2020**, *1559*, 012042. [CrossRef]
25. Bean, C.P. Magnetization of Hard Superconductors. *Phys. Rev. Lett.* **1962**, *8*, 250. [CrossRef]
26. Bean, C.P. Magnetization of High-Field Superconductors. *Rev. Mod. Phys.* **1964**, *36*, 31. [CrossRef]
27. Haberkorn, N.; Maiorov, B.; Usov, I.O.; Weigand, M.; Hirata, W.; Miyasaka, S.; Tajima, S.; Chikumoto, N.; Tanabe, K.; Civale, L. Influence of random point defects introduced by proton irradiation on critical current density and vortex dynamics of $Ba(Fe_{0.925}Co_{0.075})_2As_2$ single crystals. *Phys. Rev. B* **2012**, *82*, 180520.
28. Civale, L.; Maiorov, B.; Serquis, A.; Willis, J.O.; Coulter, J.Y.; Wang, H.; Jia, Q.X.; Arendt, P.N.; MacManus-Driscoll, J.L.; Maley, M.P.; et al. Angular-dependent vortex pinning mechanisms in $YBa_2Cu_3O_7$ coated conductors and thin films. *Appl. Phys. Lett.* **2004**, *84*, 2121. [CrossRef]
29. Iida, K.; Hänisch, J.; Schulze, M.; Aswartham, S.; Wurmehl, S.; Büchner, B.; Schultz, L.; Holzapfel, B. Generic Fe buffer layers for Fe-based superconductors: Epitaxial $FeSe_{1-x}Te_x$ thin films. *Appl. Phys. Lett.* **2011**, *99*, 202503. [CrossRef]
30. Yuan, P.; Xu, Z.; Ma, Y.; Sun, Y.; Tamegai, T. Angular-dependent vortex pinning mechanism and magneto-optical characterizations of $FeSe_{0.5}Te_{0.5}$ thin films grown on CaF_2 substrates. *Supercond. Sci. Technol.* **2016**, *29*, 035013. [CrossRef]
31. Braccini, V.; Kawale, S.; Reich, E.; Bellingeri, E.; Pellegrino, L.; Sala, A.; Putti, M.; Higashikawa, K.; Kiss, T.; Holzapfel, B.; et al. Highly effective and isotropic pinning in epitaxial Fe(Se, Te) thin films grown on CaF_2 substrates. *Appl. Phys. Lett.* **2013**, *103*, 172601. [CrossRef]
32. Spechta, E.D.; Goyal, A.; Li, J.; Martin, P.M.; Li, X.; Rupich, M.W. Stacking faults in $YBa_2Cu_3O_{7-x}$: Measurement using x-ray diffraction and effects on critical current. *Appl. Phys. Lett.* **2006**, *89*, 162510. [CrossRef]
33. Civale, L.; Maiorov, B.; MacManus-Driscoll, J.L.; Wang, H.; Holesinger, T.G.; Foltyn, S.R.; Serquis, A.; Arendt, P.N. Identification of Intrinsic ab-Plane Pinning in $YBa_2Cu_3O_7$ Thin Films and Coated Conductors. *IEEE Trans. Appl. Supercond.* **2005**, *15*, 2808. [CrossRef]

34. Iida, K.; Hänisch, J.; Reich, E.; Kurth, F.; Hühne, R.; Schultz, L.; Holzapfel, B. Intrinsic pinning and the critical current scaling of clean epitaxial Fe(Se, Te) thin films. *Phys. Rev. B* **2013**, *87*, 104510. [CrossRef]
35. Grimaldi, G.; Leo, A.; Nigro, A.; Pace, S.; Braccini, V.; Bellingeri, E.; Ferdeghini, C. Angular dependence of vortex instability in a layered superconductor: The case study of Fe(Se, Te) material. *Sci. Rep.* **2018**, *8*, 4150. [CrossRef] [PubMed]

quantum beam science

Article

Irradiation Hardening Behavior of He-Irradiated V–Cr–Ti Alloys with Low Ti Addition

Ken-ichi Fukumoto [1,*], Yoshiki Kitamura [1], Shuichiro Miura [1], Kouji Fujita [1], Ryoya Ishigami [2] and Takuya Nagasaka [3]

[1] Research Institute of Nuclear Engineering, University of Fukui, Tsuruga, Fukui 914-0055, Japan; ha170518@u-fukui.ac.jp (Y.K.); jn190083@u-fukui.ac.jp (S.M.); jn180098@u-fukui.ac.jp (K.F.)
[2] The Wakasa Wan Energy Research Center, Tsuruga, Fukui 914-0192, Japan; rishigami@werc.or.jp
[3] National Institute of Fusion Science, Toki, Gifu 509-5292, Japan; nagasaka@nifs.ac.jp
* Correspondence: fukumoto@u-fukui.ac.jp

Abstract: A set of V–(4–8)Cr–(0–4)Ti alloys was fabricated to survey an optimum composition to reduce the radioactivity of V–Cr–Ti alloys. These alloys were subjected to nano-indenter tests before and after 2-MeV He-ion irradiation at 500 °C and 700 °C with 0.5 dpa at peak damage to investigate the effect of Cr and Ti addition and gas impurities for irradiation hardening behavior in V–Cr–Ti alloys. Cr and Ti addition to V–Cr–Ti alloys for solid–solution hardening remains small in the unirradiated V–(4–8)Cr–(0–4)Ti alloys. Irradiation hardening occurred for all V–Cr–Ti alloys. The V–4Cr–1Ti alloy shows the highest irradiation hardening among all V–Cr–Ti alloys and the gas impurity was enhanced to increase the irradiation hardening. These results may arise from the formation of Ti(CON) precipitate that was produced by He-ion irradiation. Irradiation hardening of V–Cr–1Ti did not depend significantly on Cr addition. Consequently, for irradiation hardening and void-swelling suppression, the optimum composition of V–Cr–Ti alloys for structural materials of fusion reactor engineering is proposed to be a highly purified V–(6–8)Cr–2Ti alloy.

Keywords: vanadium alloy; ion irradiation; irradiation hardening; radiation damage

Citation: Fukumoto, K.-i.; Kitamura, Y.; Miura, S.; Fujita, K.; Ishigami, R.; Nagasaka, T. Irradiation Hardening Behavior of He-Irradiated V–Cr–Ti Alloys with Low Ti Addition. *Quantum Beam Sci.* **2021**, *5*, 1. https://doi.org/10.3390/qubs5010001

Received: 29 November 2020
Accepted: 29 December 2020
Published: 31 December 2020

Publisher's Note: MDPI stays neutral with regard to jurisdictional claims in published maps and institutional affiliations.

1. Introduction

Vanadium alloys are attractive blanket structural materials in fusion power systems because of their low induced activation characteristics, high-temperature strength, good compatibility with a liquid lithium environment and high thermal stress [1–3]. Critical issues of vanadium alloys, such as corrosion and oxidation have been resolved by Cr and Ti addition to the vanadium matrix, and recent efforts have focused on developing the V–4Cr–4Ti alloy as a candidate alloy for Li-blanket systems in fusion reactors [4,5]. The susceptibility of V–4Cr–4Ti alloys to low-temperature embrittlement during neutron irradiation may limit their application in low-temperature (<400 °C) regimes [6]. To improve this drawback, highly purified V–4Cr–4Ti alloys, such as NIFS-HEAT-1 and -2, have been developed by the National Institute for Fusion Science (NIFS) [7–10]. The NIFS-HEAT-2 has shown significantly lower radioactivity for full remote recycle over 25 years of cooling after being used in the first wall of a fusion commercial reactor. A reduction in the cooling period provides economical and safety benefits to reduce the amount of radiation waste material [11].

Nagasaka suggested that the cooling period of V–Cr–Ti alloy for full remote recycling can be reduced within 10 years by reducing the Ti addition from a V–4Cr–4Ti alloy [12]. The reduction in Ti addition reduces the cooling time and results in swelling and a loss of mechanical strength under neutron irradiation.

To optimize the Ti and Cr addition to V–Cr–Ti alloys to balance the irradiation hardening and swelling behavior, a set of V–Cr–Ti alloys with a lower Ti content and higher Cr content was produced from V–4Cr–4Ti alloys. In this study, nano-indenter tests were performed to investigate irradiation hardening in a new set of V–Cr–Ti alloys after He-ion irradiation to survey the optimum composition to compensate between radioactivity reduction and irradiation hardening behavior in V–Cr–Ti alloys.

2. Experimental Procedure

In total, 15 types of V–(4–8)Cr–(0–4)Ti ternary alloys were fabricated by arc melting. Table 1 shows the chemical composition of the V–Cr–Ti alloys that were used in this study. Two impurity levels of each alloy were prepared to investigate the effect of interstitial impurity for irradiation hardening during He-ion irradiation. The highly purified alloy was marked as "(h)". The V–Cr–Ti alloys from conventional fabrication contain ~500 ppm of C+N+O interstitial gas impurity. In contrast, highly purified V–Cr–Ti alloys, which are marked as "(h)", contain approximately half the interstitial gas impurity in conventional alloys by using highly purified vanadium ingots in fabrication [11]. Thin specimen plates of 10 mm × 2 mm × 0.2 mm were cut out and annealed for 2 h at 1000 °C in a vacuum ($\sim 2 \times 10^{-4}$ Pa). The samples were irradiated with 2-MeV ^{4}He ions using a tandem accelerator at the Wakasa Wan Energy Research Center. Sectional shapes of the ^{4}He^{2+} beams existed in a 9-mm-diameter circle or ellipse with a major axis of 10 mm and a minor axis of 6 mm. These beams were scanned to irradiate the samples uniformly. The horizontal and vertical widths of the scanned beams were 13 mm × 13 mm for the former beam and 13 mm × 17 mm for the latter beam. Time-averaged current densities were 0.4 μA/cm^2 and 0.9 μA/cm^2, respectively. During the irradiation, the sample stage was heated on a Mo holder with a ceramic heater. The temperature was maintained within ± 5 °C during ion irradiation. Specimens were irradiated at 500 °C and 700 °C up to doses of 0.5 dpa at a peak position and 3.4 μm depth.

Table 1. Chemical composition of V–Cr–Ti alloys. Marks (h) indicate the type of highly purified alloy from the original alloy.

Composition (wt.%)	Cr	Ti	C	N	O	Mo	Al	Si
V-4Cr	3.80	0.002	0.004	0.005	0.036	(<0.001)	0.005	(0.02)
V-4Cr (h)	3.90	0.002	0.009	0.003	0.018	(<0.001)	0.011	(0.02)
V-4Cr-0.1Ti	3.88	0.09	0.005	0.005	0.038	(<0.001)	0.012	(0.02)
V-4Cr-0.1Ti (h)	3.90	0.09	0.007	0.003	0.017	(<0.001)	0.011	(0.02)
V-4Cr-1Ti	3.86	0.96	0.005	0.006	0.035	<0.001	0.006	0.016
V-4Cr-1Ti (h)	4.02	0.96	0.008	0.004	0.016	(<0.001)	0.009	(0.02)
V-4Cr-2Ti	3.94	1.93	0.005	0.005	0.037	(<0.001)	0.005	(0.02)
V-4Cr-2Ti (h)	3.89	1.92	0.008	0.003	0.015	(<0.001)	0.006	(0.02)
V-4Cr-3Ti (h)	3.92	2.99	0.009	0.003	0.016	(<0.001)	0.007	(0.02)
V-4Cr-4Ti	3.93	3.91	0.006	0.006	0.036	<0.001	0.009	0.016
V-4Cr-4Ti (h)	4.11	3.89	0.008	0.003	0.018	(<0.001)	0.018	(0.02)
V-6Cr-1Ti	5.97	0.96	0.006	0.006	0.036	<0.001	0.006	0.016
V-6Cr-1Ti (h)	5.95	0.95	0.010	0.003	0.015	(<0.001)	0.016	(0.02)
V-8Cr-1Ti	7.89	0.99	0.006	0.006	0.038	<0.001	0.008	0.016
V-8Cr-1Ti (h)	7.83	1.00	0.008	0.003	0.015	(<0.001)	0.016	(0.02)

Figure 1 shows the damage profile and He ion range in vanadium as calculated by using the SRIM-code (the stopping and range of ions in matter). After He ion irradiation, a nano-indentation test was examined at room temperature by using a Elionix ENT-1100a (Elionix Inc., Tokyo, Japan) nano-indenter with a Berkovich diamond indenter tip and a direction of indentation parallel to the ion beam axis, which is normal to the irradiated surface. The nano-indenter test was carried out with an indentation depth of 500 nm. The indentation depth was determined from the effective depth where the plastic and elastically deformed area was expanded during the nano-indenter test that corresponds to five times the length of the nano-indenter depth. To avoid uncertainty about the specimen surface condition, a scanning electron microscopy (SEM) observation with electron backscattered diffraction measurement was carried out on the indentation surface after the indentation test to ascertain that the indentation test was not on/near grain boundary on the specimen surface. For each sample, 12 tests were conducted, and 10 of the measured data points were included in the analysis, rejecting the maximum and minimum measured values. In this study He ions were used as the projectile because the penetration depth is sufficient to assess mechanical properties as irradiation hardening through nano-indentation hardness measurements [13]; it should be noted that the first wall and divertor of a fusion reactor are expected to be subject to high fluxes of moderate to low-energy helium ions created as fusion reaction products and as a result He bubbles may form around the He ion penetration range which was 3.6 μm in this study.

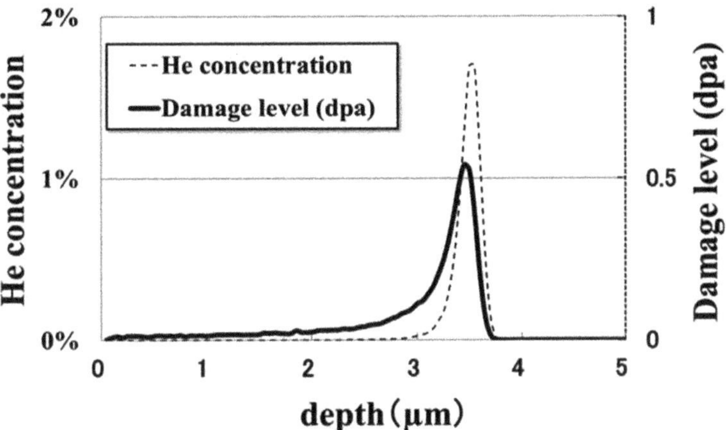

Figure 1. Calculation results of ion range profile of implanted He ion and vacancy concentration profile in V by 2-MeV He$^+$ ion irradiation.

3. Results

Figure 2 shows the results from the nano-indenter test for unirradiated V–4Cr–xTi alloys and He-irradiated V–4Cr–xTi alloys at 500 °C. In the unirradiated V–4Cr–xTi alloys, the effects of Ti addition and interstitial gas impurity on the nano-indentation hardness were not apparent. The hardness of the V–4Cr–xTi and V–4Cr–xTi(h) alloys did not change at ~1800 N/mm^2. The He-irradiated V–4Cr–xTi alloys showed a large irradiation hardening. V–4Cr–1Ti showed the largest irradiation hardening of the alloys and the hardness reached 5100 N/mm^2, which was approximately 2.8 times larger than the hardness of unirradiated V-4Cr-1Ti alloys. For the V–4Cr–xTi with 2% to 4% Ti addition, the irradiation hardening decreased and did not change as much with increasing Ti addition. Therefore 1% Ti addition is most effective to produce irradiation hardening in V–4Cr–xTi (x = 0 to 4). From 0.1% to 1%Ti addition in V–4Cr–xTi alloys, the alloys that contained more interstitial gas impurity showed a larger increase in irradiation hardening. V–4Cr–xTi alloys that contained more than 2% Ti addition did not change the irradiation hardening increase even though the amount of gas interstitial impurity increased. This result suggests that the interstitial gas impurity does not contribute to irradiation hardening in V–4Cr–xTi alloys with more than 2% Ti addition.

Figure 3 shows the irradiation hardening increase of V–4Cr–xTi alloys with the conventional impurity level irradiated at 500 °C and 700 °C. The He-irradiated V–4Cr–xTi alloys at 700 °C showed irradiation hardening but the amount of irradiation hardening increase at 700 °C was smaller than that at 500 °C. This result indicates that damaged microstructures that formed at 700 °C irradiation may be coarser than those formed at 500 °C irradiation, and irradiation hardening at 700 °C irradiation is smaller than that at 500 °C irradiation. These microstructural features depending on irradiation temperature and damage level have been reported elsewhere [5,14,15].

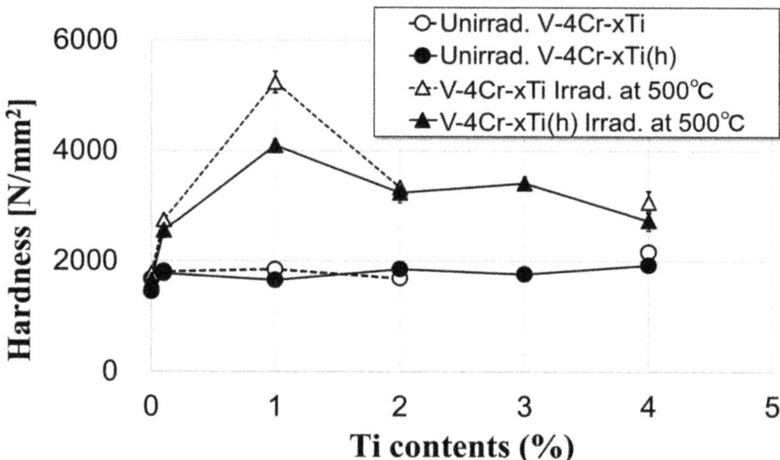

Figure 2. Ti content dependence of nano-indentation hardening for unirradiated V–4Cr–xTi alloys and He-irradiated V–4Cr–xTi alloys at 500 °C.

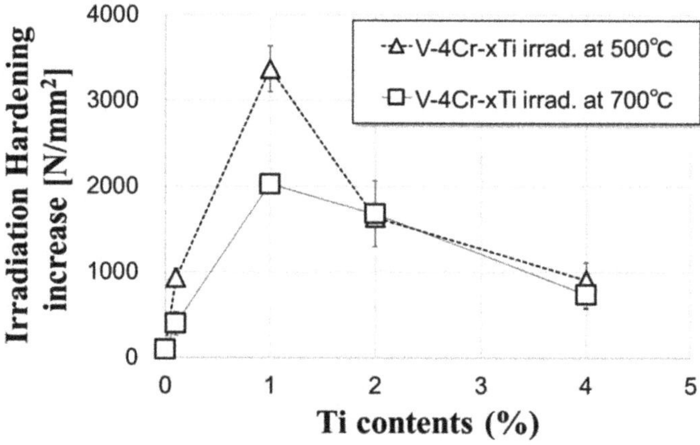

Figure 3. Irradiation hardening increase of V–4Cr–xTi alloys with conventional impurity level irradiated at 500 °C and 700 °C as a function of Ti content.

Figure 4 shows the nano-indenter test results for unirradiated and He-irradiated V–yCr–1Ti alloys with a conventional impurity level irradiated at 500 °C and 700 °C. In the unirradiated V–yCr–1Ti alloys, the nano-indentation hardness of the V–yCr–1Ti alloys increased slightly with an increase in Cr addition, which was caused by solution hardening because of Cr addition in V–Cr–Ti alloys. The interstitial gas impurity in the V–yCr–1Ti alloys did not affect the hardening behavior much. Irradiation hardening of He-irradiated V–yCr–1Ti was apparent in all alloys. Significant irradiation hardening occurred among all alloys, but the irradiation hardening decreased with an increase in Cr addition. The significant reduction in irradiation hardening in the V–8Cr–1Ti alloys was apparent. The effect of interstitial gas impurity on the irradiation-hardening behavior of V–yCr–1Ti showed that the conventional fabricated alloys had a larger irradiation hardening increase than the highly purified alloys. The irradiation hardening at 500 °C irradiation was larger than that at 700 °C irradiation in all V–yCr–1Ti alloys. At 500 °C irradiation, the irradiation hardening decreased with an increase in Cr addition. The irradiation hardening of the V–yCr–1Ti alloys that were irradiated at 700 °C increased with an increase in Cr addition.

From the result of heat treatment of V-4Cr-4Ti alloys at high temperatures, it has been reported that fine precipitate forms above 700 °C of annealing temperature [16]. Hence, it is concluded that the hardness increase at 500 °C in this study is not solely caused by the precipitation induced by the irradiation temperature at least.

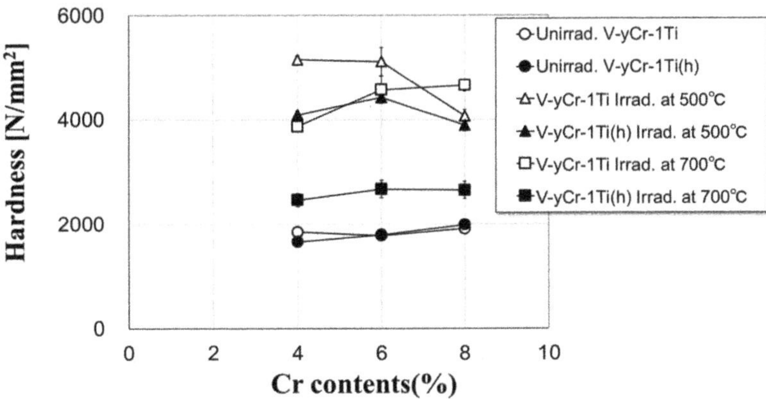

Figure 4. Cr content dependence of nano-indentation hardening for unirradiated V–yCr–1Ti alloys and He-irradiated V–yCr–1Ti alloys at 500 °C and 700 °C.

4. Discussion

Previous work on the thermal creep deformation of V–Ti alloy indicated that 3% Ti addition to V resulted in the lowest activation energy for creep strain behavior in the V–xTi alloys (x = 0% to 20%) [17–19]. The low activation energy of creep deformation reduces the creep strain rate at a given temperature in the material and leads to a high creep strength. The creep strength is strengthened by Ti addition to V in the V–Ti system alloy from the increase in yield and tensile strengths because of a solid–solution hardening mechanism and/or a dispersed particle-strengthening mechanism. The main reason why the V–3Ti alloy exhibits the strongest creep strength in the V–Ti system alloy (0% to 20% Ti addition) cannot be explained by the solid–solution hardening mechanism because solid–solution hardening due to Cr and Ti addition to V–Cr–Ti alloys was not apparent in this work as determined from the nano-indentation test of unirradiated V–Cr–Ti alloys as shown in Figure 2. In the thermal creep deformation in V–Ti alloy at an elevated temperature, the formation of a dislocation-network as well as formation of a titanium-oxycarbonitride precipitate, Ti(CON) have been observed as thermal vacancy migration and dislocation slip motion. Figure 5 provides an example of precipitate formation in V–4Cr–4Ti alloys in creep deformation at 600 °C with a stress of 200 MPa for 2800 h in a liquid Na environment [20]. Small precipitates formed near the grain boundary or large bulk precipitates and the precipitate was Ti(CON) titanium-oxycarbonitride precipitate on a habit pane of {100}. The nature of the Ti(CON) precipitate has been reported by Impagnatiello [21–23]. In the thermal creep test, it is likely that Ti(CON) precipitates are formed along with defect sinks such as grain boundaries and bulk precipitates where the thermal vacancy is absorbed preferentially. It is expected that the Ti(CON) precipitate is more likely to form in an environment where that migration of Ti is enhanced due to vacancy flux to sinks such as grain boundaries through the Kirkendall effect for a long period and the nucleation of Ti(CON) precipitate occurs around sinks such as grain boundaries and large bulk precipitates due to the thermal heat treatment. Since He irradiation at high temperature also provides a large defect flux of not only excess vacancies but also displaced Ti atoms produced by He ion bombardment in the matrix, this kind of irradiation may lead to the formation of Ti(CON) precipitates in the matrix. Previous transmission electron microscopy (TEM) work for the microstructural observation of neutron- and ion-irradiated V–Cr–Ti and V–Ti alloys reported that Ti(CON) precipitate formation occurred above 350 °C with more than 0.1 dpa of damage level [24–28]. It is deduced that the significant irradiation hardening in V–Cr–Ti alloys in this work was caused by irradiation-induced Ti(CON) precipitation during 500 °C and 700°C irradiation, even though microstructural observation by TEM work remained unexamined in this work. The high density of small Ti(CON) precipitate formed in V–4Cr–1Ti alloys and V–4Cr–1Ti shows the highest obstacle resistance against dislocation slip based on the Orowan equation termed as dispersed barrier hardening in [29];

$$\Delta\sigma\_y = M\alpha\mu b\sqrt{ND} \tag{1}$$

where M is the Taylor factor (3.06 for fcc polycrystals), μ the shear modulus of the matrix, b the magnitude of the Burgers vector of the moving dislocation, and $\Delta\sigma_y$ represents the increment in yield strength because of the obstacles of size D, number density, N and barrier strength α. It is assumed that the density of small precipitate in V–4Cr–1Ti will be highest among all V-4Cr–xTi and shows a significant irradiation hardening based on the dispersed precipitate hardening produced by He-ion irradiation.

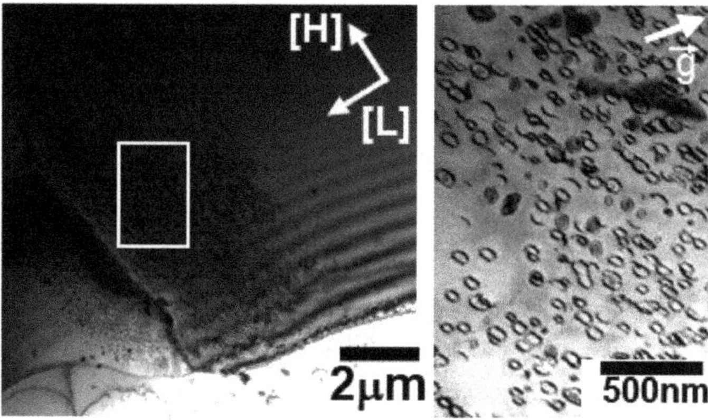

Figure 5. TEM micrographs of thermal-creep-deformed V–4Cr–4Ti alloy at 600 °C and 150 MPa for 2800 h. The left side of the low-magnification image shows that precipitates gathered along a grain boundary and the right side of the high magnification image shows that precipitates on {100} habit plane were formed. [L] and [H] indicate the direction of the longitudinal direction and horizontal direction of the creep tube, respectively. (after Fukumoto et al. [20]).

The nano-indentation test results from the V–Cr–Ti alloys show that the unirradiated V–6Cr–1Ti and V–8Cr–1Ti alloy hardness did not change much compared with that of the V–4Cr–1Ti alloys. Therefore, Cr addition may be ineffective for solid–solution hardening in V–Cr–Ti alloys. The He-irradiated V–yCr–1Ti alloy results show that the effect of Cr addition for irradiation hardening is independent of Cr addition in the V–(4–8)Cr–Ti alloys.

The effect of gas impurity level for irradiation hardening appears in irradiation hardening of V–4Cr–1Ti alloys irradiated at 500 °C and 700 °C, and the highly purified V–Cr–Ti alloys reduce the irradiation hardening as shown in Figures 2 and 4. This reduction of irradiation hardening in highly purified V–Cr–Ti alloys may be caused by the formation of Ti(CON) precipitates during He-ion irradiation. The nucleation and growth of Ti(CON) precipitate should be rate-limited to the concentration of gas impurities of C, N and O, and Ti atoms and the nucleation rate is proportional to the product $C_{imp} \cdot C_{Ti}$ of the gas impurity concentration C_{imp} and Ti concentration C_{Ti} from the kinetics of the point defect reaction [30], when the nuclei of Ti(CON) are assumed to be TiO- or TiC-type [31]. The reduction in gas impurities in the V–Cr–Ti alloy matrix is connected with the nucleation rate of Ti(CON) precipitate and a decrease of irradiation hardening in the V–Cr–Ti alloys.

Microstructural observation by TEM is required to clarify the correlation between irradiation hardening and microstructural evolution, and especially Ti(CON) formation in the future.

Because V–4Cr–1Ti alloys show significant irradiation hardening and more than 2% Ti addition to V–4Cr–xTi alloys results in a lower irradiation hardening than the V–4Cr–1Ti alloy, the optimum amount of Ti addition to candidate alloys for structural materials of fusion reactor application should be 2% to avoid surplus irradiation hardening at low-temperature irradiation. In terms of the swelling behavior of V–Cr–Ti alloys, 1% Ti addition to V–Cr–Ti and V–Fe–Ti alloys is enough to suppress void swelling from 400 to 600 °C with heavy damage levels to 30 dpa [32]. Hence, 2% Ti addition to V–Cr–Ti alloys helps to suppress void swelling at a high temperature with heavy damage. Consequently, from the viewpoints of the suppression of surplus irradiation hardening and void swelling, an optimum composition of V–Cr–Ti alloys for structural materials of fusion reactor engineering is proposed to be a highly purified V–(6–8)Cr–2Ti.

5. Conclusions

To survey an optimum composition to reduce the radioactivity in V–Cr–Ti alloys, 15 types of V–(4–8)Cr–(0–4)Ti alloys were fabricated. These alloys were subjected to 2-MeV He-ion irradiation using a tandem accelerator at the Wakasa Bay Energy Research Center. Specimens were irradiated at 500 °C and 700 °C up to doses of 0.5 dpa at peak positions in 3.6 μm depth. To investigate the effect of Cr and Ti addition and gas impurities for irradiation hardening behavior in the V–Cr–Ti alloys, nano-indentation tests were examined at room temperature. Cr and Ti addition to V–Cr–Ti alloys for solid–solution hardening is low in the unirradiated V–(4–8)Cr–(0–4)Ti alloys. Irradiation hardening could be observed among all V–Cr–Ti alloys irradiated at 500 °C and 700 °C. The V–4Cr–1Ti alloy irradiated at 500 °C shows the highest irradiation hardening among all V–Cr–Ti alloys and the irradiation hardening was 2.8 times larger than the hardness of unirradiated V–4Cr–1Ti alloy. The gas impurity increased the irradiation hardening in V–4Cr–xTi alloys. The Cr addition dependence on irradiation hardening to V–yCr–1Ti was not remarkable. The effect of interstitial gas impurity for irradiation hardening of V–yCr–Ti showed that the conventional fabricated alloys had a larger irradiation hardening increase than the highly purified alloys. The significant irradiation hardening in V–4Cr–1Ti is caused by the formation of Ti(CON) precipitate from He-ion irradiation. Because the thermal creep strength of the V–3Ti is highest among all V–Ti system alloys and small precipitates formed in thermal-creep deformed V–4Cr–4Ti alloys in previous studies, the formation of Ti(CON) precipitate results from hardening during irradiation. Consequently, to suppress irradiation hardening and void swelling, the optimum composition of V–Cr–Ti alloys for structural materials of fusion reactor engineering is proposed to be a highly purified V–(6–8)Cr–2Ti alloy.

Author Contributions: Conceptualization, K.-i.F.; TEM observation, K.i.F. and T.N.; Sample preparation, T.N., S.M., and K.F.; Ion irradiation, R.I. and S.M.; Nano indentation, S.M., K.F. and Y.K.; Manuscript writing, K.F. All authors have read and agreed to the published version of the manuscript.

Funding: This study was partly supported by the NIFS Budget, Code NIFS17KEMF098. This work was supported by a JSPS Grant-in-Aid for Scientific Research (A) 20H00144.

Institutional Review Board Statement: Not applicable.

Informed Consent Statement : Not applicable.

Data Availability Statement: Data available in a publicly accessible repository.

Acknowledgments: The authors are grateful to the technical staffs of the tandem accelerator at the Wakasa-Wan Energy Research center for supplying high-quality ion beams.

Conflicts of Interest: The authors declare no conflict of interest.

References

1. Matsui, H.; Fukumoto, K.; Smith, D.; Chung, H.M.; Van Witzenburg, W.; Votinov, S. Status of vanadium alloys for fusion reactors. *J. Nucl. Mater.* **1996**, *233*, 92–99. [CrossRef]
2. Kurtz, R.J.; Abe, K.; Chernov, V.M.; Hoelzer, D.T.; Matsui, H.; Muroga, T.; Odette, G.R. Vanadium alloys for fusion blanket ap-plications. *J. Nucl. Mater* **2004**, *329–333*, 47–55. [CrossRef]
3. Muroga, T.; Chen, J.; Chernov, V.; Kurtz, R.; Le Flem, M. Present status of vanadium alloys for fusion applications. *J. Nucl. Mater.* **2014**, *455*, 263–268. [CrossRef]
4. Smith, D.L.; Chung, H.M.; Loomis, B.A.; Tsai, H.-C. Reference vanadium alloy V-4Cr-4Ti for fusion application. *J. Nucl. Mater.* **1996**, *233*, 356–363. [CrossRef]
5. Chung, H.M.; Loomis, B.A.; Smith, D.L. Properties of V-4Cr-4Ti for application as fusion reactor structural components. *Fusion Eng. Des.* **1995**, *29*, 455–464.
6. Zinkle, S.J.; Matsui, H.; Smith, D.; Rowcliffe, A.; Van Osch, E.; Abe, K.; Kazakov, V. Research and development on vanadium alloys for fusion applications. *J. Nucl. Mater.* **1998**, *258*, 205–214. [CrossRef]
7. Nagasaka, T.; Muroga, T.; Imamura, M.; Tomiyama, S.; Sakata, M. Fabrication of high-purity V-4Cr-4Ti low activation alloy products. *Fusion Technol.* **2001**, *39*, 659–663. [CrossRef]
8. Muroga, T.; Nagasaka, T.; Abe, K.; Chernov, V.M.; Matsui, H.; Smith, D.L.; Xu, Z.-Y.; Zinkle, S.J. Vanadium alloys—Overview and recent results. *J. Nucl. Mater.* **2002**, *307*, 547–554. [CrossRef]
9. Nagasaka, T.; Muroga, T.; Yican, W.; Zengyu, X.; Imamura, M. Low activation characteristics of several heats of V-4Cr-4Ti ingots. *J. Plasma Fusion Res. Ser.* **2002**, *5*, 545–550.
10. Muroga, T. Vanadium alloys for fusion blanket application. *Mater. Trans.* **2005**, *46*, 405–411. [CrossRef]
11. Tanaka, T.; Nagasaka, T.; Muroga, T.; Yamazaki, M.; Toyama, T. Activation analysis for the reference low-activation vanadium alloy NIFS-HEAT-2. *Nucl. Mater. Eng.* **2020**, *25*, 100782. [CrossRef]
12. Nagasaka, T. Unpublished Work in Grant-in-Aid for Scientific Research (A) 20H00144, 2020, Japan. Available online: https://kaken.nii.ac.jp/grant/KAKENHI-PROJECT-20H00144/ (accessed on 31 December 2020).
13. Saleh, M.; Xu, A.; Hurt, C.; Ionescu, M.; Daniels, J.E.; Munroe, P.; Edwards, L.; Bhattacharyya, D. Oblique cross-section nanoindentation for determining the hardness change in ion-irradiated steel. *Int. J. Plast.* **2019**, *112*, 242–256. [CrossRef]
14. Fukumoto, K.; Matsui, H.; Chung, H.; Gazda, J.; Smith, D. Helium behavior in vanadium-based alloys irradiated in the dynamic helium charging experiments. *Sci. Rep. Res. Inst. Tohoku Univ. A Phys. Chem. Met.* **1997**, *45*, 149–155.
15. Fukumoto, K.; Onitsuka, T.; Narui, M. Dose dependence of irradiation hardening of neutron irradiated vanadium alloys by using temperature control rig in JMTR. *Nucl. Mater. Energy* **2016**, *9*, 441–446. [CrossRef]
16. Heo, N.J.; Nagasaka, T.; Muroga, T. Effect of impurity levels on precipitation behavior in the low-activation V–4Cr–4Ti alloys. *J. Nucl. Mater.* **2002**, *307–311*, 620–624. [CrossRef]
17. Yaggee, F.L.; Gilbert, E.R.; Styles, J.W. Thermal expansivities, thermal conductivities, and densities of vanadium, titanium, chromium and some vanadium-base alloys: A comparison with austenitic stainless steel. *J. Less Common Met.* **1969**, *19*, 39–51. [CrossRef]
18. Börm, H.; Schirra, M. Zeitstand- und kriechverhalten von Vanadin-Titan und Vanadin-Titan-Niob-legierungen. *Z Metallkde* **1968**, *59*, 715–723.
19. Börm, H.; Schirra, M. Untersuchungen über das zeitstand- und kreichverhalten binärer und ternärer vanadin-legierungen. *J. Less Common Met.* **1967**, *12*, 280–293.
20. Fukumoto, K.; Matsui, H. Precipitation behavior of vanadium alloys during creep deformation in a liquid sodium environment. *Mater. Jpn.* **2008**, *47*, 611. (In Japanese) [CrossRef]
21. Impagnatiello, A.; Shubeita, S.M.; Wady, P.T. Monolayer-thick TiO precipitation in V-4Cr-4Ti alloy induced by proton irradiation. *Scr. Mater.* **2017**, *130*, 174–177. [CrossRef]
22. Impagnatiello, A.; Toyama, T.; Jimenez-Melero, E. Ti-rich precipitate evolution in vanadium-based alloys during annealing above 400 °C. *J. Nucl. Mater.* **2017**, *485*, 122–128. [CrossRef]
23. Impagnatiello, A.; Hernandez-Maldonado, D.; Bertali, G. Atomically resolved chemical ordering at the nm-thick TiO precipitate/matrix interface in V-4Ti-4Cr alloy. *Scr. Mater.* **2017**, *126*, 50–54. [CrossRef]
24. Chung, H.M.; Loomis, B.A.; Smith, D.L. Creep properties of vanadium-base alloys. *J. Nucl. Mater.* **1994**, *212–215*, 772–777. [CrossRef]
25. Fukumoto, K.; Matsui, H.; Candra, Y.; Takahashi, K.; Sasanuma, H.; Nagata, S.; Takahiro, K. Radiation-induced precipitation in V–(Cr,Fe)–Ti alloys irradiated at low temperature with low dose during neutron or ion irradiation. *J. Nucl. Mater.* **2000**, *283–287*, 535–540. [CrossRef]
26. Fukumoto, K.; Iwasaki, M. A replica technique for extracting precipitates from neutron-irradiated or thermal-aged vanadium alloys for TEM analysis. *J. Nucl. Mater.* **2014**, *449*, 315–319. [CrossRef]
27. Watanabe, H.; Muroga, T.; Nagasaka, T. Effects of Irradiation Environment on V-4Cr-4Ti Alloys. *Plasma Fusion Res.* **2017**, *12*, 2405011. [CrossRef]
28. Rice, P.M.; Zinkle, S.J. Temperature dependence of the radiation damage microstructure in V–4Cr–4Ti neutron irradiated to low dose. *J. Nucl. Mater.* **1998**, *258–263*, 1414–1418. [CrossRef]

29. Seeger, A. *Proceedings of 2nd United Nations International Conference on the Peaceful Uses of Atomic Energy, Geneva*; United Nations: New York, NY, USA, 1958; Volume 6, p. 250.
30. Kiritani, M. Microstructure evolution during irradiation. *J. Nucl. Mater.* **1994**, *216*, 220–264. [CrossRef]
31. Kinoshita, C.; Fukumoto, K.; Nakai, K. Radiation-induced microstructural change in ceramic materials. *Ann. Chim. Fr.* **1991**, *16*, 379.
32. Fukumoto, K.; Kimura, A.; Matsui, H. Swelling behavior of V–Fe binary and V–Fe–Ti ternary alloys. *J. Nucl. Mater.* **1998**, *258–263*, 1431–1436. [CrossRef]

quantum beam science

MDPI

Article

Fabrication of Size- and Shape-Controlled Platinum Cones by Ion-Track Etching and Electrodeposition Techniques for Electrocatalytic Applications

Yuma Sato [1], Hiroshi Koshikawa [2], Shunya Yamamoto [2], Masaki Sugimoto [2], Shin-ichi Sawada [2] and Tetsuya Yamaki [1,2,*]

1   Graduate School of Science and Technology, Gunma University, Kiryu 376-8515, Gunma, Japan; gummama310@gmail.com
2   Takasaki Advanced Radiation Research Institute, National Institutes for Quantum and Radiological Science and Technology, Takasaki 370-1292, Gunma, Japan; koshikawa.hiroshi@qst.go.jp (H.K.); yamamoto.shunya@qst.go.jp (S.Y.); sugimoto.masaki@qst.go.jp (M.S.); sawada.shinnichi@qst.go.jp (S.-i.S.)
*   Correspondence: yamaki.tetsuya@qst.go.jp

Abstract: The micro/nanocone structures of noble metals play a critical role as heterogeneous electrocatalysts that provide excellent activity. We successfully fabricated platinum cones by electrodeposition using non-penetrated porous membranes as templates. This method involved the preparation of template membranes by the swift-heavy-ion irradiation of commercially available polycarbonate films and subsequent chemical etching in an aqueous NaOH solution. The surface diameter, depth, aspect ratio and cone angle of the resulting conical pores were controlled in the ranges of approximately 70–1500 nm, 0.7–11 μm, 4–12 and 5–13°, respectively, by varying the etching conditions, which finally produced size- and shape-controlled platinum cones with nanotips. In order to demonstrate the electrocatalytic activity, electrochemical measurements were performed for the ethanol oxidation reaction. The oxidation activity was found to be up to 3.2 times higher for the platinum cone arrays than for the platinum plate. Ion-track etching combined with electrodeposition has the potential to be an effective method for the fabrication of micro/nanocones with high electrocatalytic performance.

Keywords: ion-track etching; electrodeposition; micro/nano-sized metal cones; template synthesis; electrocatalyst

check for **updates**

Citation: Sato, Y.; Koshikawa, H.; Yamamoto, S.; Sugimoto, M.; Sawada, S.-i.; Yamaki, T. Fabrication of Size- and Shape-Controlled Platinum Cones by Ion-Track Etching and Electrodeposition Techniques for Electrocatalytic Applications. *Quantum Beam Sci.* **2021**, *5*, 21. https://doi.org/10.3390/qubs5030021

Academic Editors: Akihiro Iwase and Alberto Bacci

Received: 6 April 2021
Accepted: 24 June 2021
Published: 1 July 2021

Publisher's Note: MDPI stays neutral with regard to jurisdictional claims in published maps and institutional affiliations.

1. Introduction

Micro/nano-sized wire and cone structures of platinum directly integrated with a conducting substrate have several advantages for electrocatalytic performance [1–3]. First, the large surface areas accelerate the surface reaction. Second, the open spaces among the nanostructures lead to efficient mass transfer. Third, the direct contact with the electrolyte as well as with the substrate makes mixing with a polymer binder unnecessary in electrode production. The resulting binder-free architecture is expected to maintain a high electric conductivity and effectively avoid blocking the active sites. Among various nanostructures, the cone shape enhances the mechanical stability because it has not only a fine tip but also a large base.

One of the methods to prepare metal micro/nanocones is the deposition of metal in a template with conically shaped pores; different techniques using electrodeposition or electroless deposition can be employed. Electroless deposition does not require conducting substrates; however, the surface must be pretreated with Sn^{2+}, Ag, or Pd via a sensitization and activation processes [4]. Consequently, these metal ions and atoms are left as contaminants on the surface of the cones and thus affect their electrocatalytic performance [5,6] On the other hand, electrodeposition requires no pretreatment steps and therefore enables the fabrication of metal-contaminant-free platinum cones.

The most commonly used templates are anodic porous alumina and track-etched membranes [7,8]. Our particular interest is focused on track-etched membranes with nanometer-to-micrometer pores, which are prepared by the swift-heavy-ion irradiation of polymer films followed by chemical etching. This is because the ion-track technique offers flexible porous templates with large areas and enables the adjustment of the shape, size, orientation and density of the pores independently by varying the conditions of the irradiation and chemical etching.

To date, studies have been conducted on the fabrication of track-etched membranes with various shapes and configurations. For example, Rauber et al. reported a modified template fabrication method for the preparation of 13–44.2-nm-thick platinum wire networks, which involved ion-irradiation in several steps from different directions [9]. Cylindrical, conical and double-cone-shaped pores were prepared with diameters of approximately 500–1500 nm by a multi-step etching technique, penetrating through a polyethylene terephthalate (PET) membrane of a 12 μm thickness [10]. In order to make the conically shaped pores as a template, chemical etching is usually performed with an etchant on one side of the membrane and a stop solution on the other [11]. For the subsequent electrodeposition of the desired metal, one side of the track-etched membranes is then sputter-coated with a thin metal layer to prepare the cathode [11,12]. In this conventional method using penetrating pores (also called through-holes), the length of the fabricated micro/nanostructures is always equal to the thickness of the track-etched membranes and is never controlled in principle. In contrast, a track-etched membrane with non-penetrating pores can be used for the preparation of cones with a controlled length, thereby enhancing the degree of freedom in the electrocatalyst's design and production. Until the present time, gold cones were fabricated by electroless deposition using a track-etched membrane template with non-penetrating pores, which had diameters and depths in the range of 1–8 μm and 1–11 μm, respectively [13].

In this study, platinum cones were fabricated by an electrodeposition method using track-etched membranes with non-penetrating pores. The size and shape of the template pores were controlled by the etching time and etchant concentration, thereby enabling the preparation of platinum cones with different base diameters and lengths. Finally, we demonstrated their higher electrocatalytic activity by the electro-oxidation of ethanol.

2. Experimental

A stack of two polymer films, a polyimide (PI) film (Kapton, DuPont-Toray Co., Ltd., Tokyo, Japan) with a thickness of 12 μm on a polycarbonate (PC) film (Panlite, Teijin Ltd., Tokyo, Japan) of a 50 μm thickness, was irradiated with ^{40}Ar ions at an energy of 150 MeV using the azimuthally varying field (AVF) cyclotron at Takasaki Ion Accelerators for Advanced Radiation Application (TIARA). Figure 1 depicts the depth profile of the linear energy transfer (LET) of the ^{40}Ar ion calculated by the Stopping and Range of Ions in Matter (SRIM) code [14], together with the configuration for the irradiation of the first 12 μm-thick PI and the second 50 μm-thick PC films. The stopping depth was 54 μm from the front surface, meaning that the impinging ion penetrated through the first PI film and stopped in the second PC film. Consequently, we employed the present irradiation configuration. The track etching of the PC film irradiated in this way started from one side, yielding non-penetrating conical pores. The number of irradiated ions per 1 cm^2, namely the ion fluence, was fixed at 3.0×10^7 and 1.0×10^8 ions/cm^2.

Figure 2 shows the preparation procedure, which involved filling the non-penetrating pores of track-etched membranes with electroplated platinum. After the ^{40}Ar irradiation, the PC films were etched in 2.0, 4.0 and 6.0 mol/dm^3 aqueous NaOH solutions at 60 °C. The etched films were then washed with water and dried at room temperature. The surface and cross-section of the track-etched membranes were observed by field emission scanning electron microscopy (FE-SEM) (JSM-6700F, JEOL Ltd., Tokyo, Japan). Before the observation, a thin gold layer was deposited on the sample surface to improve its conductivity. In order to create the fine cross-section of the etched pores, the membranes were embrittled by exposure to UV light [10].

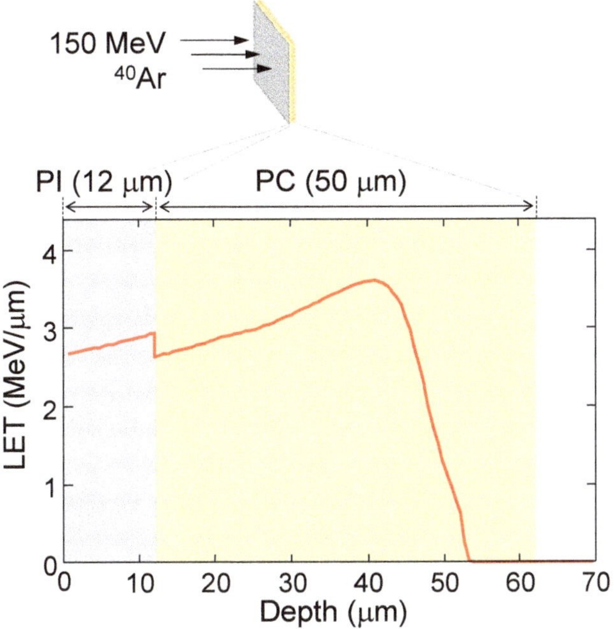

Figure 1. Depth profile of the LET of 150 MeV ^{40}Ar ions for a 12 µm-thick PI overlaid on 50 µm-thick PC, which was calculated by our SRIM simulation.

For the electrodeposition, the track-etched membrane was physically coated with a thin platinum layer from its pore-open side; the current and time of the vapor deposition were 20 mA and 520 s, respectively. This layer reached a thickness of approximately 40 nm and worked as a cathode for the electrodeposition of the platinum cones in a commercial platinum electrolyte (PRECIOUSFAB Pt3000, Tanaka Kikinzoku Kogyo K.K., Tokyo, Japan) at 50 °C. The applied voltage was repeatedly pulsed at 1.8 V for 1 s and at 0 V for 2 s; the repetition number was 200 with a total deposition time of 10 min. After filling the non-penetrating pores with electroplated platinum from the bottom and reinforcement with a thick electroplated gold substrate, the track-etched membranes were dissolved in a 6.0 mol/dm^3 aqueous NaOH solution at 60 °C, leaving the free-standing platinum cone array.

The obtained platinum cones were observed by FE-SEM and transmission electron microscopy (TEM) (JEM-2100F, JEOL Ltd., Tokyo, Japan). The elemental composition was investigated using an energy-dispersive X-ray (EDX) analyzer (X-Max, HORIBA Ltd., Kyoto, Japan). The electrochemical measurements were performed using a three-electrode cell at room temperature. The platinum cones were used directly as the working electrodes. A platinum foil and Ag/AgCl (KCl sat.) were used as the counter and reference electrodes, respectively. All of the potentials were converted to the reversible hydrogen electrode (RHE) scale according to the Nernst equation: E(RHE) = E(Ag/AgCl) + 0.059 × pH + 0.197. The electrochemically active surface area (ECSA) of the platinum was measured by an established procedure using cyclic voltammetry (CV) [15]. The potential of the working electrode was scanned from 0.02 to 1.17 V vs. RHE at a scan rate of 50 mV/s in a N$_2$-saturated 0.5 mol/dm^3 aqueous H$_2$SO$_4$ solution. The coulombic charge corresponding to the adsorption peak of atomic hydrogen, Q_{Pt-H}, was estimated by integrating the CV curve in the hydrogen underpotential deposition region. Assuming a coulombic charge of

0.21 mC/cm^2 for hydrogen adsorption on a smooth polycrystalline platinum surface, we used the equation ECSA (cm^2) = Q_{Pt-H}/0.21. Subsequently, in order to measure the ethanol oxidation reaction activity, the CV was performed in a N$_2$-saturated aqueous solution containing 0.5 mol/dm^3 ethanol and 0.5 mol/dm^3 H$_2$SO$_4$. The electrode was cycled in the potential range from 0 to 1.2 V vs. RHE at a scan rate of 20 mV/s.

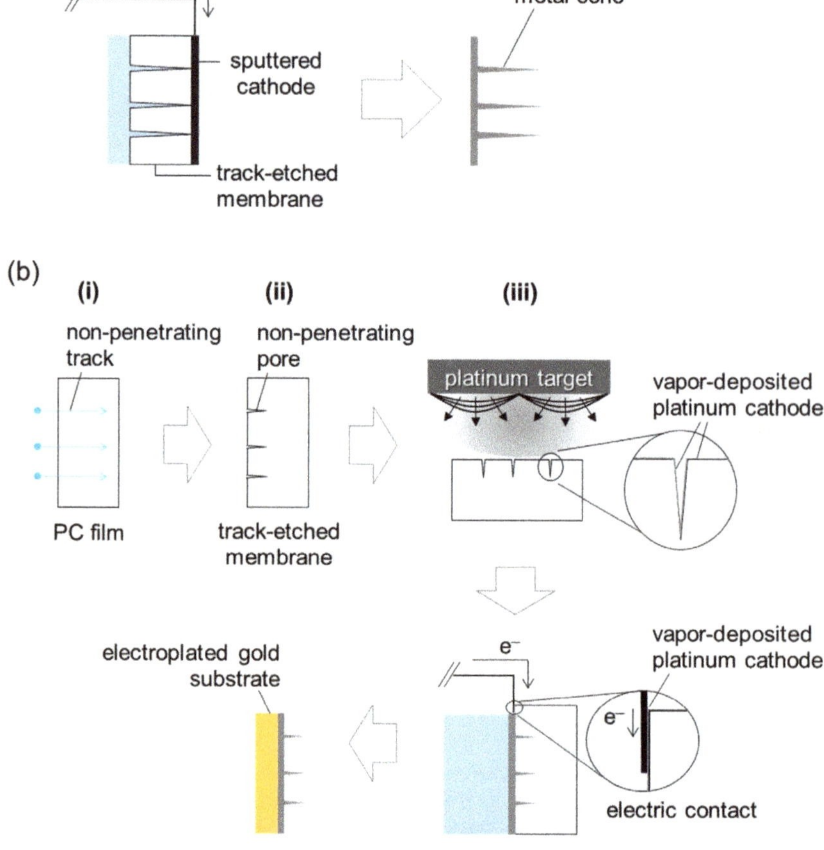

Figure 2. (a) Metal cones prepared using the track-etched membranes with penetrating pores as a template, as was the case in most of the previous studies. The conventional method involves sputter-coating with a thin metal layer on one side of track-etched membranes in order to prepare the cathode for electrodeposition. (b) Schematics of the main steps in the fabrication of the platinum cones based on a combination of ion-track etching and electrodeposition techniques: (i) ion irradiation of a PC film to obtain non-penetrating tracks; (ii) asymmetric etching to form non-penetrating conical pores; (iii) vapor deposition of a thin platinum layer from the pore-open side as a cathode; (iv) electrochemical deposition of platinum from the bottom of the non-penetrating pores, and from the surface of the track-etched membranes; (v) free-standing platinum cone array after reinforcement with a thick electroplated gold substrate and the removal of the template.

3. Results and Discussion

Figure 3 shows the FE-SEM images of the surface and cross-section of the PC membrane etched for 30 min in a 4.0 mol/dm^3 aqueous NaOH solution. The membranes were found to have uniform conical pores. The surface diameter of the pores was roughly estimated to be 540 nm by taking the average of the neighboring 50 pores. On the other hand, the depth seemed scattered, which may have been because, in the sample preparation for our cross-sectional SEM observations, the membrane was not necessarily cracked exactly in the middle of the pore. Thus, we determined the maximum depth at which the resulting cut plane was assumed to pass through the close vicinity of the conical vertex. Figure 3b presents the image of the best cut plane for the four pores, where the diameter of the pore opening was almost the same as the diameter measured on the surface (Figure 3a). Thus, the tip would have been included. This resulted in a depth of 3.8 µm.

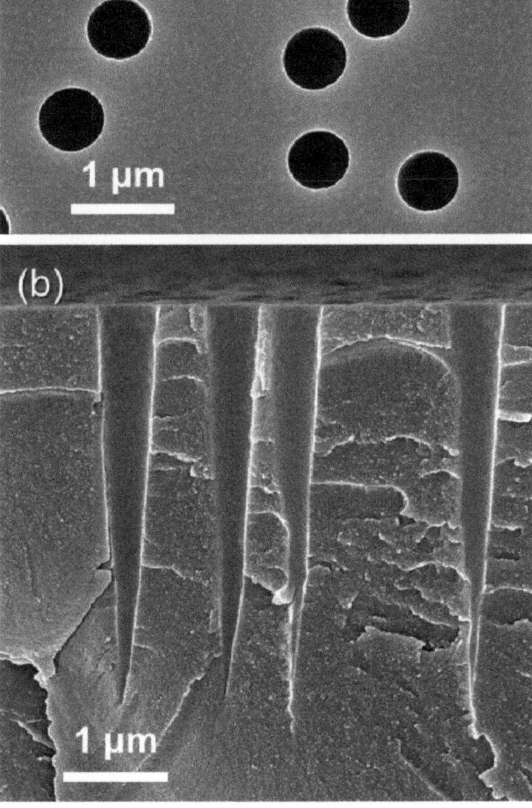

Figure 3. FE-SEM images of (**a**) the surface and (**b**) the cross-section of the PC membrane etched for 30 min in a 4.0 mol/dm^3 NaOH solution at 60 °C.

On the other side of the membrane, no pores were observed, confirming that the bombarded ions did not penetrate through the PC film, as calculated by the SRIM code. A previously proposed method for the fabrication of conically shaped pores that do not propagate through the entire thickness, that is, non-penetrating conical pores, involves the chemical etching of the penetrating tracks only from one side of the irradiated film and its stopping before the etchant breaks through to the opposite side [13]. In this method, a compression-sealed two-component cell must be employed to avoid the contact of another side with the etchant during the one-side etching [16]. However, we obtained non-penetrating conical pores even without such a dedicated cell and special care for the etching. This is because the tracks in the PC film did not propagate through its entire thickness. The immersion of the irradiated films in the etchant resulted in track-etched membranes with non-penetrating conical pores.

Non-penetrating conical pores were prepared by varying the etching time and etchant concentration. Figure 4 plots the pore diameter and depth as a function of the etching time for different concentrations of the NaOH solutions. In Figure 4a, the surface diameter is observed to be uniform with rather small error bars under all of the etching conditions. As mentioned above, the maximum depth was taken for at least 20 pores; thus, Figure 4b plots the depth with no error bars. The diameter and depth of the pores linearly increase with the etching time; therefore, the slope of the plots corresponding to the growth rate of the pores was estimated by least-squares regression. At concentrations of 2.0, 4.0 and 6.0 mol/dm^3, the pore diameter was enhanced at 16, 26 and 50 nm/min, while the pore depth increased more significantly at 60, 130 and 660 nm/min, respectively. Consequently, the pore diameter and depth were controlled by a combination of the etching time and the NaOH concentration.

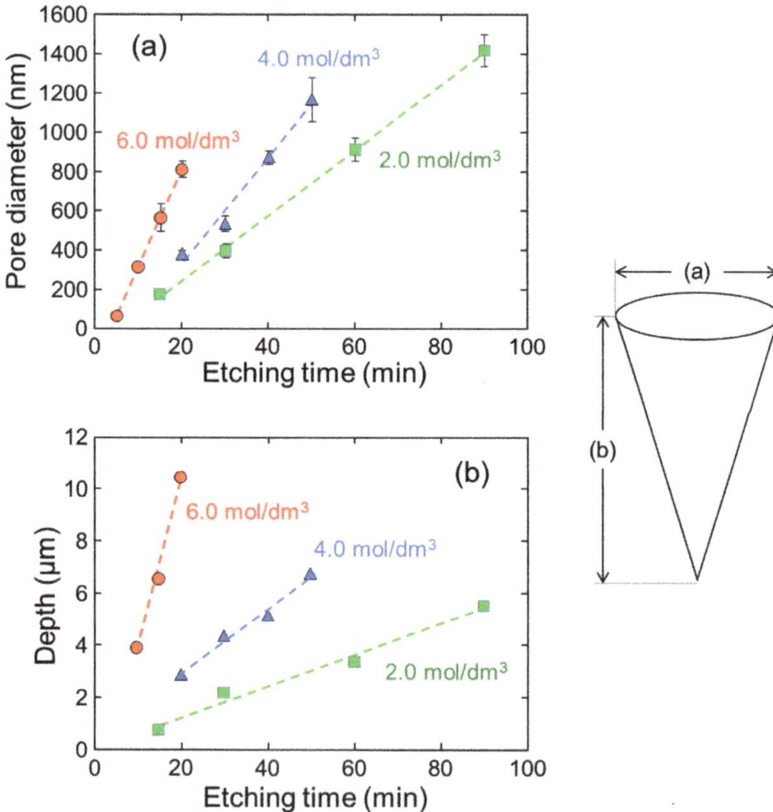

Figure 4. (a) Pore diameter and (b) depth as a function of the etching time in an aqueous NaOH solution with different concentrations at 60 °C.

Figure 5 depicts the aspect ratio (AR) [17] and the cone angle, α, [18] of the conical pores, which are defined as the ratio of the depth to the surface diameter and the angle between two generatrix lines, respectively. Based on the assumption of a perfect conical geometry, the α values were calculated by the relationship $\alpha = 2\tan^{-1}(1/AR)$. For example, these were estimated to be 6.8 and 8.4°, respectively, for the pores shown in Figure 3. For all of the etching times, we observed their average values change from 4.3 to 12.2 and from 13.3° to 4.7°, respectively, with the increase of the concentration of the aqueous NaOH solution from 2.0 to 6.0 mol/dm³. This concentration dependence can be rationalized by considering that the etch rate in the depth direction was more sensitive to the NaOH concentration than that in the transverse direction, as discussed earlier. More importantly, the AR and α, as well as the diameter and depth of the conical pores, were controlled by adjusting the etchant concentration.

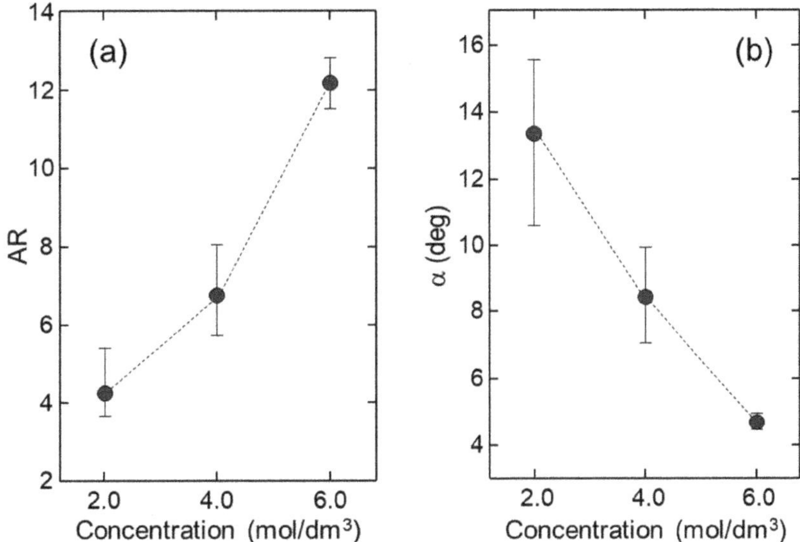

Figure 5. Dependence of (**a**) the AR and (**b**) α of the conical pores on the concentration of the aqueous NaOH solution. Both of the values were calculated using the data in Figure 4a,b, which were obtained from the experiments at all of the etching times.

Track-etched membranes with non-penetrating conical pores were used for the fabrication of the platinum cones. Figure 6a–c depicts the FE-SEM images of the platinum cone arrays obtained using the template membranes etched for 30 min in the 2.0, 4.0 and 6.0 mol/dm³ NaOH solutions, respectively. The platinum cones shown in Figure 6b have base diameters and lengths of 540 nm and 3.7 μm, respectively, both of which agree well with those of the template shown in Figure 1. The same is the case for the platinum cones shown in Figure 6a. In contrast, we obtained truncated platinum cones, as shown in Figure 6c, using the conical pores with the highest aspect ratio as a template. The vapor-deposited platinum cathode could not reach the bottom of the pores, which is likely because the platinum atoms and ions sputtered from the target arrived inside the conical pores from random directions [19,20].

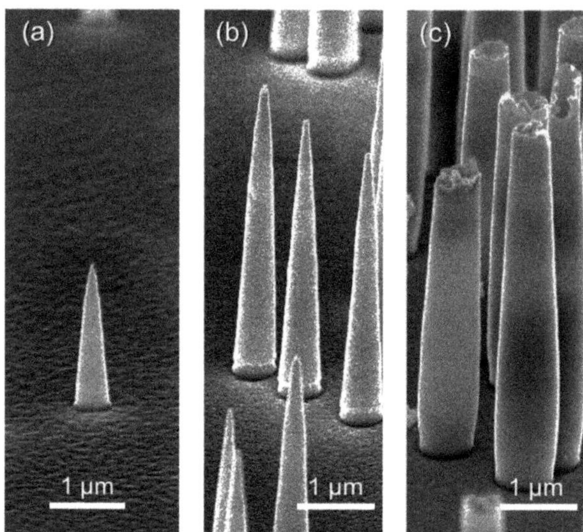

Figure 6. FE-SEM images of the platinum cones. The cones were electrochemically deposited in the conical pores prepared in (**a**) 2.0 (**b**) 4.0, and (**c**) 6.0 mol/dm^3 aqueous NaOH solutions at 60 °C.

Figure 7 depicts the EDX spectrum of the platinum cones. The platinum signals were observed at 9.42 keV (L$_\alpha$ line), 2.04 keV (M$_\alpha$ + M$_\beta$ line) and 1.59 keV (M$_\xi$ line). Meanwhile, the other emission lines were assigned to the non-metal components, likely from the electrolyte solution PRECIOUSFAB Pt3000, e.g., carbon, nitrogen, oxygen, sulfur, and chlorine. In the electroless deposition method, the other signals of the minor components, such as Sn and Ag, were observed [4]. The platinum cones without any metal contaminations were fabricated by a combination of ion-track etching with the electrodeposition technique.

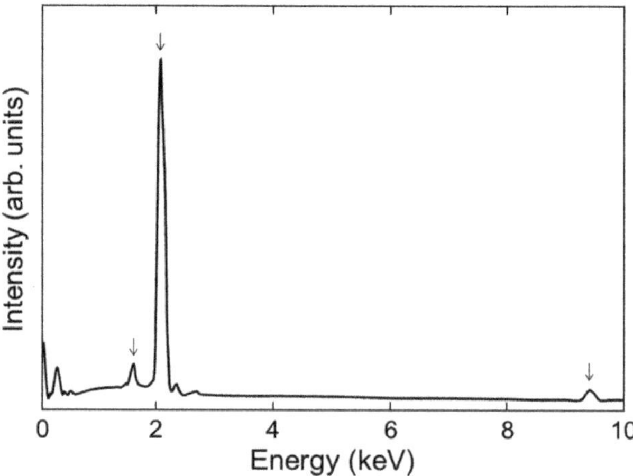

Figure 7. EDX spectrum of the obtained platinum cones. The arrows indicate the platinum signals observed at 9.42 keV (Lα line), 2.04 keV (Mα + Mβ line) and 1.59 keV (Mξ line).

The platinum cones were further analyzed using TEM. Figure 8 shows a TEM image of the tip area of the cones shown in Figure 6b. The radius of curvature of the tip was 13 nm. The selected-area electron diffraction pattern in the inset of Figure 8 exhibits concentric rings composed of bright, discrete diffraction spots that were indexed to the (111), (200), (220), and (311) crystal planes of fcc platinum, indicating the polycrystalline structure of the individual cones.

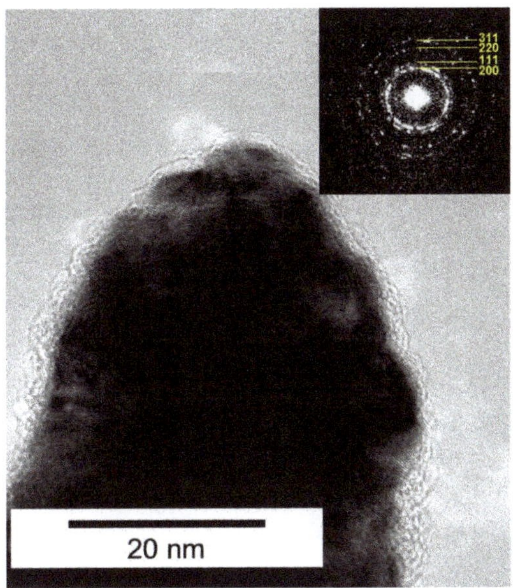

Figure 8. TEM image of the tip of the platinum cone. The inset shows the selected-area electron diffraction pattern.

Figure 9a shows the cyclic voltammograms obtained in a 0.5 mol/dm^3 aqueous H$_2$SO$_4$ solution. The base diameter, length and areal density of the platinum cones are 550 nm, 2.4 µm and 1.0×10^8/cm^2, respectively. In this areal density, nearly half of the cones could be isolated, while the remaining ones may overlap to form multiple (mainly double) cones according to the calculation using a Poisson distribution model [21]. For comparison, the platinum plate with a diameter of 1.5 cm was fabricated by electrodeposition on the PC film without any etched pores, and its circular portion (0.8 cm in diameter) was measured in the same way. Both of the samples exhibited a hydrogen adsorption and desorption region at 0.02–0.4 V vs. RHE and a double layer plateau region at 0.4–0.6 V vs. RHE with peaks for the formation and reduction of surface platinum oxide at 0.6–1.17 V vs. RHE. The ECSA for the hydrogen adsorption was 1.8 times higher for the cones than for the plate. In a previous paper [15], the surface area resulting from the double-layer capacitance was estimated for comparison with the ECSA. This was determined by the non-Faradaic double-layer charging current around 0.5 V vs. RHE. It was approximately 1.7 times higher for the platinum cones than for the plate; therefore, the increase in the ECSA could reasonably be accounted for within the allowable error.

The electrocatalytic performance for ethanol oxidation was demonstrated by CV in an aqueous solution containing 0.5 mol/dm^3 ethanol and 0.5 mol/dm^3 H$_2$SO$_4$. The CV curves of the platinum cones and platinum plate are shown in Figure 9b. The current in the forward scan exhibited the oxidation of ethanol, whereas in the backward scan, another oxidation current was observed, which was associated with the oxidation of intermediates of ethanol dissociative adsorption [22]. The ECSA-normalized current densities (called specific currents) at 0.7 V vs. RHE were extracted and compared with those of platinum on carbon (often referred to as Pt/C) and platinum black found in a study by Mao and co-workers [23]. The platinum cones and plate indicated approximately 0.06 and 0.02 mA/cm^2, respectively, whereas both of the commercial products exhibited 0.06–0.07 mA/cm^2. The only difference between our measurements and theirs was the ethanol concentration, and the CV curve was recorded in an aqueous solution containing 2 mol/cm^3 ethanol and 0.5 mol/cm^3 H$_2$SO$_4$. The twofold higher ethanol concentration would lead to an increase in the current density because it likely related to a greater amount of electroactive species in the solution [24]. Therefore, our platinum cones could exhibit the best performance among these four samples.

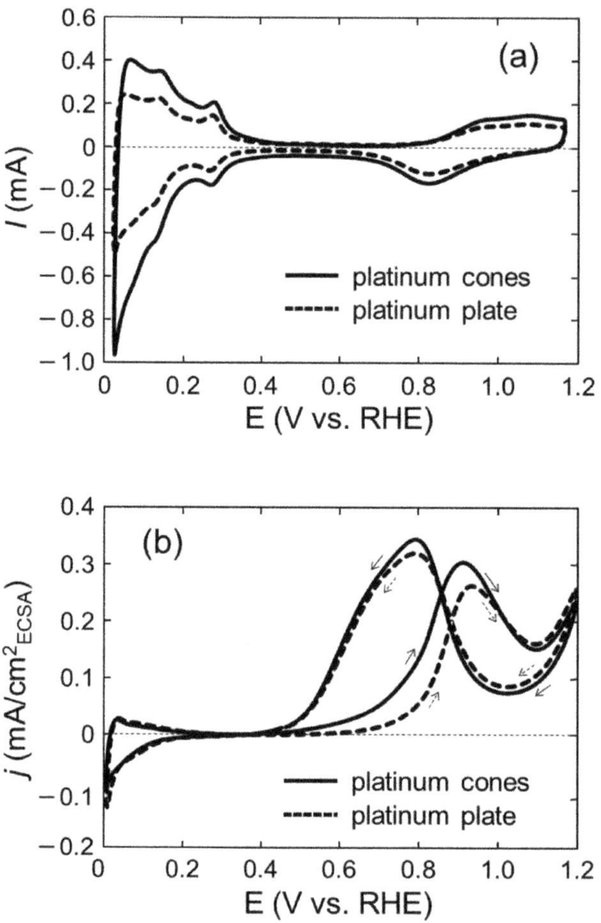

Figure 9. Cyclic voltammograms of the platinum cones and plate (**a**) in 0.5 mol/dm^3 H$_2$SO$_4$ using a scan rate of 50 mV/s and (**b**) in 0.5 mol/dm^3 ethanol + 0.5 mol/dm^3 H$_2$SO$_4$ using a scan rate of 20 mV/s.

It should be emphasized that the current density was 3.2 times higher for the platinum cones than for the platinum plate. This comparison leads to the assumption that the cone structure contributes to the improvement of the electrocatalytic activity. The high electrocatalytic activity may be exhibited by the field-induced reagent concentration [25,26]. In other words, the polar ethanol molecules are likely to approach the surface of the cone electrode because the fine tips of the cones produce high local electrical fields. Finally, we demonstrated the fabrication of platinum cones by ion-track etching and electrodeposition techniques and found them to be a promising alternative for ethanol oxidation electrocatalysts.

4. Conclusions

Platinum cones were fabricated by vapor deposition for cathode preparation followed by electrodeposition to PC track-etched membranes with non-penetrating conically shaped pores as templates. Conical pores with a surface diameter and depth of approximately 70–1500 nm and 0.7–11 µm, respectively, were obtained by varying the etching time and etchant concentration, thereby also enabling control of their aspect ratio and cone angle. The obtained platinum cones were found to be approximately the same size and geometry as the template pores, and they had a polycrystalline nanotip with no metal contaminants. In order to demonstrate the electrocatalytic activity, electrochemical measurements were performed for the ethanol oxidation reaction. The electrocatalytic activity of the platinum cones with a diameter of 550 nm, a length of 2.4 µm, and an areal density of $1.0 \times 10^8/cm^2$ was 3.2 times higher than that of the platinum plate. The combination of ion-track etching and electrodeposition is thus an effective method for the fabrication of micro/nanocones with high electrocatalytic performance.

Author Contributions: Conceptualization, T.Y.; data curation, Y.S. and M.S.; investigation, Y.S.; methodology, S.-i.S., S.Y. and H.K.; project administration, T.Y.; resources, S.Y. and H.K.; supervision, T.Y.; visualization, Y.S. and S.-i.S.; writing—original draft, Y.S. and T.Y.; writing—review & editing, T.Y. and Y.S. All authors have read and agreed to the published version of the manuscript.

Funding: This research received no external funding.

Conflicts of Interest: The authors declare no conflict of interest.

References

1. Zhao, G.-Y.; Xu, C.-L.; Guo, D.-J.; Li, H.; Li, H.-L. Template Preparation of Pt Nanowire Array Electrode on Ti/Si Substrate for Methanol Electro-Oxidation. *Appl. Surf. Sci.* **2007**, *253*, 3242–3246. [CrossRef]
2. Choi, S.M.; Kim, J.H.; Jung, J.Y.; Yoon, E.Y.; Kim, W.B. Pt Nanowires Prepared via a Polymer Template Method: Its Promise toward High Pt-Loaded Electrocatalysts for Methanol Oxidation. *Electrochim. Acta* **2008**, *53*, 5804–5811. [CrossRef]
3. Li, Y.; Hasin, P.; Wu, Y. $Ni_xCo_{3-x}O4$ Nanowire Arrays for Electrocatalytic Oxygen Evolution. *Adv. Mater.* **2010**, *22*, 1926–1929. [CrossRef]
4. Muench, F.; Oezaslan, M.; Seidl, T.; Lauterbach, S.; Strasser, P.; Kleebe, H.J.; Ensinger, W. Multiple Activation of Ion Track Etched Polycarbonate for the Electroless Synthesis of Metal Nanotubes. *Appl. Phys. A* **2011**, *105*, 847–854. [CrossRef]
5. Haruta, M. New Generation of Gold Catalysts: Nanoporous Foams and Tubes—Is Unsupported Gold Catalytically Active? *ChemPhysChem* **2007**, *8*, 1911–1913. [CrossRef]
6. Byeon, J.H.; Kim, Y.W. Simple Fabrication of a Pd–P Film on a Polymer Membrane and Its Catalytic Applications. *ACS Appl. Mater. Interfaces* **2011**, *3*, 2912–2918. [CrossRef] [PubMed]
7. Molares, M.E.T. Characterization and Properties of Micro- and Nanowires of Controlled Size, Composition, and Geometry Fabricated by Electrodeposition and Ion-Track Technology. *Beilstein J. Nanotechnol.* **2012**, *3*, 860–883. [CrossRef] [PubMed]
8. Brzózka, A.; Szeliga, D.; Tabor, E.K.; Sulka, G.D. Synthesis of Copper Nanocone Array Electrodes and Its Electrocatalytic Properties Toward Hydrogen Peroxide Reduction. *Mater. Lett.* **2016**, *174*, 66–70. [CrossRef]
9. Rauber, M.; Alber, I.; Müller, S.; Neumann, R.; Picht, O.; Roth, C.; Schökel, A.; Molares, M.E.T.; Ensinger, W. Highly-Ordered Supportless Three-Dimensional Nanowire Networks with Tunable Complexity and Interwire Connectivity for Device Integration. *Nano Lett.* **2011**, *11*, 2304–2310. [CrossRef]
10. Mo, D.; Liu, J.D.; Duan, J.L.; Yao, H.J.; Latif, H.; Cao, D.L.; Chen, Y.H.; Zhang, S.X.; Zhai, P.F.; Liu, J. Fabrication of Different Pore Shapes by Multi-Step Etching Technique in Ion-Irradiated PET Membranes. *Nucl. Instr. Methods Phys. Res. B* **2014**, *333*, 58–63. [CrossRef]
11. Duan, J.L.; Lei, D.Y.; Chen, F.; Lau, S.P.; Milne, W.I.; Molares, M.E.T.; Trautmann, C.; Lie, J. Vertically-Aligned Single-Crystal Nanocone Arrays: Controlled Fabrication and Enhanced Field Emission. *ACS Appl. Mater. Interfaces* **2016**, *8*, 472–479. [CrossRef] [PubMed]
12. Karim, S.; Ensinger, W.; Mujahid, S.A.; Maaz, K.; Khan, E.U. Effect of Etching Conditions on Pore Shape in Etched Ion-Track Polycarbonate Membranes. *Radiat. Meas.* **2009**, *44*, 779–782. [CrossRef]
13. Mukaibo, H.; Horne, L.P.; Park, D.; Martin, C.R. Controlling the Length of Conical Pores Etched in Ion-Tracked Poly(ethylene terephthalate) Membranes. *Small* **2009**, *5*, 2474–2479. [CrossRef]
14. Ziegler, J.F.; Ziegler, M.D.; Biersack, J.P. SRIM–The Stopping and Range of Ions in Matter (2010). *Nucl. Instr. Methods Phys. Res. B* **2010**, *268*, 1818–1823. [CrossRef]
15. Łukaszewski, M.; Soszko, M.; Czerwiński, A. Electrochemical Methods of Real Surface Area Determination of Noble Metal Electrodes—An Overview. *Int. J. Electrochem. Sci.* **2016**, *11*, 4442–4469. [CrossRef]

16. Apel, P.Y.; Korchev, Y.E.; Siwy, Z.; Spohr, R.; Yoshida, M. Diode-Like Single-Ion Track Membrane Prepared by Electro-Stopping. *Nucl. Instr. Methods Phys. Res. B* **2001**, *184*, 337–346. [CrossRef]
17. Hadley, A.; Notthoff, C.; Mota-Santiago, P.; Dutt, S.; Mudie, S.; Carrillo-Solano, M.A.; Toimil-Molares, M.E.; Trautmann, C.; Kluth, P. Analysis of Nanometer-Sized Aligned Conical Pores Using Small-Angle X-Ray Scattering. *Phys. Rev. Mater.* **2020**, *4*, 056003. [CrossRef]
18. Bakouei, M.; Abdorahimzadeh, S.; Taghipoor, M. Effects of Cone Angle and Length of Nanopores on the Resistive Pulse Quality. *Phys. Chem. Chem. Phys.* **2020**, *22*, 25306–25314. [CrossRef]
19. Durney, A.R.; Frenette, L.C.; Hodvedt, E.C.; Krauss, T.D.; Mukaibo, H. Fabrication of Tapered Microtube Arrays and Their Application as a Microalgal Injection Platform. *ACS Appl. Mater. Interfaces* **2016**, *8*, 34198–34208. [CrossRef] [PubMed]
20. Kakitani, K.; Koshikawa, H.; Yamaki, T.; Yamamoto, S.; Sato, Y.; Sugimoto, M.; Sawada, S. Preparation of Conductive Layer on Polyimide Ion-Track Membrane by Ar Ion Implantation. *Surf. Coat. Technol.* **2018**, *355*, 181–185. [CrossRef]
21. Spohr, R. *Ion Tracks and Microtechnology, Principles and Applications*; Vieweg & Sohn Verlagsgesellschaft mbH: Braunschweig, Germany, 1990; p. 173.
22. Fu, S.; Zhu, C.; Du, D.; Liu, Y. Facile One-Step Synthesis of Three-Dimensional Pd–Ag Bimetallic Alloy Networks and Their Electrocatalytic Activity toward Ethanol Oxidation. *ACS Appl. Mater. Interface* **2015**, *7*, 13842–13848. [CrossRef]
23. Mao, J.; Chen, W.; He, D.; Wan, J.; Pei, J.; Dong, J.; Wang, Y.; An, P.; Jin, Z.; Xing, W.; et al. Design of Ultrathin Pt-Mo-Ni Nanowire Catalysts for Ethanol Electrooxidation. *Sci. Adv.* **2017**, *3*, e1603068. [CrossRef] [PubMed]
24. Huang, J.-J.; Hwang, W.-S.; Weng, Y.-C.; Chou, T.-C. Electrochemistry of Ethanol Oxidation on Ni-Pt Alloy Electrodes in KOH Solutions. *Mater. Trans.* **2009**, *50*, 1139–1147. [CrossRef]
25. Liu, M.; Pang, Y.; Zhang, B.; Luna, P.D.; Voznyy, O.; Xu, J.; Zheng, X.; Dinh, C.T.; Fan, F.; Cao, C.; et al. Enhanced Electrocatalytic CO_2 Reduction via Field-Induced Reagent Concentration. *Nature* **2016**, *567*, 328–386. [CrossRef] [PubMed]
26. Safaei, T.S.; Mepham, A.; Zheng, X.; Pang, Y.; Dinh, C.T.; Liu, M.; Sinton, D.; Kelley, S.O.; Sargent, E.H. High-Density Nanosharp Microstructures Enable Efficient CO_2 Electroreduction. *Nano. Lett.* **2016**, *16*, 7224–7228. [CrossRef] [PubMed]

quantum beam science

MDPI

Article

Effect of Irradiation on Corrosion Behavior of 316L Steel in Lead-Bismuth Eutectic with Different Oxygen Concentrations

Nariaki Okubo *, Yuki Fujimura and Masakatsu Tomobe

Nuclear Science and Engineering Center, Japan Atomic Energy Agency (JAEA), Tokai, Ibaraki 319-1195, Japan; fujimura.yuki@jaea.go.jp (Y.F.); tomobe.masakatsu@jaea.go.jp (M.T.)
* Correspondence: okubo.nariaki@jaea.go.jp; Tel.: +81-29-282-6212

Abstract: In an accelerator-driven system (ADS), the beam window material of the spallation neutron target is heavily irradiated under severe conditions, in which the radiation damage and corrosion co-occur because of high-energy neutron and/or proton irradiation in the lead–bismuth flow. The materials used in ADSs must be compatible with the liquid metal (lead–bismuth eutectic (LBE)) to prevent issues such as liquid metal embrittlement (LME) and liquid metal corrosion (LMC). This study considers the LMC behavior after ion irradiation of 316L austenitic steel for self-ion irradiations followed by the corrosion tests in LBE with critical oxygen concentration. The 316L samples were irradiated by 10.5 MeV-Fe^{3+} ions at a temperature of 450 °C, up to 50 displacements per atom (dpa). After the corrosion test performed at 450 °C in LBE with low oxygen concentration, a surface of the nonirradiated area was not oxidized but appeared with locally corrosive morphology, Ni depletion, whereas an iron/chromium oxide layer fully covered the irradiated area. In the case of the corrosion surface with high oxygen concentration in LBE, the surface of the nonirradiated area was covered by an iron oxide layer only, whereas the irradiated area was covered by the duplex layers comprising iron and iron/chromium oxides. It is suggested that irradiation can enhance the oxide layer formation because of the enhancement of Fe and/or oxygen diffusion induced by the radiation defects in 316L steel.

Keywords: accelerator-driven system (ADS); liquid metal corrosion (LMC); lead–bismuth eutectic (LBE); self-ion irradiation; oxygen concentration in LBE; irradiation effect on corrosion behavior

Citation: Okubo, N.; Fujimura, Y.; Tomobe, M. Effect of Irradiation on Corrosion Behavior of 316L Steel in Lead-Bismuth Eutectic with Different Oxygen Concentrations. *Quantum Beam Sci.* **2021**, *5*, 27. https://doi.org/10.3390/qubs5030027

Academic Editor: Akihiro Iwase

Received: 6 April 2021
Accepted: 30 August 2021
Published: 31 August 2021

Publisher's Note: MDPI stays neutral with regard to jurisdictional claims in published maps and institutional affiliations.

1. Introduction

Decreasing the risk of spent nuclear fuel elements has become a major concern, especially in Japan after Fukushima's first nuclear power plant accident. An accelerator-driven system (ADS) is an important concept to realize partitioning and transmutation [1] to reduce the hazards associated with the spent fuel. In an ADS, the beam window, which is the boundary between a high-energy accelerator for protons in a vacuum and a spallation target of lead–bismuth eutectic (LBE), is irradiated under severe conditions to sufficiently transmute the minor actinides in the fuel cladding. The ADS irradiation conditions, which induce considerable displacement damage with high concentrations of helium (He) and hydrogen (H) atoms in the materials, are produced by the high-energy proton and spallation neutron irradiation. Degradation of the mechanical and corrosion properties after irradiation in the LBE at system temperatures, e.g., from 350 to 550 °C, should be maintained within a permissible range for a good system design [2]. High-fluence neutron irradiation experiments with up to about 20 and 100 displacements per atom (dpa, the parameter for indicating the radiation damage level) are the estimated upper limits of the irradiation damage tolerated by the beam window and cladding materials in ADS [2], respectively; however, these are practically challenging to execute using an experimental nuclear reactor because of the required long irradiation time.

An oxygen control for the LBE coolant is critical in ADSs, and the control system needs to be properly installed in the LBE coolant system. The oxygen concentration of LBE must be controlled below the upper concentration of lead oxide (PbO) formation because PbO results in stuck valves and plugs the narrow flow channels of LBE during long-term operation. In steel materials, the surface-protective layer of iron and/or chromium oxide dissolves when the oxygen concentration is low, reducing the metal oxide. The acceptable oxygen concentration depends on the LBE temperature and ranges from 10^{-5} to 10^{-7} wt.% at 450 °C [3]. These upper and lower oxygen concentrations are critical values of the oxygen concentration boundary and need to be controlled in the system for the LBE in ADS.

The material corrosion behavior in LBE has been studied for approximately two decades [4,5], and the corrosion kinetics of steels, such as austenitic steels, 316L, and ferritic/martensitic steels, T91, has been discussed [6–8]. In the case of 316L steels, it is well-known that Ni, which is one of the main austenite formers, can dissolve in the LBE, because of the high solubility in LBE. The depleted area of Ni appears to be ferritic phase [9], followed by the substitutive Pb and/or Bi penetration [10], especially at a higher temperature of approximately 500 °C, under flowing LBE [5] and long-time corrosion test over 3000 h, even at a relatively low temperature, such as 450 °C [11].

The few studies [7,8] that have investigated the effect of neutron irradiation on the material properties in LBE have been conducted only from the viewpoint of liquid metal embrittlement (LME); thus, reports concerning liquid metal corrosion (LMC) of irradiated materials are limited. However, ion irradiation is a powerful technique to simulate ADS irradiation conditions with accurate temperature control, making it appropriate to investigate the mechanism of microstructural evolution under irradiation and further select candidate materials before irradiation by neutrons. Firstly, the formation of oxide layers on the surface of T91 steel was reported to have improved because of the triple ion irradiation of Fe, He, and H beams, which simulated the ADS spallation neutron irradiation even after the corrosion test conducted under the low oxygen concentration in LBE [12]. As per the report, only Fe ion irradiation for T91 specimens proved incongruous to the oxide-formation enhancement. It was suggested that the vacancy defects, which were effectively induced by the triple ion irradiations in T91, played a vital role in enhancing the oxide formation. On the contrary, displacement damage can produce vacancy defects in 316L steels [13]. In this study, the LMC behavior after irradiation is considered for self-ion irradiations without simultaneous He and H irradiations, followed by the corrosion tests to study the effect of single irradiation on the corrosion of 316L steels.

2. Experiments

In an ADS, SS316L (316L) is one of the candidate materials for the in-core structure. The material composition of 316L steel is shown in Table 1. The 316L steel was solution-annealed at 1040 °C for 10 min and then cooled in water. A specimen with a length of 6.0 mm, width of 3.0 mm, and thickness of 1.0 mm was cut from the bulk sample to mechanically polish and finish its surface via buffing, using 50 nm alumina nanoparticles for a mirror finish, thus achieving a final thickness of approximately 0.75 mm. As explained below, two such specimens were set in the irradiation specimen holder with the same irradiation condition but different corrosion conditions.

Table 1. Chemical composition of 316L steel (wt.%).

Fe	Cr	Ni	Mo	Mn	Si	P	C	S
Bal.	17.46	12.11	2.19	0.82	0.51	0.027	0.017	0.001

The 316L specimens were irradiated by 10.5 MeV-Fe^{3+} ions at a temperature of 450 °C up to about 50 dpa. Ion irradiation experiments were conducted at Takasaki Ion Accelerators for Advanced Radiation Application, Japan's National Institutes for Quantum and Radiological Science and Technology (QST). As shown in Figure 1a, two specimens were held by a 10 mmφ steel mask, and the irradiation area was covered by an aluminum foil to create a nonirradiated area on the surface of the same specimen, having almost the same experimental condition as the temperature history during ion irradiation and also immersion in LBE. Irradiation experimental conditions were determined using the Stopping and Range of Ions in Matter (SRIM) code [14]. The displacement damage depth profile is shown in Figure 1b. The schematic cross-sectional image of the irradiated specimen was also inserted. Both irradiation and corrosion temperature were controlled at 450 °C because the medium ADS component temperature is around 450 °C [2]. An infrared pyrometer (NIKON) monitored the surface temperature under irradiation and actively controlled it via electron bombardment heating, Joule heating, and beam heating to achieve the accurate irradiation temperature. Displacement damage in the corrosion test was determined to be approximately 4 and 8 dpa at the specimen surface. For example, the specimen was irradiated up to 8 dpa at its surface and 52 dpa at a depth of 2 μm, as shown in Figure 1b. In the case of 4 dpa, the depth profile of the displacement damage represented half of the whole depth. After the ion irradiation, the specimen was immersed at a lower temperature compared with the corrosion test temperature in the LBE pot, keeping the LBE at the desired oxygen concentration. The specimen was fixed in LBE using a stainless-steel wire connected to a tungsten weight. The experimental setup and further details of the corrosion test are provided in [12]. The oxygen concentration in LBE was measured using a Pt/air type-6YSZ (yttria-stabilized zirconia) oxygen sensor (fabricated by JAEA [15]) and maintained using a covering gas of premixed Ar + 5.0%H_2 for a low oxygen concentration spanning the order of 10^{-8}–10^{-9} wt.%, which was around the lower critical concentration for oxide (Fe_3O_4 and FeO) formation [3]. The corrosion test time started when the LBE temperature reached 450 ± 5 °C, shown in Figure 2 as an open triangle. Notably, during the initial 10% of the immersion period, approximately 35 h, high oxygen concentration in the range of 2.4×10^{-4}–2.3×10^{-8} wt.% was employed because of the opening of the pot to set the specimens into the LBE, as shown in Figure 2a. Consequently, the incident of this higher oxygen concentration highlighted the irradiation effect for the ion-irradiated specimen before reaching a low oxygen concentration. Before the corrosion test's completion under lower oxygen concentration, an unforeseen leakage of approximately 6.0×10^{-8} wt.% was observed, but it could be considered insignificant compared with the oxygen concentration at working conditions. After conducting the corrosion test at the lower oxygen concentration, another test was conducted under a saturated oxygen concentration of 3.6×10^{-4} wt.% at 450 °C, as shown in Figure 2b. In a previous study [12], the corrosion time of 1000 h was too long to observe the effect of irradiation on the specimen, especially at low oxygen concentration at approximately 10^{-8} and 10^{-9} wt.%, when comparing the radiation damage depth range of about 2 μm to the oxide thickness. Hence, we chose to operate at a third of the 1000 h, i.e., 330 h. After completing the corrosion test in 330 h, the LBE on the surface of the specimen was removed using silicone oil at 200 °C to melt the LBE into the oil bath. The silicone oil exhibited a high affinity for the corroded surface, thus penetrating the oxide layer with ease, as partially shown in the latter energy-dispersive X-ray spectroscopy (EDS) spectra. Removing the silicone oil was difficult and required care to avoid unexpected distortions of the corrosion behavior. To identify the oxide layer, X-ray diffraction (XRD; MAC Science, MXP3 with a Cu Kα X-ray source) was performed after the corrosion test. The surface observation before LBE removal and the cross-sectional corrosion behavior were assessed using a field-emission scanning electron microscope (high-resolution FE-SEM; Zeiss, Sigma) at an acceleration voltage of 30 kV. Two types of SEM detector modes of normal secondary electron and AsB (angular selective backscattering electron, showing material contrast and topographical information) were used for cross-sectional observation after fixing in electrically conductive resin via hot pressing, polishing to a mirror surface, and depositing a thin Os coating (thickness of a few nanometers) to avoid charging up under SEM observations. EDS was also performed to obtain line analytical data and mapping images of material elements in the specimens after corrosion.

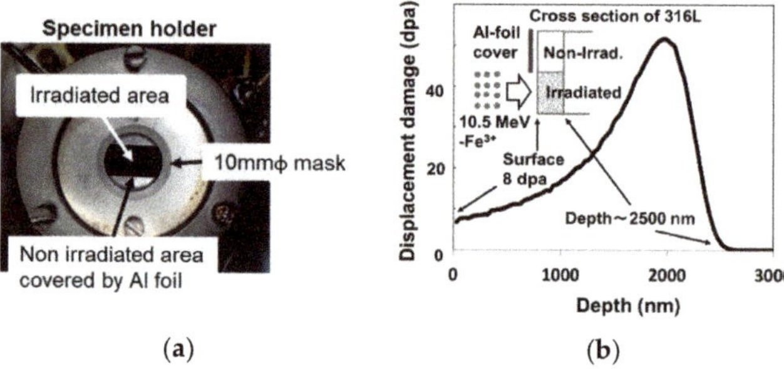

(**a**)   (**b**)

Figure 1. (**a**) Irradiation specimen holder and (**b**) depth distribution of displacement damage of ion irradiation. The experimental dpa was determined based on the depth distribution.

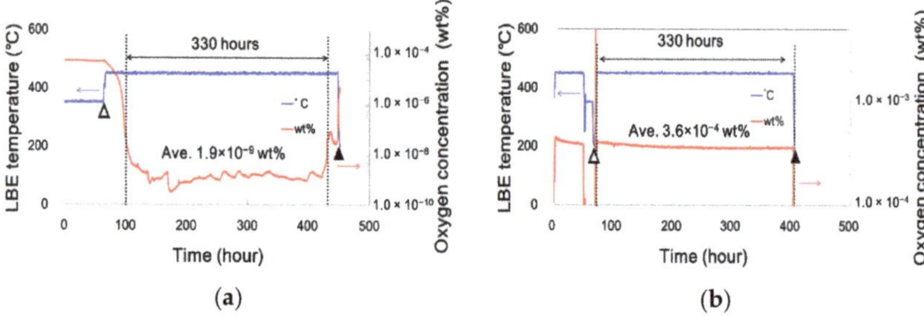

(**a**)   (**b**)

Figure 2. LBE temperature and oxygen concentration under corrosion tests for the 316L specimens. The corrosion tests were conducted for 330 h at (**a**) low oxygen concentration and (**b**) high (saturated) oxygen concentration in LBE at 450 °C. The triangles denote the insertion and extraction moments of the specimens as the open and solid triangles, respectively.

3. Results

After the corrosion test at 450 °C in LBE with a sufficiently low oxygen concentration, a surface within the nonirradiated area was left unoxidized, but a locally corrosive morphology was observed; however, an iron/chromium oxide layer covered the irradiated area. Ion irradiation of 10.5 MeV-Fe^{3+} up to 8 dpa at the specimen surface was conducted at 450 °C, followed by a corrosion test in LBE at 450 °C for 330 h under the low oxygen concentration (in the range of 10^{-8}–10^{-9} wt.%, as shown in Figure 2a). Figure 3 shows typical surface SEM images observed on the boundary area between the nonirradiated (masked) and irradiated areas after removing LBE. The irradiation displacement damage was 4 and 8 dpa for the specimen shown in Figure 3a,b, respectively. The boundary lines between nonirradiated and irradiated areas clearly appeared in both upper images. In the 4 dpa irradiation, the irradiated surface showed a patchy pattern texture, which is nonuniform, whereas the covered surface showed a uniform texture in the 8 dpa irradiation. These surface SEM images were observed before cleaning the LBE with silicone oil. The lower images are a high-magnification version of both the nonirradiated and irradiated areas. As shown in the bottom left images of Figure 3a,b, the nonirradiated areas of both specimens exhibited similar local corrosion, observed as black dots. The area irradiated to 8 dpa became fully covered by the rough texture, as it was in the case of 4 dpa, indicated as a round shape with white contrast on the upper part of Figure 3a. The white contrast textures in the Figure 3a upper section comprised the elements of iron, chromium, and oxygen, as confirmed by EDS measurements (not shown in this report). The uniform texture in Figure 3b seen in the upper image comprises iron, chromium, nickel, oxygen, lead, and bismuth. Comparing the surface images among nonirradiated, irradiated, 4 dpa, and 8 dpa irradiated areas suggested that the irradiation affects the corrosion behavior of 316L in LBE post-ion irradiation. Figure 4 shows the cross-sectional SEM images of nonirradiated and irradiated areas on the surface after a corrosion test conducted at 450 °C for 330 h. These cross-sectional SEM images correspond to the surface SEM images of identical specimens shown in Figure 3, but the surface LBE was removed using silicone oil. The upper and lower images denote: (a), (c) nonirradiated area and (b), (d) irradiated area of 4 and 8 dpa specimens, respectively. The oxygen concentration was low in LBE at 450 °C. In the cases of nonirradiated sites for both specimens, local corrosion was observed on both surfaces as a weak contrast revealed by the AsB detector, which could show the compositional difference, as indicated in Figure 4a,c. The local corrosion size was estimated to be approximately 100–300 nm in diameter and about 100–150 nm in depth, measured to a higher magnification via high-resolution FE-SEM. This local corrosion observed in the cross-sectional image corresponds to the black dots on the surface, as shown in Figure 3. The Ni component of steel is considered to have dissolved in the LBE at low oxygen concentration because the oxygen concentration is close to or lower than that necessary for a Fe and/or Cr surface oxide formation. Although the nonirradiated area appeared to have almost no oxide formation, rather local corrosion, an oxide layer of about 140 nm thickness was formed on the 4 dpa irradiated area by efficiently reacting with the small amount of oxygen present within the initial 10% immersion period, as shown in Figure 2a. The oxide area occupied approximately 70% of the total irradiated area in the case of 4 dpa, and some other residual areas appeared to onset the ferritization via Ni-depletion. Unfortunately, the EDS line spectra for all the images in Figure 4 did not exhibit the distinct reduction of Ni concentration. It is obvious to consider the initial state of ferritization by Ni-depletion, as depicted in Figure 4. However, in the 8 dpa, fully irradiated areas were covered by an approximately 230 mm thick oxide layer. The irradiation enhanced the oxide formation even though the oxygen concentration was relatively low.

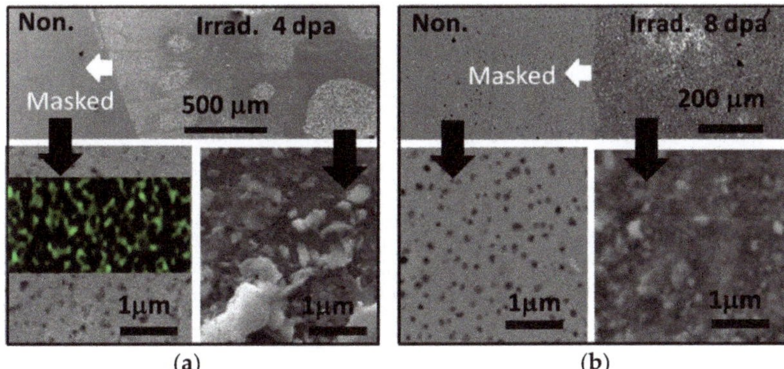

Figure 3. Surface SEM images after corrosion tests for the irradiated 316L specimens. Upper images were observed on the boundary area irradiated at (**a**) 4 dpa and (**b**) 8 dpa. The mapping image from Ni is inserted in the masked area shown in (**a**). The black dots correspond to the Ni-depleted areas and behave similarly for both (**a**,**b**).

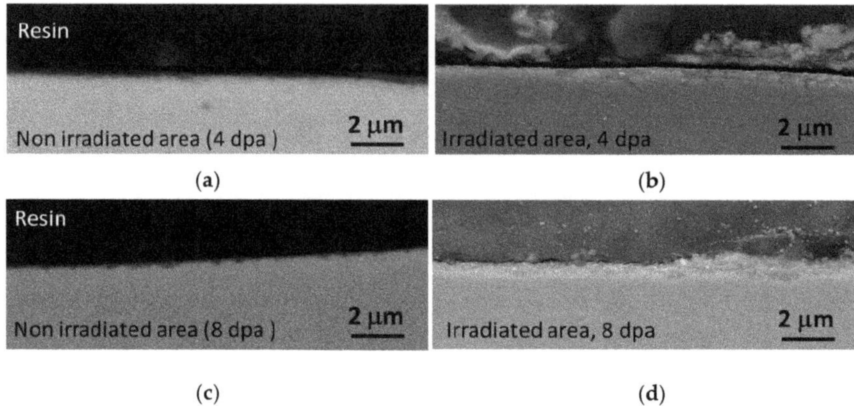

Figure 4. Cross-sectional SEM images after conducting corrosion tests for the irradiated 316L specimens. Upper and lower images denote: (**a,c**) nonirradiated area and (**b,d**) irradiated area of specimens irradiated at 4 and 8 dpa, respectively. The oxygen concentration was in the order of 10^{-9} wt.% in LBE. The images (**a,c**) and (**b,d**) were observed by AsB and SE detectors respectively, for convenience.

In the case of high oxygen concentration in LBE, the surface of the nonirradiated area reacted to form an iron oxide layer, whereas the irradiated area developed duplex layers of iron and iron/chromium oxides as coverings. A corrosion test for the other pieces of the irradiated specimens was conducted for 330 h in LBE at a saturated oxygen concentration of 2.4×10^{-4} wt.%. SEM images and EDS line spectra of the nonirradiated and irradiated areas are shown in Figure 5. The EDS spectra from the resin included Si and O signals from silicone oil and Fe. In contrast, the other elements of steel came from the residues in the oil because the acetone cleaning was insufficient to clean the silicone oil of its residues. Even though both areas were subjected to the same preparation process, the morphologies of the surface in the two regions were different. The surface of the nonirradiated area was rough and appeared to have a thin white contrast, whereas that of the irradiated area was coarse and contained numerous cracks. The EDS line spectra in the regions are also shown in Figure 5b,c. An oxide layer with a thickness of half a micrometer, observed as a thin white layer, was formed on the surface of the nonirradiated region. In contrast, a distinct oxide layer with a thickness of approximately 1 µm, which is twice that in the nonirradiated area, covered the surface of the irradiated area. The oxygen diffused into the 1 µm depth from the original surface, which was considered to be approximately the half depth of two dotted lines, as shown in Figure 5c. The breakage of the oxide layer was caused by polishing in the irradiated region. XRD analysis revealed that the oxide layers of the nonirradiated and irradiated areas mainly comprised Fe_3O_4 (magnetite) and small signals of $FeCr_2O_4$ (spinel). The boundary is not very clear for the irradiated area, but it is somewhat distinguishable, as shown in Figure 5c, upper SEM image. In the case of 316L steel after immersion in LBE under a saturated oxygen concentration of 2.4×10^{-4} wt.%, the surface of the irradiated area was coated with duplex oxide layers, as shown in Figure 5c. For corrosion in LBE with a sufficiently high oxygen concentration (in the range of 10^{-4}–10^{-6} wt.%) and/or a long corrosion time (typically over 1000 h), a duplex oxide layer forms on the steel surface [5,16]. In this duplex, the outer layer is Fe_3O_4 and the inner layer is $FeCr_2O_4$. In this study, the duplex layer could not be observed in the SEM images of the irradiated area in Figure 4 and the nonirradiated site in Figure 5b because the oxygen concentration was too low and corrosion time too short for the formation of the duplex layer. Nevertheless, ion irradiation up to 8 dpa enhanced the oxidation reaction by effectively using the oxygen introduced into LBE at the beginning of the corrosion test, as shown in Figure 2a, even at a low oxygen concentration, as shown in Figure 4d. The original surface is assumed to be located between the outer and inner layers of the duplex, indicating that the inner layer of the irradiated region includes radiation damage.

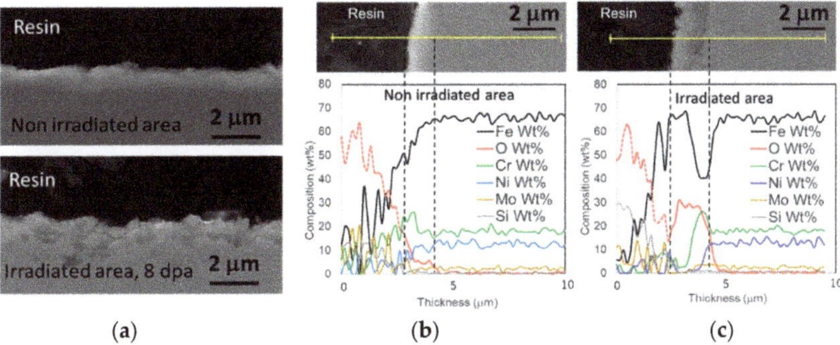

Figure 5. Cross-sectional SEM images and EDS line spectra after conducting corrosion tests for the specimen irradiated up to 8 dpa. Upper and lower images of (**a**) correspond to that of nonirradiated and irradiated areas, respectively. Figures (**b**,**c**) denote the compositions of nonirradiated and irradiated areas respectively, measured by an EDS line analysis, which was shown in the upper image. The oxygen concentration was 2.4×10^{-4} wt.% in LBE at 450 °C. The red dotted and gray lines are signals come from the compositions of silicone oil.

In 316L steel, after conducting the corrosion test in LBE, a cross-sectional SEM image captured at a region exposed to ion irradiation showed a thicker oxide layer than the nonirradiated surface, even at a low oxygen concentration in LBE. The mean thicknesses of the local corrosion (depletion of Ni) oxide layers in nonirradiated and irradiated areas are shown in Figure 6. The total thickness of the duplex layer was also measured from each image in the case of the saturated oxygen concentration. In the case of the low oxygen concentration of an order of 10^{-9} wt.%, the irradiation enhances the oxide formation by a factor of two at maximum, and the irradiation effect on the oxidization increases with increasing displacement damage up to 8 dpa. In the case of a saturated oxygen concentration, the irradiation enhances the oxide formation by a factor of two, and the irradiation is undoubtedly effective in the surface oxidization; however, it appears to increase the displacement damage up to 8 dpa.

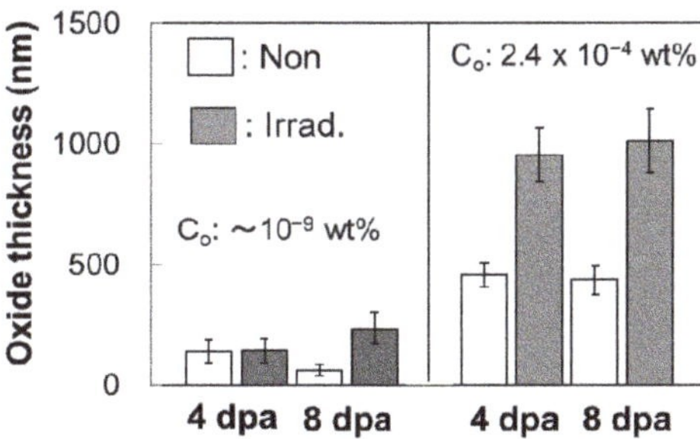

Figure 6. Oxide thickness of non-irradiated and irradiated areas after corrosion tests at low and saturated oxygen concentrations for 316L specimens irradiated at 4 and 8 dpa.

4. Discussion

The process of inward diffusion of oxygen from the LBE side and outward diffusion of Fe from the steel matrix mutually governed the oxide formation at the boundary between the steel surface and LBE. The cold working is a good simulation of the microstructural evolution under radiation damage in dislocation defects and interstitial atom types of defects. In the case of the corrosion behavior exhibited by the cold-worked austenitic stainless-steels in LBE, the cold working accelerated the formation of the duplex oxide layer and the ferritization via Ni dissolution [10] at 500–550 °C in 1000–3000 h. As shown in Figure 5c, even at 450 °C and in 330 h experimental conditions, Ni dissolution was observed in the irradiated region and not in the nonirradiated region. As shown in Figure 6, oxide formation was enhanced via ion irradiation in LBE even at a low oxygen concentration by the formation of radiation defects. In the nonirradiated region, the oxide layer can grow in LBE with an oxygen concentration dissolved in LBE under thermal equilibrium. The thermal equilibrium state is also applicable to the irradiated region because oxidation reactions occur after ion irradiation. However, the number of Fe atoms diffusing outward might increase through site changes with vacancies, which might be trapped by impurities or the strain field induced via ion irradiation, leading to more Fe self-diffusion than that found under normal thermal activity. The activation energy of Fe self-diffusion in α-Fe is 2.87 eV, in which the vacancy formation energy (E_{vf}) and migration energy (E_{vm}) are 1.61 and 1.3 eV, respectively [17]. The diffusion coefficient (D) is calculated by $D = D_0\exp(-(E_{vf} + E_{vm})/kT$, where D_0 is 8.0×10^{-5} m^2/s, k is 8.62×10^{-5} eV/K, and T is 723 K. The thermal diffusion length (L_d) is described by $L_d = (6Dt)^{1/2}$, and L_d of Fe was calculated to be 1.4 nm for the specimen subjected to the corrosion test performed at 450 °C for 330 h. Therefore, Fe self-diffusion does not affect oxide formation at the specimen surface. However, after ion irradiation, assuming all vacancies exist even at RT, when $E_{vf} = 0$, which means the neighbor of the diffusive atom is a vacancy, D exhibits its maximum value and L_d is 580 μm. The actual L_d is smaller than this maximum value because of the void formation and residual vacancies trapped by the strain field induced by radiation damage. From an SRIM calculation shown in Figure 1b, the total number of vacancies was estimated to be 1.5×10^{19} m^{-2} around 1 μm depth, induced by the displacement damage of 20 dpa, which is considered average damage through 2 μm depth. This is just the case of $E_{vf} = 0$, then, L_d shows a maximum value of 580 μm.

On the contrary, using the void size of 1.3 nm and the density of 3×10^{24} m^{-3} for 316L irradiated by 4 MeV-Au ions at 450 °C up to about 20 dpa from Reference [13], the number of vacancies included in the total voids was estimated to be 4.0×10^{17} m^{-2}, assuming that the one void had 100 vacancies. From these rough estimations, the number ratio of residual (invisible) vacancies is 0.0267 ($4.0 \times 10^{17}/1.5 \times 10^{19}$). Then, a mean free path (MFP) of vacancy diffusion is approximately 15 μm. This value is reasonable because recombining with interstitials and/or disappearance into sink sites reduces the MFP to be several micrometers. However, this might be a limitation of the simulation method that employed (ex situ) the corrosion test after ion irradiation.

In contrast, oxygen atoms diffuse inward from the surface as interstitial atoms. The migration energy of an interstitial atom (E_{im}) is 0.89 eV [14]. Here, $D = D_0\exp(-E_{im})/kT$, where D_0 is 1.79×10^{-7} m^2/s, k is 8.62×10^{-5} eV/K, and T is 723 K. The oxygen diffusion length (L_d) was calculated to be 910 μm according to $L_d = (6Dt)1/2$. Therefore, oxygen can sufficiently diffuse into the material in this corrosion experiment in LBE. Comparing the EDS line spectra shown in Figure 5b,c, the inward oxygen diffusion lengths of the nonirradiated and irradiated regions are comparable. However, in the irradiation region, the oxygen concentration increased about six times more than that of the nonirradiated case. From these rough estimations, two reasons for the enhanced oxide formation in the case of ion irradiation in 316L are considered: (1) enhanced Fe diffusion caused by vacancy diffusion after ion irradiation and (2) enhancement of O interstitial diffusion induced via radiation damage. This enhances the oxidation reaction between Fe, Cr, and O. Based on the results of this study, it is suggested that radiation-induced diffusion during irradiation enhances the oxidation much more than that after irradiation.

5. Conclusions

After ion irradiation of 316L austenitic steel, the LMC behavior was studied for self-ion irradiations, followed by the corrosion tests in LBE with different oxygen concentrations.

The 316L specimens were irradiated by 10.5 MeV-Fe^{3+} ions at a temperature of 450 °C, up to 8 dpa at the surface. After conducting the corrosion test at 450 °C in LBE with low oxygen concentration, a surface of the nonirradiated area was not oxidized, but corrosive morphology appeared, whereas an iron/chromium oxide layer covered the irradiated area. In the case of a high oxygen concentration in LBE, the surface of the nonirradiated area was oxidized to an iron oxide layer, whereas the irradiated area was covered by the duplex layers of iron and iron/chromium oxides. It is suggested that irradiation can advance the oxide layer formation because of the enhancement of Fe and/or oxygen diffusion induced by the radiation defects in the 316L steel.

Author Contributions: Conceptualization, N.O.; SEM observation, Y.F. and M.T.; formal analysis, N.O.; investigation, N.O.; writing—original draft preparation, N.O. All authors have read and agreed to the published version of the manuscript.

Funding: This research received no external funding.

Acknowledgments: These ion irradiation experiments were performed under the Shared Use Program of QST Facilities. The authors are grateful to S. Ukai for useful comments.

Conflicts of Interest: The authors declare no conflict of interest.

References

1. Tsujimoto, K.; Oigawa, H.; Kikuchi, K.; Kurata, Y.; Mizumoto, M.; Sasa, T.; Saito, S.; Nishihara, K.; Umeno, M.; Takei, H. Feasibility of lead-bismuth-cooled accelerator-driven system for minor-actinide transmutation. *Nucl. Tech.* **2008**, *161*, 315–328. [CrossRef]
2. Tsujimoto, K.; Nishihara, K.; Takei, H.; Sugawara, T.; Kurata, Y.; Saito, S.; Obayashi, H.; Sasa, T.; Kikuchi, K.; Tezuka, M.; et al. Feasibility study for transmutation system using lead-bismuth cooled accelerator-driven system. *JAEA-Research* **2010** , 2010–2012.
3. Müller, G.; Heinzel, A.; Schumacher, G.; Weisenburger, A. Control of oxygen concentration in liquid lead and lead–bismuth. *J. Nucl. Mater.* **2003**, *321*, 256–262. [CrossRef]
4. Müller, G.; Schumacher, G.; Zimmermann, F. Investigation on oxygen controlled liquid lead corrosion of surface treated steels. *J. Nucl. Mater.* **2000**, *278*, 85–95. [CrossRef]
5. Tsisar, V.; Schroer, C.; Wedemeyer, O.; Skrypnik, A.; Konys, J. Characterization of corrosion phenomena and kinetics on T91 ferritic/martensitic steel exposed at 450 and 550 °C to flowing Pb-Bi eutectic with 10^{-7} mass% dissolved oxygen. *J. Nucl. Mater.* **2017**, *494*, 422–438. [CrossRef]
6. Schroer, C.; Wedemeyer, O.; Skrypnik, A.; Novotny, J.; Konys, J. Corrosion kinetics of Steel T91 in flowing oxygen-containing lead–bismuth eutectic at 450 °C. *J. Nucl. Mater.* **2012**, *431*, 105–112. [CrossRef]
7. Stergar, E.; Eremin, S.G.; Gavrilov, S.; Lambrecht, M.; Makarov, O.; Iakovlev, V. Influence of LBE long term exposure and simultaneous fast neutron irradiation on the mechanical properties of T91 and 316L. *J. Nucl. Mater.* **2016**, *473*, 28–34. [CrossRef]
8. Magielsen, A.J.; Jong, M.; Bakker, T.; Luzginova, N.V.; Mutnuru, R.K.; Ketema, D.J.; Fedorov, A.V. Irradiation of structural materials in contact with lead bismuth eutectic in the high flux reactor. *J. Nucl. Mater.* **2011**, *415*, 311–315. [CrossRef]
9. Deloffre, P.; Terlain, A.; Barbier, F. Corrosion and deposition of ferrous alloys in molten lead–bismuth. *J. Nucl. Mater.* **2002**, *301*, 35–39. [CrossRef]
10. Kurata, Y. Corrosion behavior of cold-worked austenitic stainless steels in liquid lead–bis muth eutectic. *J. Nucl. Mater.* **2014**, *448*, 239–249. [CrossRef]
11. Heinzel, A.; Weisenburger, A.; Müller, G. Corrosion behavior of austenitic steels in liquid leadbismuth containing 10-6 wt% and 10-8 wt% oxygen at 400–500 °C. *J. Nucl. Mater.* **2014**, *448*, 163–171. [CrossRef]
12. Okubo, N.; Fujimura, Y. Irradiation Influence on Swelling and Corrosion Behavior of ADS Beam Window Materials, T91 Steels, in Lead Bismuth. In Proceedings of the 14th International Workshop on Spallation Materials Technology, JPS Conference Proceedings, Fukushima, Japan, 11–16 November 2018. [CrossRef]
13. Jublot-Leclerca, S.; Li, X.; Legras, L.; Lescoat, M.-L.; Fortuna, F.; Gentilsa, A. Microstructure of Au-ion irradiated 316L and FeNiCr austenitic stainless steels. *J. Nucl. Mater.* **2016**, *480*, 436–446. [CrossRef]
14. Ziegler, J.F. SRIM-2003. *Nucl. Instr. Meth. B* **2004**, *219–220*, 1027–1036. [CrossRef]
15. Sugawara, T.; Yamaguchi, K. Measurement Experiment of Oxygen Concentration in Liquid Lead-Bismuth Eutectic; Fabrication of Oxygen Sensor and Measurement under Static Condition. 2015. Available online: https://inis.iaea.org/search/search.aspx?orig_q=RN:47070968 (accessed on 6 April 2021).
16. *LBE Handbook*; NEA No. 7268; OECD/NEA: Boulogne-Billancourt, France, 2015.
17. Watanabe, Y.; Morishita, K.; Nakasuji, T.; Ando, M.; Tanigawa, H. Helium effects on microstructural change in RAFM steel under irradiation: Reaction rate theory modeling. *Nucl. Instr. Meth. Phys. Res. B* **2015**, *352*, 115–120. [CrossRef]

quantum beam science

MDPI

Article

Desorption of Implanted Deuterium in Heavy Ion-Irradiated Zry-2

Hideo Watanabe [1,*], Yoshiki Saita [2], Katsuhito Takahashi [2,3] and Kazufumi Yasunaga [4]

[1] Research Institute for Applied Mechanics, Kyushu University, Kasuga-kouenn, Kasugashi, Fukuoka 816-8580, Japan
[2] Interdisciplinary Graduate School of Engineering Sciences, Kyushu University, Kasuga-kouenn, Kasugashi, Fukuoka 816-8580, Japan; koga@archkyushu-u.ac.jp (Y.S.); katsuhito.takahashi.hy@hitachi.com (K.T.)
[3] Hitachi, Co., Ltd., Omika, Hitachshi, Ibaraki 319-1292, Japan
[4] The Wakasa Wan Energy Research Center, Nagatani, Tsurugashi, Fukui 914-0192, Japan; kyasunaga@werc.or.jp
* Correspondence: watanabe@riam.kyushu-u.ac.jp

Abstract: To understand the degradation behavior of light water reactor (LWR) fuel-cladding tubes under neutron irradiation, a detailed mechanism of hydrogen pickup related to the point defect formation (i.e., a-component and c-component dislocation loops) and to the dissolution of precipitates must be elucidated. In this study, 3.2 MeV Ni^{3+} ion irradiation was conducted on Zircaloy-2 samples at room temperature. Thermal desorption spectroscopy is used to evaluate the deuterium desorption with and without Ni^{3+} ion irradiation. A conventional transmission electron microscope and a spherical aberration-corrected high-resolution analytical electron microscope are used to observe the microstructure. The experimental results indicate that radiation-induced dislocation loops and hydrides form in Zircaloy-2 and act as major trapping sites at lower (400–600 °C) and higher (700–900 °C)-temperature regions, respectively. These results show that the detailed microstructural changes related to the hydrogen pickup at the defect sinks formed by irradiation are necessary for the degradation of LWR fuel-cladding tubes during operation.

Keywords: light water reactor; zirconium alloys; nuclear fuel cladding; thermal desorption spectroscopy; transmission electron microscopy

check for **updates**

Citation: Watanabe, H.; Saita, Y.; Takahashi, K.; Yasunaga, K. Desorption of Implanted Deuterium in Heavy Ion-Irradiated Zry-2. *Quantum Beam Sci.* **2021**, 5, 9. https://doi.org/10.3390/qubs5020009

Academic Editor: Akihiro Iwase

Received: 22 March 2021
Accepted: 23 April 2021
Published: 26 April 2021

Publisher's Note: MDPI stays neutral with regard to jurisdictional claims in published maps and institutional affiliations.

1. Introduction

Hydrogen embrittlement of Zircaloy-2 is one of the main factors that limits the life of fuel rods in light water reactor (LWR) fuel-cladding tubes. Because the hydrogen centration exceeds the solid solubility of the metal, some hydrogen atoms are precipitated as hydride, causing hydride embrittlement. To understand how hydrogen pickup during operation influences the growth acceleration of the fuel rod, the effect of hydrogen on the neutron irradiation-induced microstructure should be studied. Hydrogen diffusivity under neuron irradiation is generally controlled by the trapping and de-trapping processes of hydrogen on the material. A dislocation loop is a radiation-induced defect cluster formed by the irradiation of neutrons [1,2], electrons [3,4], or charged particles [5–10]. Second-phase particles (SPPs) also form in Zircaloy-2 [11]. The size distribution of the SPPs and the chemical composition of the cladding tube affect the tube's corrosion rate inside boiling water reactors (BWRs) [12–17]. However, these SPPs are unstable during neutron irradiation and undergo an amorphous transformation resulting in the decomposition and redistribution of the constituent elements the precipitates into defect sinks [15–17]. Ion irradiation has been used on nuclear materials in multiple studies because, unlike neutron irradiation, the displacement per atom (dpa) level, irradiation temperature, and other irradiation conditions can be precisely controlled [5–10].

In this study, 3.2 MeV Ni^{3+} ion irradiation is applied to Zircaloy-2 samples. The samples are injected with 5.0–30 keV D_2^+ ions to understand the trapping and de-trapping process of hydrogen on N^{3+} ion irradiated Zrycaloy-2. Thermal desorption spectroscopy (TDS) and conventional transmission electron microscopy (C-TEM) are conducted following the irradiation to evaluate the details of retention and desorption of the implanted deuterium and to identify the responsible traps.

2. Experimental Procedures

The Zircaloy-2 specimens were annealed at 630 °C for 2 h and subsequently air-cooled. Table 1 shows the results of the chemical analyses and the measurements of the hydrogen impurity levels of these specimens. The samples were irradiated at room temperature with 5.0–30 keV D_2^+ ions and 3.2 MeV Ni^{3+} ions by an ion implanter and a tandem accelerator at Kyushu University. Figure 1a shows the ion irradiation chamber (with a duo-plasma ion gun) used for the D_2^+ irradiation, and Figure 1b shows the specimen holder used for the D_2^+ and Ni^{3+} ion irradiation. Figure 2a shows the depth profile of the damage rate, and Figure 2b shows the concentration of D_2^+ ions irradiated at each accelerating voltage. The damage estimation was obtained from the Stopping and Range of Ions in Matter (SRIM) calculation [18], for which the threshold energy for displacement was assumed to be 40 eV. After irradiation at room temperature at a flux of 1.0×10^{18} ions/m^2s, the samples were transferred into the vacuum chamber of the TDS apparatus. After they were evacuated, the specimens were held at room temperature for a period of time (less than 2 h) and TDS measurements were conducted. During the heating, with a ramping rate of 1 °C/s up to 900 °C, the thermal desorption of HD (mass = 3) and D_2^+ (mass = 4) were measured using quadrupole mass spectroscopy. The desorption rate was calibrated by a He standard leak and corrected the relative ionization efficiency.

Table 1. Chemical composition of the materials used in the present study (wt%).

	Sn	Fe	Cr	Ni	H (wtppm)	Zr
Zircaloy-2	1.38	0.15	0.09	0.05	46	Bal.

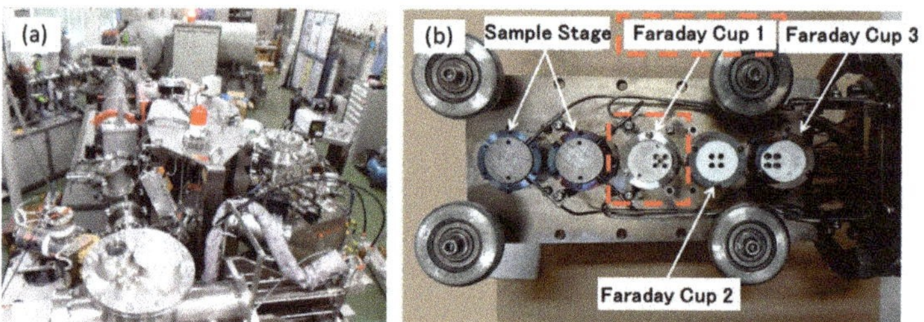

Figure 1. The triple ion beam facilities at RIAM Kyushu University: (**a**) tandem-type accelerator attached with two ion guns: terminal voltage 1.0 MeV and (**b**) specimen holder for ion irradiation.

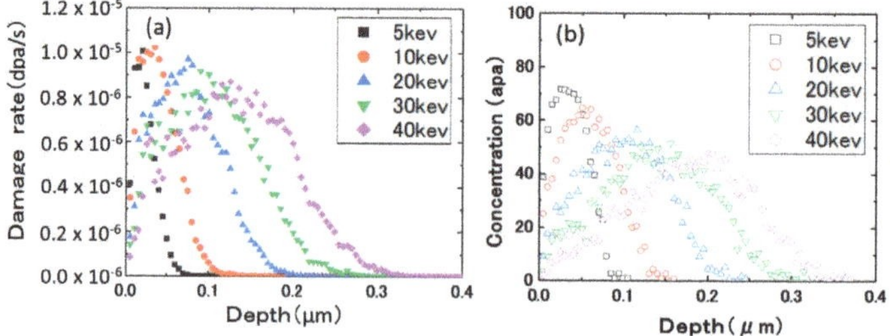

Figure 2. The depth profile of the damage rate (**a**) and the concentration of D^+ at 3.0×10^{21} ions/m^2 (**b**) in the case of each accelerating voltage.

Figure 3 shows the damage distribution of the 3.2 MeV Ni^{3+} ions and the impurity concentration (Ni^{3+} ions) in the pure Zr following irradiation. The damage estimation was also obtained by the SRIM calculation. The samples for microscopy were thinned by the electropolishing method in which the electrolyte was a mixture of 50 mL perchloric acid and 950 mL acetic acid. During the thinning process, the electrolyte was held at $-40\,°C$ and the potential applied was 25 V.

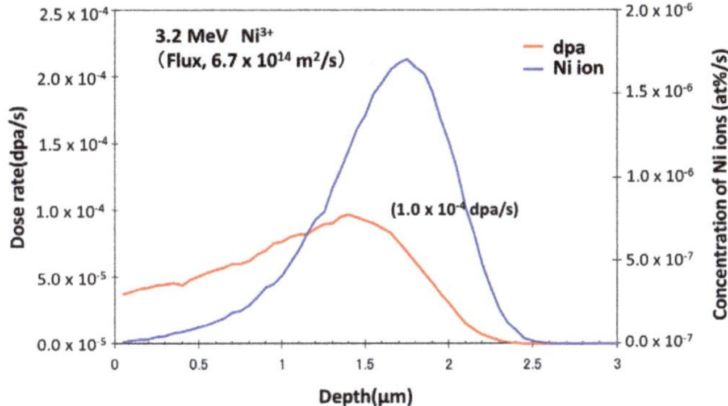

Figure 3. The SRIM calculation [18,19] of the damage distribution and the concentration of Ni^{3+} atoms in Zr irradiated by Ni^{3+} ions at 3.2 MeV. The values are estimated in the case of 6.7×10^{14} ions/m²s.

The microstructure was observed before and after irradiation via C-TEM and using a spherical aberration (Cs)-corrected high-resolution analytical electron microscope (JEOL ARM200FC) operated at a voltage of 200 kV in a radiation-controlled area at Kyushu University.

3. Results and Discussions

3.1. Ion Dose and Energy Dependance of Thermal Desorption Behavior

Figure 4 shows the dose dependance of the thermal desorption spectrum after 5.0 keV D_2^+ ion irradiation at room temperature. Figure 4a–c show the cases of unirradiated samples, samples radiated with 3.0×10^{21} ions/m², and samples irradiated with 10×10^{21} ions/m², respectively. As shown in Figure 4a, the desorption stages of HD (mass = 3) and D_2 (mass = 4) were not detected prior to irradiation, but, after the D_2^+ ion irradiation, two major peaks at approximately 600 °C (mass = 3) and 800 °C (mass = 4) were detected. As demonstrated, the desorption stages were designated Peak A and Peak B on the lower temperature side. By increasing the irradiation dose from 3.0 to 10×10^{21} ions/m², the total desorption of HD and D_2 increased from 1.8 to 5.5×10^{19} ions/m², and from 2.5 to 6.6×10^{18} ions/m², respectively. The increasing rate was 3.1 for HD and 3.8 for D_2. These values were nearly consistent with the increasing dose level values, particularly 3.3.

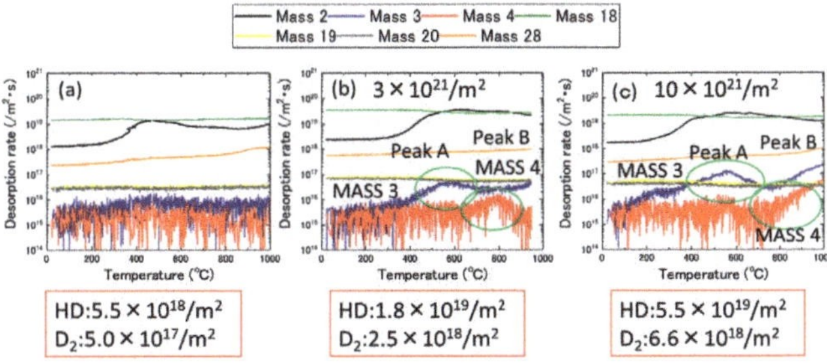

Figure 4. The dose dependance of the thermal desorption spectrum after 5.0 keV D_2^+ ion irradiation at room temperature: (**a**) before irradiation, (**b**) 3.0×10^{21} ions/m², and (**c**) 10×10^{21} ions/m².

Figure 5a,b shows the ion energy dependance of the thermal desorption spectrum after the administration of 5.0 keV D_2^+ ions and 30 keV D_2^+ ions with a dose of 3.0×10^{21} ions/m^2, respectively. Figure 5b, shows that Peak B became prominent when the ion energy increased to 30 keV. In this figure, the total desorption of deuterium after 30 keV ion irradiation was much higher than that of the sample after 5.0 keV. Since the implanted deuterium atoms do not stay at the same position which is calculated in Figure 2, they diffuse to the thick region of the sample during irradiation and hydrides are formed. In the case of 5.0 keV irradiation, more deuterium atoms were released from the specimen surface than at 30 keV. Detailed estimation of deuterium atom diffusion during irradiation and desorption is needed.

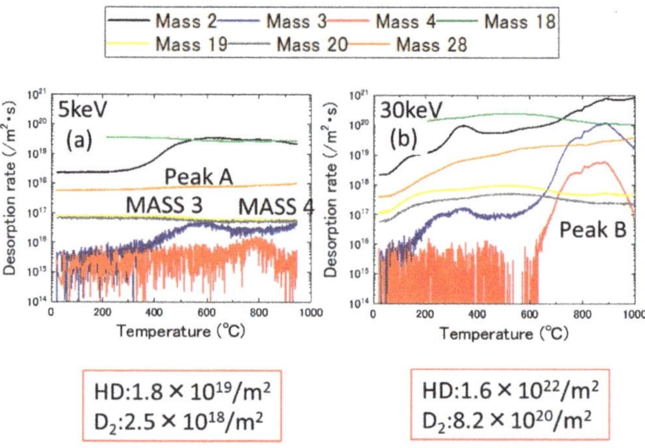

Figure 5. The energy dependance of the thermal desorption spectrum after the ion irradiation with a dose of 3.0×10^{21} ions/m^2 at room temperature: (**a**) 5.0 keV and (**b**) 30 keV.

Figure 6a,b shows the microstructures of these samples. Hydrides were observed in the specimen's edge region following irradiation at 5.0 keV (Figure 6a). However, hydride formation was only detected in the thick region of the C-TEM samples following irradiation at 30 keV (Figure 6b).

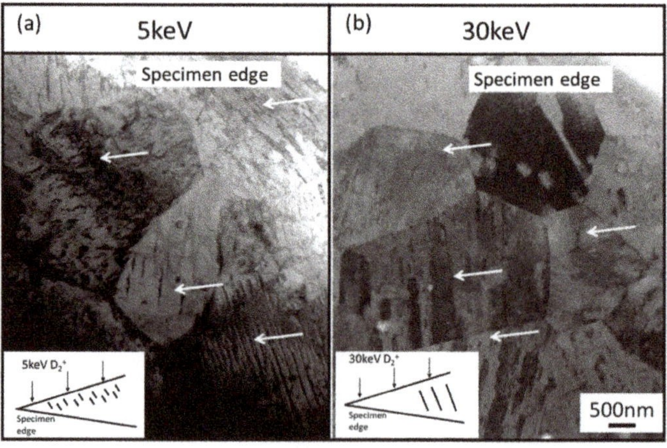

Figure 6. The energy dependance of the microstructure after the ion irradiation with a dose of 3.0×10^{21} ions/m^2 at room temperature: (**a**) 5.0 keV and (**b**) 30 keV (the arrows show the hydrides formed by irradiation).

In our previous study of Zircaloy-2 [19], it was concluded that hydrides formed by annealing were stable up to 700 °C. The cross-sectional view of the sample showed that large hydrides were formed in the thick region of the samples. However, as Figure 6a shows, small hydrides formed in the specimen surface region by 5.0 keV D_2^+ ions irradiation were not stable in a higher temperature region (Peak B in Figure 5b). To investigate the thermal stability of these small hydrides in the surface region of the samples, the TEM samples was annealed at each temperature for 30 min, and the observation was conducted using a heating TEM holder. Figure 7 shows the thermal stability of these small hydrides, and they disappeared around 400 °C (in Peak A).

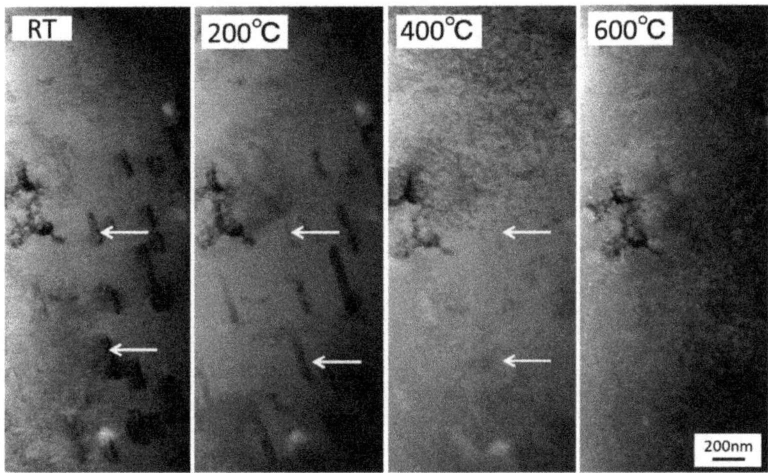

Figure 7. The thermal stability of the small hydrides formed in the surface region (30 keV, 3.0×10^{21} ions/m^2 at room temperature). The hydrides (shown by arrows) disappeared at 400 °C.

3.2. Effects of Nickel Ion Irradiation

Figure 8a,b shows the microstructure after Ni^{3+} ion irradiation at room temperature. After the irradiation (up to 3.0 dpa), high-density dislocation loops (approximately 4.1×10^{20} m^{-3}) were detected. In the Zr alloys, interstitial-type dislocation loops (a-loops) and vacancy-type dislocation loops (c-loops) are known to form. Nakamichi et al. investigated the formation and growth process of a-loops in Zry-2 under electron irradiation using a high voltage electron irradiation (HVEM) [20]. Estimated migration energy for interstitial and vacancy were 0.17 eV and 1.0 eV, respectively. These dislocations formed in this study were identified as a-loops because of vacancy mobility at room temperature. A-loop formation was already saturated, and the loops were connected. Figure 9a,b shows the desorption spectrum before and after the 3.2 MeV Ni^{3+} ion irradiation at room temperature, respectively. A new stage appeared in the position close to Peak A because of the Ni^{3+} ion irradiation (up to 3.0 dpa).

3.3. Effects of Dislocations Formed by Cold Work

To determine the effects of the dislocations on the desorption spectrum, TDS experiments were conducted on cold-worked specimens in which dense dislocations were introduced. Figure 10a–d shows the desorption spectrum for the cold-worked specimens after irradiation with 5.0 keV D^{2+} ion to the fluence of 3.0×10^{21} ions/m^2 at room temperature. In Figure 10a,b, the desorption stage A shifted to the lower temperature side when the level of cold work increased. However, the desorption stage B remained consistent when the level of cold work increased. Table 2 summarizes the total amounts of HD and D^2. These values became saturated at 5% cold work and did not show any dependance on the level of cold work. Table 3 summarizes the desorption temperature range and their responsible trapping site for deuterium obtained by the present study.

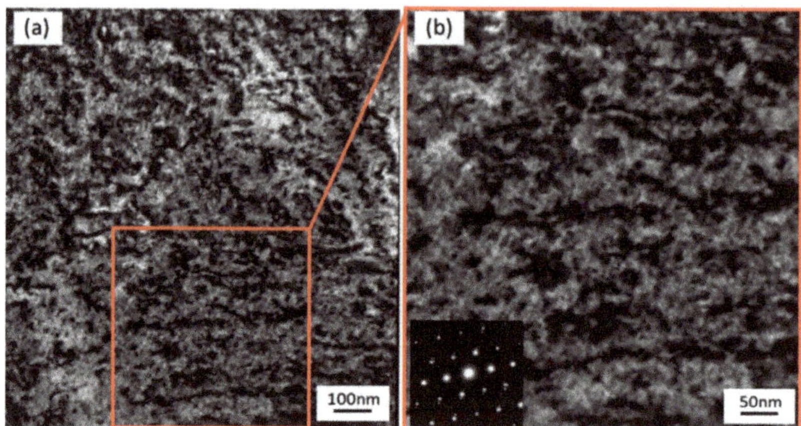

Figure 8. (**a**,**b**) The dislocation loops formed after 3.2 MeV Ni^{3+} ion irradiation at room temperature. The irradiation dose was 3.0 dpa. Ion irradiation and TEM observation were conducted perpendicular to the <C> direction.

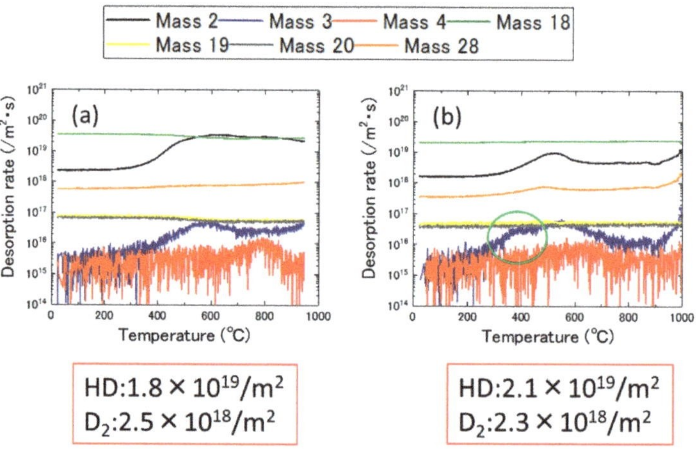

Figure 9. The thermal desorption spectra (**a**) before and (**b**) after 3.2 MeV Ni^{3+} ion irradiation with a dose of 3.0 dpa at room temperature.

As was discussed in Section 3.2, Ni^{3+} ion irradiation at room temperature induced a-loops into the samples. By increasing the level of cold work, stage A moved to the lower temperature side. Stage A corresponded to the recovery stage for weak trap sites as dislocation loops and hydrides formed in the surface region. Tangled dislocations (formed by cold work) and hydrides (formed in the thick region) contributed to stage B in the higher-temperature region.

In the Zircaloy-2 samples, a relatively large number (approximately 30 nm) of $Zr_2(Fe,Ni)$ precipitates were formed in the matrix [19]. Small $Zr(Fe,Cr)_2$ precipitates were also found in the vicinity of the $Zr_2(Fe,Ni)$ precipitates. The number density of the $Zr(Fe,Cr)_2$ precipitates was 2.0×10^{19} m^{-3} and that of the $Zr_2(Fe,Ni)$ precipitates was 3.2×10^{18} m^{-3}. The stability of these SPPs is necessary at higher dose levels because it is at these levels that the formation of c-loops and the dissolution of the SPPs are known to occur simultaneously [17,19].

Among these SPPs, the $Zr(Fe,Cr)_2$ precipitates are unstable during irradiation and undergo an amorphous transformation resulting in the decomposition and redistribution of other precipitates into defect sinks [15–17]. In this study, low-dose irradiation was chosen where phase stability of the SPPs was not essential. The irradiation dose was estimated to be approximately 18 dpa [21] at a burn-up of 50 GWd/t. This corresponded with the end of the life of the fuel rods for BWRs. The role of the redistributed precipitates and the number of c-loops formed at higher dose levels for hydrogen pickup are essential for the degradation of LWR fuel-cladding tubes during operation.

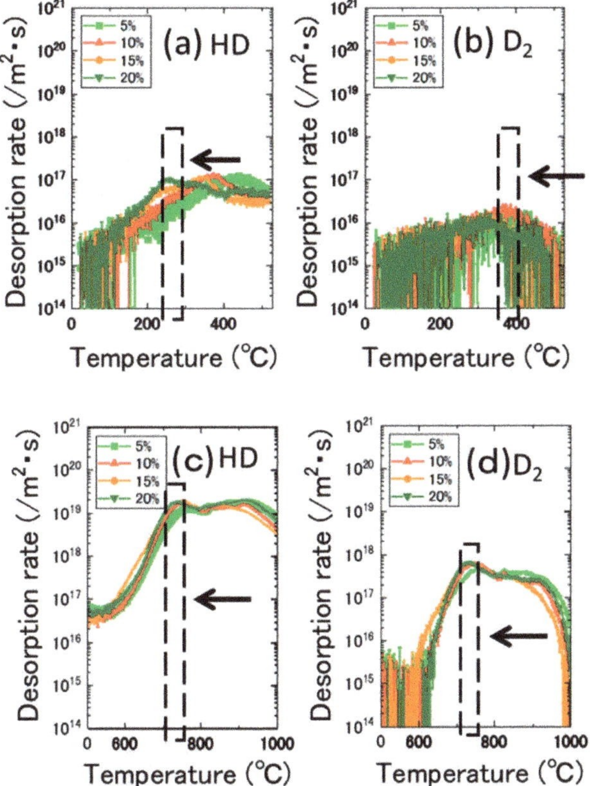

Figure 10. The amount of cold work dependance of the thermal desorption spectrum after 5.0 keV D_2^+ ion irradiation at room temperature: (**a,b**) lower temperature region and (**c,d**) higher-temperature region.

Table 2. Total desorption of HD and D_2 obtained in cold-worked samples.

Cold Work (%)	HD (m^{-2})	D_2(m^{-2})
0	1.8×10^{19}	2.5×10^{18}
5	4.0×10^{21}	1.0×10^{20}
10	3.8×10^{21}	9.7×10^{19}
15	4.2×10^{21}	9.5×10^{19}
20	4.8×10^{21}	9.9×10^{19}
25	4.0×10^{21}	1.3×10^{20}

Table 3. Desorption temperature range and their responsible trapping site for deuterium.

Stage	Temp. (°C)	Type of Trapping
Stage A	400–600	Weak Trap (dislocation loop, hydrides formed in surface region)
Stage B	700–900	Strong Trap (tangled dislocation, hydrides formed in thick region)

4. Summary

In this study, the details of retention and desorption of implanted deuterium were investigated and the responsible traps in irradiated Zircaloy-2 were identified. The thermal desorption of deuterium was conducted on Ni^{3+} ion-irradiated samples and cold-worked samples. The following conclusions have been drawn:

(1)   The desorption spectrum shows two major desorption stages (named Peak A and Peak B) in the temperature ranges of 400–600 °C and 700–900 °C, respectively.

(2)   As the thermal annealing experiments of ion-irradiated samples and cold-worked samples demonstrated, stage A corresponded to the recovery stage of weak trapping sites (i.e., the dislocation loops formed by irradiation and the hydrides that formed in the surface region of the specimens).

(3)   Stage B corresponds to the strong trapping sites explained by the tangled dislocations and hydrides that formed in the thick region of the specimens.

Author Contributions: Conceptualization, H.W.; TDS, Y.S. and K.T.; TEM, K.Y.; project administration, H.W.; funding acquisition, H.W., All authors have read and agreed to the published version of the manuscript.

Funding: This research received no external funding.

Acknowledgments: This work was supported by the Collaborative Research Program of Research Institute for Applied Mechanics, Kyushu University.

Conflicts of Interest: The authors declare no conflict of interest.

References

1.   Jonstsons, A.; Kelly, P.M.; Blake, R.G.; Farrell, K. Neutron Irradiation-Induced Defect Structure in Zirconium. *ASTM STP* **1979**, *683*, 46–61.

2.   Griffiths, M. A Review of Microstructure Evolution in Zirconium Alloys During Irradiation. *J. Nucl. Mater.* **1988**, *159*, 190–218. [CrossRef]

3.   Griffiths, M.; Loretto, M.H.; Smallman, R.E. Electron Damage in Zirconium. *J. Nucl. Mater.* **1983**, *115*, 313–332. [CrossRef]

4.   Griffiths, M.; Styles, R.C.; Woo, C.H.; Pillipp, F.; Frank, W. Study of Point Defect Mobilities in Zirconium During Electron Irradiation in a High-Voltage Electron Microscope. *J. Nucl. Mater.* **1944**, *208*, 324–334. [CrossRef]

5.   Lee, D.; Koch, E.F. Irradiation Damage in Zircaloy-2 Produced by High Dose Ion Bombardment. *J. Nucl. Mater.* **1974**, *50*, 162–174. [CrossRef]

6.   Tournadre, L.; Onimus, F.; Bechade, J.-L.; Gilbon, D.; Cloue, J.-M.; Mardon, J.-P.; Feaugas, X.; Toader, O.; Bachelet, C. Experimental Study of the Nucleation and Growth of C-Component Loops Under Charge Particle Irradiation of Recrystallized Zircaloy-4. *J. Nucl. Mater.* **2012**, *425*, 76–82. [CrossRef]

7.   Tournadre, L.; Onimus, F.; Bechade, J.-L.; Gilbon, D.; Cloue, J.-M.; Mardon, J.-P.; Feaugas, X. Toward a Better Understanding of the Hydrogen Impact on the Radiation Induced Growth of Zirconium Alloys. *J. Nucl. Mater.* **2013**, *441*, 222–231. [CrossRef]

8.   Yamada, S.; Kameyama, T. Observation of C-Component Dislocation Structures Formed in Pure Zr and Zr-base Alloy by Self-Ion Accelerator Irradiation. *J. Nucl. Mater.* **2012**, *422*, 167–172. [CrossRef]

9.   Shinohara, Y.; Abe, H.; Iwai, T.; Sekimura, N.; Kido, T.; Yamamoto, H.; Taguchi, T. In situ TEM Observation of Growth Process of Zirconium Hydride in Zircaloy-4 During Hydrogen Ion Implantation. *J. Nucl. Sci. Technol.* **2009**, *46*, 564–571. [CrossRef]

10.   Idress, Y.; Yao, Z.; Kirk, M.A.; Daymond, M.R. In situ Study of Defect Accumulation in Zirconium Under Heavy Ion Irradiation. *J. Nucl. Mater.* **2013**, *433*, 95–107. [CrossRef]

11.   Chemelle, P.; Knorr, D.B.; Van der Sand, J.B.; Pelloux, R.M. Morphology and Composition of Second Phase Precipitates in Zircaloy-2. *J. Nucl Mater.* **1983**, *113*, 58–64. [CrossRef]

12.   Taegtstrom, P.; LimBaeck, M.; Dahlbaeck, M.; Andersson, T.; Pettersson, H. Effects of Hydrogen Pickup and Second Phase Particle Dissolution on the in-Reactor Corrosion Performance of BWR Cladding. *ASTM STP* **2002**, *1423*, 96–118.

13. Valizadeh, S.; Ledergerber, G.; Abolhassani, S.; Jaedernaes, D.; Dahlbaeck, M.; Mader, E.V.; Zhou, G.; Wright, J.; Hallstadius, L. Effects of Secondary Phase Particle Dissolution on the In-Reactor Performance of BWR Cladding. *STM STP* **2011**, *1529*, 729–753.
14. Sundell, G.; Thuvander, M.; Tejland, P.; Dahlbaeck, M.; Hallstadius, L.; Andren, H.-O. Redistribution of Alloying Elements in Zircaloy-2 After In-Reactor Exposure. *J. Nucl. Mater.* **2014**, *454*, 178–185. [CrossRef]
15. Griffiths, M.; Gilbert, R.W.; Carpenter, G.J.C. Phase Instability Decomposition and Redistribution on Intermetallic precipitates Zircaloy-2 and -4 During Neutron Irradiation. *J. Nucl. Mater.* **1987**, *150*, 53–66. [CrossRef]
16. Griffiths, M.; Gilbert, R.W.; Fidleris, V.; Tucker, R.P.; Adamson, R.B. Neutron Damage in Zirconium Alloys Irradiated at 644 to 710K. *J. Nucl. Mater.* **1987**, *150*, 159–168. [CrossRef]
17. Griffiths, M.; Gilbert, R.W. The Formation of c-Component defects in Zirconium Alloys During Neutron Irradiation. *J. Nucl. Mater.* **1987**, *150*, 169–181. [CrossRef]
18. Ziegler, Z.F. SRIM–The Stopping and Ranges of Ions in Matter. Available online: http://www.strim.org (accessed on 13 November 2019).
19. Watanabe, H.; Takahashi, K.; Yasunaga, K.; Wan, Y.; Aono, Y.; Maruno, Y.; Hashizume, K. Effects of an Alloying Element on a C-Component Loop Formation and Precipitate Resolution in Zr Alloys During Ion Irradiation. *J. Nucl. Sci. Technol.* **2018**, *55*, 1212–1224. [CrossRef]
20. Nakamichi, M.; Kinoshita, C.; Yasuda, K.; Fukada, S. Formation and Growth Process of Dislocation Loops in Zircaloys under Electron Irradiation. *J. Nucl. Sci. Technol.* **1987**, *34*, 1079–1086. [CrossRef]
21. Sonoda, T.; Sawabe, T.; Kitajima, S.; Nishikawa, N. Radiation Damage Accumulation and Dissolution of Second Phase Particles in Zircaloy-2 by means of Ion Accelerator. In Proceedings of the WRFPM 2014, Sendai, Japan, 14–17 September 2014; p. 100091.

quantum beam science

MDPI

Article

Modification of SiO$_2$, ZnO, Fe$_2$O$_3$ and TiN Films by Electronic Excitation under High Energy Ion Impact

Noriaki Matsunami [1,*]**, Masao Sataka** [2]**, Satoru Okayasu** [2] **and Bun Tsuchiya** [1]

[1] Faculty of Science and Technology, Meijo University, Nagoya 468-8502, Japan; btsuchiya@meijo-u.ac.jp
[2] Japan Atomic Energy Agency (JAEA), Tokai 319-1195, Japan; sataka@tac.tsukuba.ac.jp (M.S.);
    okayasu.satoru@jaea.go.jp (S.O.)
* Correspondence: atsu20matsu@gmail.com

Abstract: It has been known that the modification of non-metallic solid materials (oxides, nitrides, etc.), e.g., the formation of tracks, sputtering representing atomic displacement near the surface and lattice disordering are induced by electronic excitation under high-energy ion impact. We have investigated lattice disordering by the X-ray diffraction (XRD) of SiO$_2$, ZnO, Fe$_2$O$_3$ and TiN films and have also measured the sputtering yields of TiN for a comparison of lattice disordering with sputtering. We find that both the degradation of the XRD intensity per unit ion fluence and the sputtering yields follow the power-law of the electronic stopping power and that these exponents are larger than unity. The exponents for the XRD degradation and sputtering are found to be comparable. These results imply that similar mechanisms are responsible for the lattice disordering and electronic sputtering. A mechanism of electron–lattice coupling, i.e., the energy transfer from the electronic system into the lattice, is discussed based on a crude estimation of atomic displacement due to Coulomb repulsion during the short neutralization time (~fs) in the ionized region. The bandgap scheme or exciton model is examined.

Keywords: electronic excitation; lattice disordering; sputtering; electron–lattice coupling

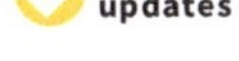
check for **updates**

Citation: Matsunami, N.; Sataka, M.; Okayasu, S.; Tsuchiya, B. Modification of SiO$_2$, ZnO, Fe$_2$O$_3$ and TiN Films by Electronic Excitation under High Energy Ion Impact. *Quantum Beam Sci.* **2021**, *5*, 30. https://doi.org/10.3390/qubs5040030

Academic Editors: Akihiro Iwase and Rozaliya Barabash

Received: 30 August 2021
Accepted: 17 October 2021
Published: 27 October 2021

Publisher's Note: MDPI stays neutral with regard to jurisdictional claims in published maps and institutional affiliations.

1. Introduction

Material modification induced by electronic excitation under high-energy (> 0.1 MeV/u) ion impact has been observed for many non-metallic solids since the late 1950's; for example, the formation of tracks (each track is characterized by a long cylindrical disordered region or amorphous phase in crystalline solids) in LiF crystal (photographic observation after chemical etching) by Young [1], in mica (a direct observation using transmission electron microscopy, TEM, without chemical etching, and often called a latent track) by Silk et al. [2], in SiO$_2$-quartz, crystalline mica, amorphous P-doped V$_2$O$_5$, etc. (TEM) by Fleischer et al. [3,4], in oxides (SiO$_2$-quartz, Al$_2$O$_3$, ZrSi$_2$O$_4$, Y$_3$Fe$_5$O$_{12}$, high-Tc superconducting copper oxides, etc.) (TEM) by Meftah et al. [5] and Toulemonde et al. [6], in Al$_2$O$_3$ crystal (atomic force microscopy, AFM) by Ramos et al. [7], in Al$_2$O$_3$ and MgO crystals (TEM and AFM) by Skuratov et al. [8], in Al$_2$O$_3$ crystal (AFM) by Khalfaoui et al. [9], in Al$_2$O$_3$ crystal (high resolution TEM) by O'Connell et al. [10], in amorphous SiO$_2$ (small angle X-ray scattering (SAXS)) by Kluth et al. [11], in amorphous SiO$_2$ (TEM) by Benyagoub et al. [12], in polycrystalline Si$_3$N$_4$ (TEM) by Zinkle et al. [13] and by Vuuren et al. [14], in amorphous Si$_{3.55}$N$_4$ (TEM) by Kitayama et al. [15], in amorphous SiN$_{0.95}$:H and SiO$_{1.85}$:H (SAXS) by Mota-Santiago et al. [16], in epilayer GaN (TEM) by Kucheyev et al. [17], in epilayer GaN (AFM) by Mansouri et al. [18], in epilayer GaN and InP (TEM) by Sall et al. [19], in epilayer GaN (TEM) by Moisy et al. [20], in InN single crystal (TEM) by Kamarou et al. [21], in SiC crystal (AFM) by Ochedowski et al. [22] and in crystalline mica (AFM) by Alencar et al. [23]. Amorphization has been observed for crystalline SiO$_2$ [5] and the Al$_2$O$_3$ surface at a high ion fluence (though the XRD peak remains) by Ohkubo et al. [24] and Grygiel et al. [25]. The counter process, i.e., the recrystallization of the amorphous or disordered regions, has been reported for SiO$_2$ by Dhar et al. [26], Al$_2$O$_3$ by Rymzhanov [27] and InP, etc., by Williams [28]. Density modification, i.e., a lower density in the track core surrounded by a shell with a higher density, has been observed for Al_2O_3 [10], amorphous SiO_2 [11], Si_3N_4 [14] and amorphous $SiN_{0.95}$:H and $SiO_{1.85}$:H [16]. Interestingly, an electrically conducting track formation in tetrahedral-amorphous carbon (sp^3 into sp^2 bond transformation) has been observed by Gupta et al. [29]. The track radius, hillock height and diameter characterizing the surface morphology modification associated with the track are well described in terms of the electronic stopping power S_e (defined as the energy loss due to electronic excitation and ionization per unit path length), and the velocity effect has been noticed [12]. The threshold of S_e for the track formation has been reported [3,6,8,9,12,13,30] and the data appear to scatter, and it seems that the threshold S_e depends on the observation method of the track [12]. No track formation by monatomic ions has been observed in AlN [19].

Moreover, electronic sputtering (the erosion of solid materials caused by electronic energy deposition) has been observed for various compound solids: UO_2 by thermal-neutron-induced ^{235}U fission fragments by Rogers [31,32] and by Nilsson [33], UO_2 by energetic ions by Meins et al. [34], Bouffard et al. [35] and Schlutig [36], H_2O ice by Brown et al. [37,38], Bottiger et al. [39], Baragiola et al. [40], Dartois et al. [41] and Galli et al. [42], frozen gas films of Xe, CO_2 and SF_6 [39], those of CO, Ar and N_2 by Brown et al. [43], CO_2 ice by Mejia et al. [44], SiO_2 by Qui et al. [45], Sugden et. al. [46], Matsunami et al. [47,48], Arnold-bik et al. [49] and Toulemonde et al. [50,51], $MgAl_2O_4$ [48], UF_4 ([34], by Griffith et al. [52] and Toulemonde et al. [53]), $LiNbO_3$ [45], Al_2O_3 ([45] and by Matsunami et al. [54]), various oxides by Matsunami et al. ($SrTiO_3$ and $SrCeO_3$ [47,54], CeO_2, MgO, TiO_2 and ZnO [54], Y_2O_3 and ZrO_2 [55], Cu_2O [56,57], WO_3 [58], CuO [59], Fe_2O_3 [60]), Si_3N_4 [45], Si_3N_4 and AlN by Matsunami et al. [55], Cu_3N by Matsunami et al. [56,61], LiF ([50], by Assmann et al. [62] and Toulemonde et al. [63]), KBr [56], NaCl [63], CaF_2 [53] and SiC [56]. The sputtering of frozen Xe films has been observed for low energy electron impact, against the anticipation of no atomic displacement [39], and the result confirms that the sputtering is caused by electronic excitation. Mechanisms of electronic excitation leading to atomic displacement will be discussed in Section 4.

As mentioned above, electronic sputtering has been observed for a variety of non-metallic materials, indicating that it seems to be a general phenomenon for non-metallic solids by high-energy ion impact. In many cases, ions with an equilibrium charge have been employed, which is usually attained by inserting thin foils, such as carbon, metals, etc., before impact on samples, and sputtered atoms are collected in carbon, metals, etc., followed by neutron activation and ion beam analysis to obtain sputtering yields. This article concerns the equilibrium charge incidence, though charge-state effects for non-equilibrium charge incidence have been observed and discussed ([23,29,34,49,52,58,62,64]). The electronic energy deposition or electronic stopping power (S_e) at the equilibrium charge can be calculated using a TRIM or SRIM code by Ziegler et al. [65,66], a CasP code by Grande et al. [67] and the nuclear stopping power (S_n, defined as the energy loss due to elastic collisions per unit path length) [65,66]. Characteristic features of the electronic sputtering by high-energy ions are as follows:

(a). Electronic sputtering yields (Y_{SP}) are found to be larger by $10\text{--}10^3$ than nuclear sputtering yields due to elastic collision cascades, which can be estimated assuming linear dependence on S_n;

(b). Y_{SP} super linearly depends on S_e and is approximated by the power-law fit: $Y_{SP} = (B_{SP}S_e)^{N_{SP}}$ with $1 \leq N_{SP} \leq 4$ for most cases, with B_{SP} being a material dependent constant.

Stoichiometric sputtering has been observed for many materials, whereas a considerable deviation from the stoichiometric sputtering has been reported for $YBa_2Cu_3O_7$ [68], $Gd_3Ga_5O_{12}$ and $Y_3Fe_5O_{12}$ [69] and CaF_2 and UF_4 [53]. Only the heavy element of U [31–36] and the light element of O [49] have been detected.

Besides track formation and electronic sputtering, lattice disordering (the degradation of X-ray diffraction (XRD) intensity) with lattice expansion (an increase in the lattice parameter) by high-energy ions has been observed for the polycrystalline films of SiO_2 [70] and WO_3 [58], and lattice disordering with lattice compaction for those of Cu_2O [57],

CuO [59], Fe_2O_3 [60], Cu_3N [61] and Mn-doped ZnO [71]. Only lattice disordering has been observed for the ultra-thin films of WO_3 [72]. It should be noted that a comparison between high-energy and low-energy ion impact effects is important. Lattice expansion has been observed for a few keV D ion irradiation on Fe_2O_3 [73], and this can be understood by the incorporation of D into non-substitutional sites (incorporation or implantation effect). Thus, lattice expansion by medium-energy ions (100 keV Ne) on Fe_2O_3 [60] could be understood by Ne incorporation and/or interstitial-type defects generated by ion impact, with a possible stabilization by incorporating Ne in the film, whereas lattice compaction has been observed for a 100 MeV Xe ion impact on Fe_2O_3 [60]. It should be noted that the incorporation of ions in thin films does not take place for high-energy ions, since the projected range of ions (R_p) is much larger than the film thickness (e.g., R_p of 14 μm for 100 MeV Xe in SiO_2), unless the thickness is too large. The lattice expansion due to the incorporation effect has been observed for a few keV H and D irradiation at a low fluence on WNO_x with $x \approx 0.4$, whereas lattice compaction has been observed at a high fluence of D [74]. Peculiarly, lattice expansion at a low ion fluence and compaction at a high fluence, as well as disordering, have been reported for medium-energy (100 keV Ne and N) and high-energy (100 MeV Xe and 90 MeV Ni) ion impact on WNO_x [75,76]. One speculation is that the lattice compaction is due to vacancy-type defects generated by ion impact, which is to be investigated. Furthermore, a drastic increase in electrical conductivity has been observed for Cu_3N [61], Mn-doped ZnO [71] and WNO_x with $x \approx 0.4$ [75,76]. The conductivity increase is ascribed to the increase in the carrier concentration and mobility.

There are a few reports on the S_e dependence of the XRD intensity degradation per unit fluence (Y_{XD}) for SiC and KBr [56] and WO_3 [72]. Y_{XD} is found to follow the power-law fit: $Y_{XD} = (B_{XD}S_e)^{N_{XD}}$, B_{XD} being a material-dependent constant and the exponent N_{XD} being comparable with the N_{sp} of the electronic sputtering. The results imply that similar mechanisms operate for lattice disordering and electronic sputtering. It is of interest to compare the S_e dependence of lattice disordering with that of electronic sputtering for materials other than those mentioned above. In this paper, we have measured the lattice disordering of SiO_2, ZnO, Fe_2O_3 and TiN films, and the sputtering of TiN. The XRD results are compared with the sputtering. The exciton model is examined and scaling parameters are explored for representing electronic excitation effects.

2. Materials and Methods

XRD has been measured using Cu-k_α radiation. Accuracy of the XRD intensity is estimated to be approximately 10%, based on the variation of repeated measurements. Rutherford backscattering (RBS) has been performed with MeV He ions for evaluation of film thickness and composition. Similarly, accuracy of the RBS is estimated considering the variation of the repeated measurements. High-energy ion irradiation has been performed at room temperature and normal incidence. Irradiation of high-energy ion with lower incident charge than the equilibrium charge without carbon foil is often employed for the samples of XRD measurement; however, the effect of non-equilibrium charge incidence does not come into play because the length for attaining the equilibrium charge is much smaller than the film thickness, as described for each material in Section 3.

SiO_2 films have been grown by thermal oxidation of Si(001) at 1300 °C for 5 hr. According to XRD, the films are polycrystalline with diffraction peaks at ~21°, 22°, 31°, 33°, 36° and 69°, with very weak peak at 44° and 47°. The peaks at ~21°, 22°, 44° and 47° have been assigned to (100), (002), (004) and (202) diffraction of hexagonal-tridymite structure [70]. The strong peak at 69° is Si(004) and peak at 33° is possibly Si(002). Film thickness is ~1.5 μm and the composition is stoichiometric (O/Si = 2.0 $\pm$ 5%) by RBS of 1.8 MeV He. Film density is taken to be the same as that of amorphous-SiO_2 (a-SiO_2), since it has been derived to be 2.26 gcm^{-3} from XRD results, which is close to that of a-SiO_2 (2.2 gcm^{-3})

Pure ZnO films have been prepared on MgO (001) substrate by using a radio frequency magnetron sputtering (RFMS) deposition method with ZnO target, and it has been reported that the dominant growth orientation is (001) and (100) of hexagonal-wurtzite structure depending on the substrate temperature of 350 °C and 500 °C during the film growth, respectively [71,77,78]. The composition is stoichiometric, i.e., O/Zn = 1.0 ± 0.05, and film thickness is ~100 nm by He RBS. Here, the density is taken to be 4.2×10^{22} Zn cm^{-3} (5.67 gcm^{-3}).

Preparation and characterization methods of Fe_2O_3 films are described in [60]. Briefly, Fe_2O_3 films have been prepared by deposition of Fe layers on SiO_2-glass and C-plane cut Al_2O_3 (C-Al_2O_3) substrates using a RFMS deposition method with Fe target (99.99%) and Ar gas, followed by oxidation at 500 °C for 2–5 hr in air. According to RBS of 1.4–1.8 MeV He ions, the composition is stoichiometric (O/Fe = 1.5 ± 0.1) and film thickness used in this study is ~100 nm. Here, the density of 3.96×10^{22} Fe cm^{-3} (5.25 gcm^{-3}) is employed. Diffraction peaks have been observed at ~33° and 36°, and crystalline structure has been identified as hexagonal Fe_2O_3 (hematite or α-Fe_2O_3). These correspond to (104) and (110) diffraction planes.

TiN films have been prepared on SiO_2-glass, C-plane cut Al_2O_3 (C-Al_2O_3) and R-plane cut Al_2O_3 (R-Al_2O_3) substrates at 600°C using a RFMS deposition method with Ti target (99.5%) and high purity N_2 gas. RBS of 1.4–1.8 MeV He ions shows that the composition is stoichiometric (N/Ti = 1.0 ± 0.05) and that the film thickness used in this study is ~170 nm (deposition time of 1 hr). Here, the density of 5.25×10^{22} Ti cm^{-3} (5.4 gcm^{-3}) is employed. Diffraction peaks have been observed at 36.6°, 42.6° and ~77° on SiO_2 glass and C-Al_2O_3. Crystalline structure has been identified as a cubic structure and these correspond to (111), (200) and (222) diffractions [79]. Diffraction intensity of (111) is larger than that of (200) on SiO_2 glass, and diffraction of (111) on C-Al_2O_3 is very intensive. TiN on R-Al_2O_3 has preferential growth orientation of (220) of a cubic structure (diffraction angle at ~61°). Sputtered atoms are collected in the carbon foil (100 nm) and the sputtered atoms are analyzed by RBS to obtain the sputtering yields [54] (carbon collector method).

3. Results and Discussion

3.1. SiO$_2$

The XRD intensity at the diffraction angle of ~22° (the most intensive (002) diffraction of hexagonal-trydimite) normalized to that of as-grown SiO_2 films on Si(001) is shown in Figure 1 as a function of the ion fluence for 90 MeV Ni^{+10}, 100 MeV Xe^{+14} and 200 MeV Xe^{+14} ion impact. The XRD intensity of the irradiated sample normalized to that of the unirradiated sample is proportional to the ion fluence to a certain fluence. Deviation from the linear dependence for the high fluence could be due to the overlapping effect. As observed in latent track formation (e.g., [5,6]), electronic excitation effects extend to a region (approximately cylindrical) with a radius of several nm and a length of the projected range or film thickness, and thus ions may hit the ion-irradiated part for a high ion fluence (called the overlapping effect). As described below, the XRD degradation yield per unit ion fluence (Y_{XD}) is reduced at a high fluence, and this could be understood as thermal annealing and/or a reduction in the disordered regions via ion-induced defects (recrystallization [26]). The damage cross-sections (A_D obtained by RBS-channeling (RBS-C) technique and TEM [5]) are compared with Y_{XD} in Figure 2, and it appears that both agree well for $S_e > 10$ keV. A discrepancy between A_D and Y_{XD} is seen for $S_e < 10$ keV, and the reason for this is not understood. In addition, sputtering yields are often reduced, and this is unlikely to be explained by the annealing effect. Therefore, the reasons for the sputtering suppression at a high fluence remain in question. The XRD degradation yields (Y_{XD}) per unit ion fluence are obtained and given in Table 1. The film thickness has been obtained to be ~1.5 μm, using 1.8 MeV He RBS. The attenuation length (L_{XA}) of Cu-k$_\alpha$ (8.0 keV) is obtained to be 128 μm [80] and the attenuation depth ($L_{XA} \cdot \sin(22°/2)$) = 24.3 μm. The film thickness (~1.5 μm) is much smaller than the attenuation depth and thus no correction is necessary for the XRD intensity. The lattice expansion or increase in the lattice parameter of 0.5% with an estimated error of 0.2% at 1×10^{12} cm^{-2} is found to be nearly independent of the electronic stopping power.

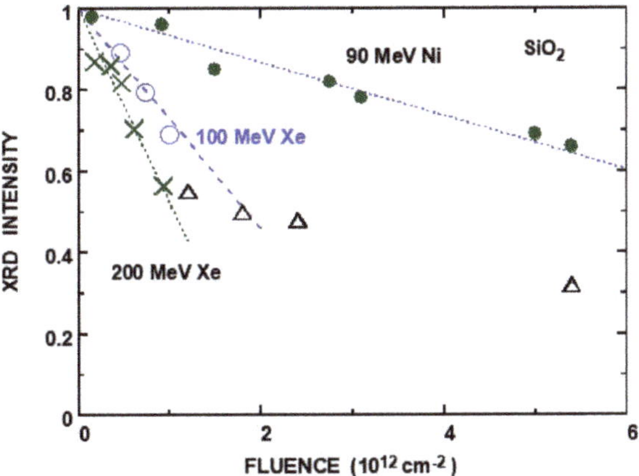

Figure 1. XRD intensity from (002) diffraction plane at ~22° normalized to as-grown films of SiO$_2$ as a function of ion fluence for 90 MeV Ni (●), 100 MeV Xe (o, Δ) and 200 MeV Xe (x) ions. Data of 90 MeV Ni (●) and 100 MeV Xe (Δ) are from [70]. Linear fit is indicated by dashed lines. An estimated error of XRD intensity is 10%.

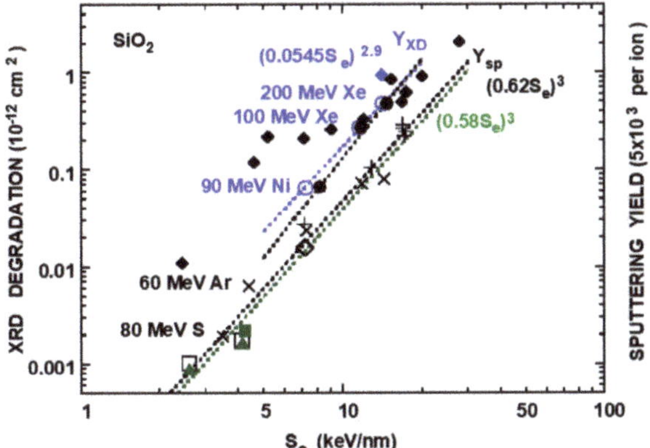

Figure 2. XRD degradation per unit fluence Y_{XD} of polycrystalline SiO$_2$ film (o, present result) and sputtering yield Ysp (x) of amorphous (or vitreous)-SiO$_2$ (□, x)) and film of SiO$_2$ (▲, ■, x, ◊, +) as a function of electronic stopping power (S$_e$). Data (□, ▲) from (Qiu et al.) [45], (■) from (Sugden et al.) [46], (x) from (Matsnami et al.) [47,48], (◊) from (Arnoldbik et al.) [49] and (+) from (Toulemonde et al.) [51]. S$_e$ is calculated using SRIM2013, and power-law fits of Y_{XD} (($0.0545S_e$)$^{2.9}$) and Y_{sp} (($0.62S_e$)$^{3.0}$) are indicated by blue and black dotted lines, respectively. Power-law fit (●) Y_{XD} (($0.055S_e$)$^{3.4}$, TRIM1997) and Y_{sp} (($0.58S_e$)$^{3.0}$, TRIM1985 through SRIM2010) from [47,48,51] are indicated by black and green dashed lines. Damage cross sections (◆) are obtained by RBS-C and (◆) by TEM from [5].

Table 1. XRD data of SiO_2 films. Ion, incident energy (E in MeV), XRD intensity degradation (Y_{XD}), appropriate E* (MeV) considering the energy loss in the film and electronic stopping power in keV/nm (S_e*) appropriate for Y_{XD} (see text). S_e from SRIM2013. The deviation ΔS_e* = (S_e*/S_e(E) − 1) × 100 is also given.

Ion	Energy	Y_{XD}	E*	S_e*	ΔS_e*
	(MeV)	(10^{-12} cm^2)	(MeV)	(keV/nm)	(%)
^{58}Ni	90	0.066	84.5	7.246	−0.32
^{136}Xe	100	0.27	91.0	11.56	−3.2
^{136}Xe	200	0.475	189	14.22	−1.3

The electronic stopping power (S_e*) appropriate for XRD intensity degradation is calculated using SRIM 2013, using a half-way approximation that the ion loses its energy for half of the film thickness (~0.75 μm), i.e., S_e* = S_e(E*) with E* = E(incidence) − S_e(E) × 0.75 μm (Table 1). The correction for the film thickness on S_e appears to be a few percent. It is noticed that the incident charge (Ni^{+10}, Xe^{+14}) differs from the equilibrium charge (+19, +25 and +30 for 90 MeV Ni, 100 MeV Xe and 200 MeV Xe, respectively (Shima et al.) [81], and +18.2, +23.9 and +29.3 (Schiwietz et al.) [82]), both being in good agreement. Following [64], the characteristic length (L_{EQ} = 1/(electron loss cross-section times N)) for attaining the equilibrium charge is estimated to be 8.7, 8.3 and 7.9 nm for 90 MeV Ni^{+10}, 100 MeV Xe^{+14} and 200 MeV Xe^{+14}, respectively, from the empirical formula of the single-electron loss cross-section σ_{1L} (10^{-16} cm^2) of 0.52 (90 MeV Ni^{+10}), 0.55 (100 MeV Xe^{+14}) and 0.57 (200 MeV Xe^{+14}) [83,84], N (2.2 × 10^{22} Si cm^{-3}) being the density, and (target atomic number)$^{2/3}$ dependence being included. Here, σ_{1L} = σ_{1L}(Si) + 2σ_{1L}(O), ionization potential I_P = 321 eV [85,86] with the number of removable electrons N_{eff} = 8 and I_P = 343 eV with N_{eff} = 12 are employed for Ni^{+10} and Xe^{+14}. L_{EQ} is much smaller than the film thickness and hence the charge-state effect is insignificant.

The sputtering yields Y_{sp} of SiO_2 (normal incidence) are summarized in Table 2 for the comparison of the S_e dependence of the XRD degradation yields Y_{XD}. There are various versions of TRIM/SRIM starting in 1985, and in this occasion, the results used the latest version of SRIM2013 are compared with those earlier versions. Firstly, the correction on the stopping power and projected range for carbon foils (20–120 nm), which have been used to achieve the equilibrium charge incidence, is less than a few %, except for low-energy Cl ions (several %). Secondly, S_e by CasP (version 5.2) differs ~30% from that by SRIM 2013. Figure 2 shows the S_e dependence of the XRD degradation yields Y_{XD} and Y_{sp}. Both Y_{XD} and Y_{sp} fit to the power-law of S_e, and the exponents of XRD degradation N_{XD} = 2.9 and N_{sp} = 3 (sputtering) are almost identical, indicating that the same mechanism is responsible for lattice disordering and sputtering. Further plotted is the sputtering yields vs. S_e calculated using earlier versions of TRIM/SRIM (TRIM1985 to SRIM2010) [45–49,51], and the plot using earlier versions give the same exponent (N_{sp} = 3) with a 6% smaller constant B_{sp} in the power-law fit (20% smaller in the sputtering yields). This means that the plot and discussion using SRIM2013 do not significantly differ from those using the earlier versions of TRIM/SRIM. One notices that no appreciable difference in sputtering yields is observed among a-SiO_2, films and single-crystal-SiO_2 (c-SiO_2) [45–48], even though the density of c-SiO_2 is larger by 20% than that of a-SiO_2, whereas much smaller yields (by a factor of three) have been observed for c-SiO_2 [51]. The discrepancy remains in question. Sputtering yields Y_{EC}, which are due to elastic collision cascades, is estimated assuming Y_{EC} is proportional to the nuclear stopping power, discarding the variation of the α-factor (order of unity) depending on the ratio of target mass over ion mass (Sigmund) [87]. The proportional constant is obtained to be 2.7 nm/keV using the sputtering yields by low-energy ions (Ar and Kr) (Betz et al.) [88]. Y_{EC} is given in Table 2 and it is shown that Y_{sp}/Y_{EC} ranges from 44 (5 MeV Cl) to 3450 (210 MeV Au).

Table 2. Sputtering data of SiO_2 (normal incidence). Ion, incident energy (E in MeV), energy (E* in MeV) corrected for the energy loss in carbon foils (see footnote), electronic stopping power (S_e), nuclear stopping power (S_n), projected range (R_p) and sputtering yield (Y_{SP}). S_e, S_n and R_p are calculated using SRIM2013. $(S_e(E^*)/S_e(E) - 1)$, $(S_n(E^*)/S_n(E) - 1)$ and $(R_p(E^*)/R_p(E) - 1)$ in % are given in the parentheses after $S_e(E^*)$, $S_n(E^*)$ and $R_p(E^*)$, respectively. Y_{SP} in the parenthesis is for SiO_2 films. $S_e(E)$ by CasP is also listed. Y_{EC} is the calculated sputtering yield due to elastic collisions.

Ion	E(E*)	S_e(E*)	S_n(E*)	R_p(E*)	Y_{SP}	S_e(CasP)	Y_{EC}
	(MeV)	(keV/nm)	(keV/nm)	(μm)		(keV/nm)	
			Qiu et al. [45]				
^{35}Cl	5 (4.6)	2.59 (−4.26)	0.0426 (6.8)	3.0 (−4.6)	5.1 (4.4)	1.87	0.12
^{35}Cl	20 (19.4)	4.15 (−0.35)	0.0134 (2.7)	7.0 (−2.0)	8.77 (8.22)	3.83	0.036
			Sugden et al. [46]				
^{35}Cl	30 (29.9)	4.24 (−2.4 × 10^{-3})	9.3×10^{-3} (0.3)	9.5 (−0.25)	11	3.99	0.025
			Matsunami et al. [47,48]				
^{58}Ni	90 (89)	7.265 (−0.055)	0.0145 (0.84)	18.3 (−0.60)	120	7.66	0.039
^{136}Xe	100 (99)	11.88 (−0.49)	0.091 (1.2)	14.4 (−0.83)	362	14.0	0.246
^{136}Xe	200 (198)	14.37 (−0.19)	0.051 (0.73)	21.9 (−0.51)	404	16.0	0.138
^{40}Ar	60 (60)	4.40	6.5×10^{-3}	16.3	32	4.19	0.018
^{32}S	80 (80)	3.49	3.3×10^{-3}	23	9.7	3.23	0.0089
			Arnoldbik et al. [49]				
^{63}Cu	50 (50)	7.17	0.027	11.6	80	7.55	0.073
			Toulemonde et al. [51]				
^{197}Au	190 (190)	16.9	0.143	20.5	1425	20.9	0.39
^{197}Au	190 (190)				1320		
^{197}Au	197 (197)	17.1	0.14	20.9	1110	21.2	0.38
^{197}Au	210 (210)	17.4	0.13	21.7	1230	21.7	0.36
^{127}I	148 (148)	12.9	0.06	20	525	15.3	0.16
^{58}Ni	69 (69)	7.15	0.018	15.5	135	7.75	0.049

Equilibrium charge has been obtained by using carbon foils of 120 nm [45], 25 nm [46], 100 nm [47,48], 20 nm [49] and 50 nm [51].

In order to obtain the stopping powers (S) for the non-metallic compounds, such as SiO_2, described above, we apply the Bragg's additive rule, e.g., $S(SiO_2) = S(Si) + 2S(O)$ and S of the constituting elements is calculated using TRIM/SRIM and CasP codes. Before moving to the discussion of the Bragg's deviation, the accuracy of S is briefly mentioned. It is estimated to be 8% (Be through U ions in Ag) near the maximum of S (~0.8 MeV/u) [66], 17% (K to U ions in Au) (Paul) [89]. Besenbacher et al. have reported no difference between solid and gas phases for 0.5–3 MeV He ion stopping in Ar with an experimental accuracy of 3% [90], and this could be understood by the fact that the binding energy (cohesive energy) of solid Ar is too small (0.08 eV (Kittel) [91]), compared with the ionization potential (I_{Bethe}) of 188 eV [92], to affect the stopping. On the other hand, Arnau et al. have reported a large deviation (~50% near the stopping power maximum at ~50 keV) for proton stopping between solid and gas phases of Zn, and the deviation reduces ~10% at ~1 MeV [93]. The cohesive energy of 1.35 eV [91] is much smaller than the mean ionization potential I_{Bethe} of 330 eV [92], and hence the small increase in I_{Bethe} cannot explain the Bragg's deviation of Zn. They have argued that the difference in the 4s into 4p transition probability and screening effect between solid and gas phases are responsible. Both TRIM/SRIM and CasP codes are based on the experiments of conveniently available solid targets and molecular gas (e.g., N_2, O_2), and thus it is anticipated that the binding effect is included to some or large extent and that the Bragg's deviation is not serious for nitrides and oxides, and is roughly 10% or less at around 1 MeV/u.

3.2. ZnO

The XRD intensity at a diffraction angle of ~34° ((001) diffraction) and 32° ((100) diffraction) normalized to those of unirradiated ZnO films on the MgO substrate is shown in Figure 3 as a function of the ion fluence for 90 MeV Ni^{+10}, 100 MeV Xe^{+14} and 200 MeV Xe^{+14} ion impact. It appears that the XRD intensity degradation is nearly independent of the diffraction planes. The XRD intensity degradation per unit fluence Y_{XD} is given in Table 3, together with sputtering yields [54], stopping powers and projected ranges (SRIM2013). The X-ray (Cu-kα) attenuation length L_{XA} is obtained to be 36.6 μm [80] and the attenuation depth is 11 and 10 μm for the diffraction angle of ~34° and 32°, respectively; thus, the X-ray attenuation correction is unnecessary. It appears that the appropriate energy for the Y_{XD} vs. S_e plot, $E - S_e\ell/2$, where ℓ = a film thickness of ~100 nm, is nearly the same as E^* for sputtering, in which the energy loss of a carbon foil of 100 nm is considered. Similarly to the case of SiO_2, the characteristic length (L_{EQ}) is estimated to be 4.6, 4.4 and 4.2 nm for 90 MeV Ni^{+10}, 100 MeV Xe^{+14} and 200 MeV Xe^{+14}, respectively, from the empirical formula of the single-electron loss cross-section $\sigma_{1L}(10^{-16} \text{ cm}^2)$ of 0.52 (90 MeV Ni^{+10}), 0.54 (100 MeV Xe^{+14}) and 0.57 (200 MeV Xe^{+14}) [83,84]. Here, $\sigma_{1L} = \sigma_{1L}(Zn) + \sigma_{1L}(O)$, and the ionization potential I_P and N_{eff} are described in Section 3.1. Again, L_{EQ} is much smaller than the film thickness and the charge-state effect is insignificant.

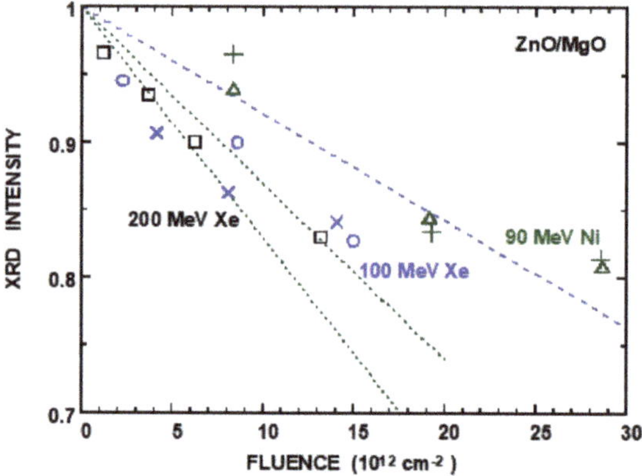

Figure 3. XRD intensity normalized to as-deposited films of ZnO as a function of ion fluence for 90 MeV Ni (Δ, +), 100 MeV Xe (o, x) and 200 MeV Xe ($\square$) ions. Diffraction planes are (002) at ~34° (Δ, o, $\square$) and (100) at ~32° (+, x). Linear fit is indicated by dashed lines. An estimated error of XRD intensity is 10%.

Table 3. XRD data of ZnO films. Ion, incident energy (E in MeV), XRD intensity degradation (Y_{XD}), $E^* = E - \Delta E$ (energy loss in carbon foil of 100 nm) (MeV) and electronic (S_e^*) and nuclear (S_n^*) stopping powers in keV/nm and projected range R_p^* (μm) at E^* calculated using SRIM2013. Sputtering yield Y_{sp} from [54]. Sputtering yield by 100 keV Ne ion is also given.

Ion	Energy (MeV)	Y_{XD} (10^{-14} cm^2)	E^* (MeV)	S_e^* (keV/nm)	S_n^* (keV/nm)	R_p^* (μm)	Y_{sp}
^{32}S	80		80	6.62	0.007	13	1.09
^{40}Ar	60		60	8.3	0.014	9.5	2.08
^{58}Ni	90	0.788	89	13.72	0.031	11	4.0
^{127}I	85		84	18.92	0.205	8.8	7.0
^{136}Xe	100	1.3	99	21.60	0.20	8.8	
^{136}Xe	200	1.7	198	27.14	0.112	13	11
^{20}Ne	0.10	0.10		0.29	0.24	0.12	0.9

Figure 4 shows the XRD intensity degradation Y_{XD} vs. electronic stopping power (S_e) (SRIM2013 and TRIM1997) together with the sputtering yields Y_{sp} vs. S_e. Both Y_{XD} and Y_{sp} follow the power-law fit and the exponent for Y_{XD} using TRIM1997 gives a slightly larger value than that using SRIM2013. The exponent of lattice disordering is nearly the same as that of sputtering. The change in the lattice parameter $\Delta\ell c$ appears to scatter, and roughly -0.2% and -0.1% with an estimated error of 0.1% are obtained for (100) and (002) diffractions by 100 MeV Xe at 10×10^{12} cm^{-2}, assuming that $\Delta\ell c$ is proportional to the ion fluence. $\Delta\ell c$ is obtained at -0.3% for (002) diffraction by 200 MeV Xe at 5×10^{12} cm^{-2}, and no appreciable change in the lattice parameter is observed by 90 MeV Ni ions at 40×10^{12} cm^{-2}; more data are desired.

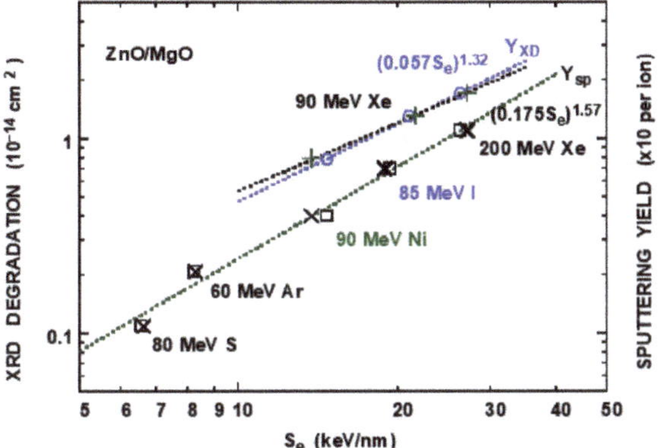

Figure 4. XRD degradation per unit fluence Y_{XD} of polycrystalline ZnO film vs. electronic stopping power S_e (TRIM1997 and SR2013). Power-law fit to $Y_{XD} = (0.057S_e)^{1.32}$ (TRIM1997) (o, blue dotted line) and $(0.0585 S_e)^{1.16}$ (SRIM2013) (+, black dotted line). Sputtering yield Ysp vs. S_e (TRIM1997, □) and S_e (SR2013, x) is also shown. Sputtering yield from [54]. Power-law fits to Ysp: $(0.175 S_e)^{1.57}$ for both S_e from TRIM1997 and SR2013 is indicated by green dotted line.

3.3. Fe$_2$O$_3$

The XRD intensity at a diffraction angle of ~33° and 36° (corresponding to diffraction planes of (104) and (110)) normalized to those of unirradiated Fe$_2$O$_3$ films on C-Al$_2$O$_3$ and SiO$_2$ glass substrates as a function of the ion fluence is shown in Figure 5 for 90 MeV Ni^{+10}, 100 MeV Xe^{+14} and 200 MeV Xe^{+14} ion impact. It appears that the XRD intensity degradation is nearly independent of the diffraction planes and substrates. The XRD intensity degradation per unit fluence Y_{XD} is given in Table 4, together with the sputtering yields [60] and stopping powers (SRIM2013). The X-ray (Cu-kα) attenuation length L_{XA} is obtained to be 8.8 μm [80] and the attenuation depth is 2.5 and 2.7 μm for the diffraction angle of ~33° and 36°, respectively, which are much larger the film thickness of ~100 nm and thus the X-ray attenuation correction is unnecessary. The appropriate energy for the XRD vs. S_e plot, using half-way approximation ($E - S_e\ell/2$) with the film thickness ℓ of ~100 nm, again gives nearly the same as E^* for sputtering, in which the energy loss of the carbon foil of 100 nm is taken into account.

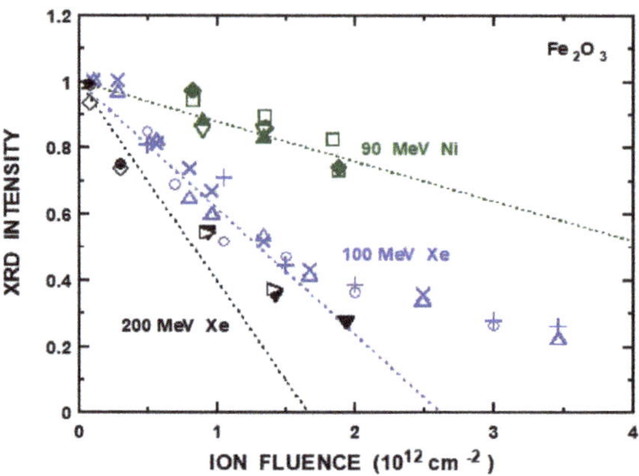

Figure 5. XRD intensity normalized to unirradiated films of Fe$_2$O$_3$ as a function of ion fluence for 90 MeV Ni ($\square$, •, ∇, $\blacktriangle$), 100 MeV Xe (o, $\triangle$, +, x) and 200 MeV Xe (•, $\Diamond$, $\blacktriangledown$, $\cap$) ions. Diffraction peaks at ~33° of Fe$_2$O$_3$ films on C-Al$_2$O$_3$ substrate ($\square$ (90 MeV Ni), o (100 MeV Xe), • (200 MeV Xe)), ~36° of films on C-Al$_2$O$_3$ (• (90 MeV Ni), + (100 MeV Xe), $\Diamond$ (200 MeV Xe)), ~33° of films on SiO$_2$ glass substrate (∇ (90 MeV Ni), $\triangle$ (100 MeV Xe), $\blacktriangledown$ (200 MeV Xe)) and ~36° of films on SiO$_2$ glass substrate ($\blacktriangle$ (90 MeV Ni), x (100 MeV Xe), $\cap$ (200 MeV Xe)). Data of 100 MeV Xe are from [60]. Linear fit is indicated by dotted lines. An estimated error of XRD intensity is 10%.

Table 4. XRD data of Fe$_2$O$_3$ films. Ion, energy (E in MeV), XRD intensity degradation (Y_{XD}), $E^* =$ E-ΔE (energy loss in carbon foil of 100 nm) (MeV) and electronic (S_e^*) and nuclear (S_n^*) stopping powers in keV/nm and projected range R_p^* (μm) calculated using SRIM2013. Sputtering yield Y_{sp} from [60]. Results by low energy (100 keV Ne) ion are also given.

Ion	Energy (MeV)	Y_{XD} (10^{-12} cm^2)	E^* (MeV)	S_e^* (keV/nm)	S_n^* (keV/nm)	R_p^* (μm)	Y_{sp}
^{58}Ni	90	0.12	89	14.28	0.030	9.8	38.3
^{136}Xe	100	0.38	99	23.25	0.19	7.9	57.9
^{136}Xe	200	0.60	198	28.27	0.11	11.7	81.7
^{20}Ne	0.10		0.10	0.354	0.258	0.12	2.3

Similarly to SiO$_2$ and ZnO, the characteristic length (L_{EQ}) is estimated to be 4.5, 4.3 and 4.1 nm for 90 MeV Ni^{+10}, 100 MeV Xe^{+14} and 200 MeV Xe^{+14}, respectively, from the empirical formula of the single-electron loss cross-section σ_{1L} (10^{-16} cm^2) of 0.56 (90 MeV Ni^{+10}), 0.59 (100 MeV Xe^{+14}) and 0.61 (200 MeV Xe^{+14}) [83,84]. Here, $\sigma_{1L} = \sigma_{1L}$(Fe) + 1.5$\sigma_{1L}$(O). L_{EQ} is much smaller than the film thickness and the charge-state effect does not come into play.

Figure 6 shows XRD intensity degradation Y_{XD} vs. electronic stopping power (S_e) (SRIM2013 and TRIM1997) together with the sputtering yields Y_{sp} vs. S_e. Both Y_{XD} and Y_{sp} follow the power-law fit and the exponent using TRIM1997 gives a slightly larger fit than those using SRIM2013. The exponent of lattice disordering is two times larger than that of sputtering (N_{sp} is exceptionally close to unity, in contrast to the SiO$_2$ and ZnO cases). The change in the lattice parameter appears to scatter depending on the substrate and diffraction planes, and is not proportional to the ion fluence. The average of the lattice parameter change in the (104) and (110) diffractions of Fe$_2$O$_3$ on C-Al$_2$O$_3$ is -0.2, -0.3% (an estimated error of 0.1%) and nearly zero at ~1×10^{12} cm^{-2} for 200 MeV Xe, 100 MeV Xe and 90 MeV Ni ion impact. The dependence of the lattice parameter change on the ion fluence and S_e is complicated, and is to be investigated.

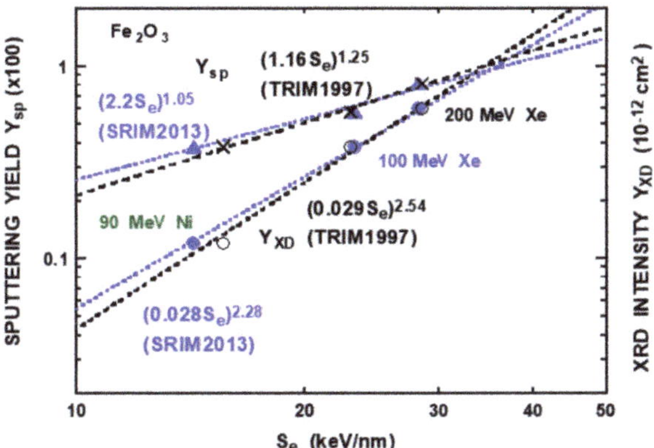

Figure 6. XRD dgradation per unit fluence Y_{XD} of polycrystalline Fe_2O_3 films ($\bullet$, o) and sputtering yield Ysp ($\blacktriangle$, x) as a function of the electronic stopping power (S_e) in keV/nm. Power-law fits are indicated by dashed lines and S_e is calculated using SRIM2013 ($\bullet$, $\blacktriangle$) and TRIM1997 (o, x): ($\bullet$) Y_{XD} = $(0.028\ S_e)^{2.28}$ (SRIM2013), (o) Y_{XD} = $(0.029 S_e)^{2.54}$ (TRIM1997), ($\blacktriangle$) Y_{sp} = $(2.2\ S_e)^{1.05}$ (SRIM2013) and (x) Y_{sp} = $(1.16\ S_e)^{1.25}$ (TRIM1997). Sputtering data and power-law fit to the sputtering yields (TRIM1997) from [60].

3.4. TiN

The XRD patterns are shown in Figure 7 for unirradiated and irradiated TiN films on the SiO_2 glass substrate. As already mentioned in Section 2, (111) and (200) diffraction peaks are observed and the XRD intensity decreases due to ion impact. Figure 8 shows XRD intensities normalized to those of unirradiated TiN films on SiO_2 glass, $C-Al_2O_3$ and $R-Al_2O_3$ substrates as a function of the ion fluence. It is seen that the XRD intensity degradation is nearly the same for the diffraction planes of (111) and (200) on SiO_2, and for (111) on $C-Al_2O_3$. The XRD intensity degradation is less sensitive to the ion impact for the diffraction plane (220) on the $R-Al_2O_3$ substrate (~30% smaller than that for (111) and (200) on SiO_2, and for (111) on $C-Al_2O_3$). The XRD intensity degradation per unit fluence Y_{XD} for (111) and (200) diffractions is given in Table 5, together with sputtering yields and stopping powers (TRIM1997 and SRIM2013). No appreciable change in the lattice parameter is observed, as shown in Figure 7. Similarly to the SiO_2, ZnO and Fe_2O_3 cases, the appropriate energy, $E - S_e\ell/2$, ℓ = film thickness of ~170 nm is taken into account, and the energy is close to that for sputtering, in which the energy loss of the carbon foil of 100 nm is considered. The X-ray (Cu-kα) attenuation length L_{XA} is obtained to be 11.8 μm [80], and the attenuation depth is 3.7, 4.3 and 6.0 μm for diffraction angles of 36.6°, ~43° and 61°, respectively; thus, the X-ray attenuation correction is insignificant.

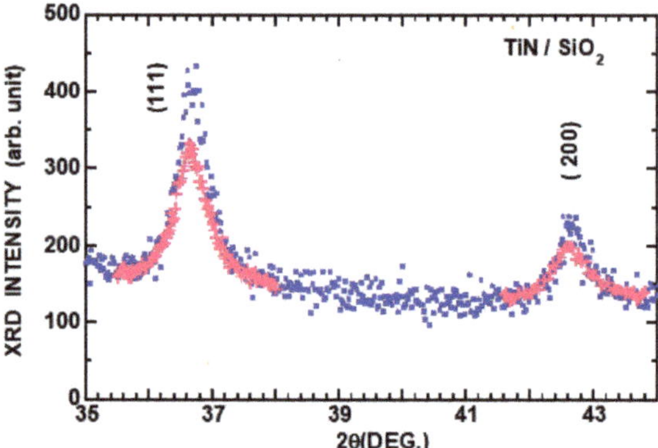

Figure 7. XRD patterns of TiN film on SiO$_2$ glass substrate: unirradiated (•) and irradiated by 100 MeV Xe at 0.72×10^{12} cm^{-2} (+).

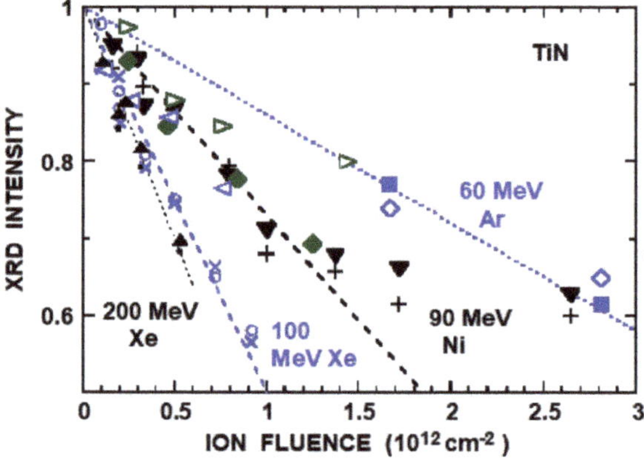

Figure 8. XRD intensity normalized to unirradiated films of TiN as a function of ion fluence for 60 MeV Ar (■, ◊), 90 MeV Ni (▼, +, ♦, ∩), 100 MeV Xe (o, x, ◁) and 200 MeV Xe (▲, •) ions. Diffraction plane (111) at diffraction angle of ~36.6° is indicated by ■, ▼, o and ▲ for SiO$_2$ substrate, (200) at ~43° by ◊, +, x and • for SiO$_2$ substrate, (111) by ♦ for C-Al$_2$O$_3$ substrate and (220) at ~61° by ∩ and ◁ for R-Al$_2$O$_3$ substrate. Linear fit is indicated by dotted lines. An estimated error of XRD intensity is 10%.

Table 5. XRD data of TiN films. Ion, energy (E in MeV), XRD intensity degradation (Y_{XD}) for (111) and (200) diffraction on SiO_2 and $C-Al_2O_3$, substrates, Y_{XD} for (220) diffraction on $R-Al_2O_3$ in the parenthesis, $E^* = E - \Delta E$ (energy loss in carbon foil of 100 nm) (MeV) and electronic (S_e^*) and nuclear (S_n^*) stopping powers in keV/nm and projected range R_p^* (µm) calculated using SRIM2013 and sputtering yield Y_{sp} of Ti. S_e^* (TRIM1997) is given in parenthesis.

Ion	Energy (MeV)	Y_{XD} (10^{-12} cm^2)	E* (MeV)	S_e^* (keV/nm)	S_n^* (keV/nm)	R_p^* (µm)	Y_{sp}(Ti)
^{40}Ar	60	0.14	60	9.41 (9.33)	0.0135	7.6	51.8
^{58}Ni	90	0.27 (0.2)	89	15.5 (16.5)	0.0305	8.6	147
^{136}Xe	100	0.50 (0.35)	99	26.7 (25.5)	0.19	6.9	380
^{136}Xe	200	0.60	198	30.85 (30.25)	0.11	10	529

The characteristic length (L_{EQ}) is estimated to be 4.5, 4.4, 4.2 and 4.0 nm for 60 MeV Ar^{+7}, 90 MeV Ni^{+10}, 100 MeV Xe^{+14} and 200 MeV Xe^{+14}, respectively, from the empirical formula of the single-electron loss cross-section σ_{1L} (10^{-16} cm^2) of 0.43 (60 MeV Ar^{+7}), 0.44 (90 MeV Ni^{+10}), 0.46 (100 MeV Xe^{+14}) and 0.48 (200 MeV Xe^{+14}) [83,84]. Here, $\sigma_{1L} = \sigma_{1L}$(Ti) + σ_{1L}(N), and the ionization potential I_P and N_{eff} are (I_P = 143 eV and N_{eff} = 1) for Ar^{+7}, with those described in Section 3.1 for Ni^{+10} and Xe^{+14}. L_{EQ} is much smaller than the film thickness, and hence the charge-state effect is insignificant.

It is found that sputtered Ti collected in the carbon foil is proportional to the ion fluence, as shown in Figure 9 for 60 MeV Ar, 90 MeV Ni, 100 MeV Xe and 200 MeV Xe ions. The sputtering yield of Ti is obtained using the collection efficiency of 0.34 in the carbon foil collector [47] and the results are given in Table 5. Sputtered N collected in the carbon foil is obtained to be 0.4×10^{14} and 0.44×10^{14} cm^{-2} with an estimated error of 20% for 200 MeV Xe at 0.22×10^{12} cm^{-2} and 60 MeV Ar at 2.8×10^{12} cm^{-2}, respectively, and this is comparable with the Ti areal density of 0.4×10^{14} cm^{-2} (200 MeV Xe) and 0.475×10^{14} cm^{-2} (60 MeV Ar). The results imply stoichiometric sputtering, due to the collection efficiency of N in the carbon foil collector of 0.35 [55], which is close to that of Ti. Thus, the total sputtering yield (Ti + N) is obtained by doubling Y_{sp} (Ti) in Table 5. The sputtering yields of TiN (Y_{EC}) due to elastic collisions can be estimated assuming that Y_{EC} is proportional to the nuclear stopping power. Here, the proportional constant is obtained to be ~1.6 nm/keV using the experimental yields of 0.527 (0.6 keV Ar) and 0.427 (0.6 keV N) [94] and 0.7 (0.5 keV Cd) [88]. Y_{sp}(TiN)/Y_{EC} ranges from 2.5×10^3 to 6×10^3. The XRD intensity degradations Y_{XD} and Y_{sp}(Ti + N) are plotted as a function of the electronic stopping power S_e in Figure 10. It appears that both fit to the power-law: $Y_{XD} = (0.0224S_e)^{1.26}$ and $Y_{sp} = (1.17S_e)^{1.95}$. The exponents are comparable for XRD intensity degradation and sputtering.

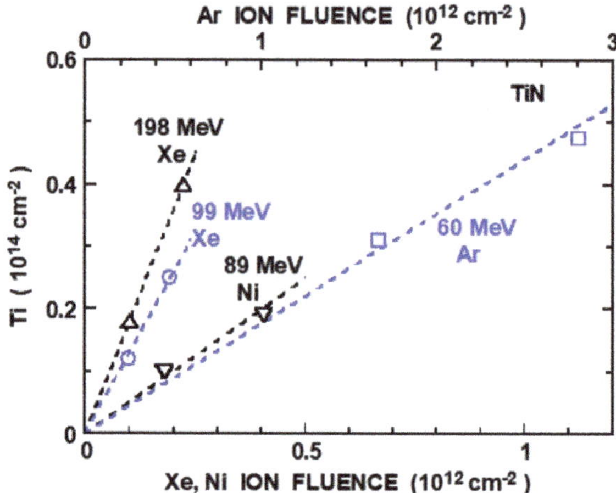

Figure 9. Areal density of sputtered Ti from TiN on SiO_2 substrate collected in carbon foil vs. ion fluence for 60 MeV Ar ($\square$), 89 MeV Ni (∇), 99 MeV Xe (o) and 198 MeV Xe (Δ) ions. An estimated error of areal density is 20%.

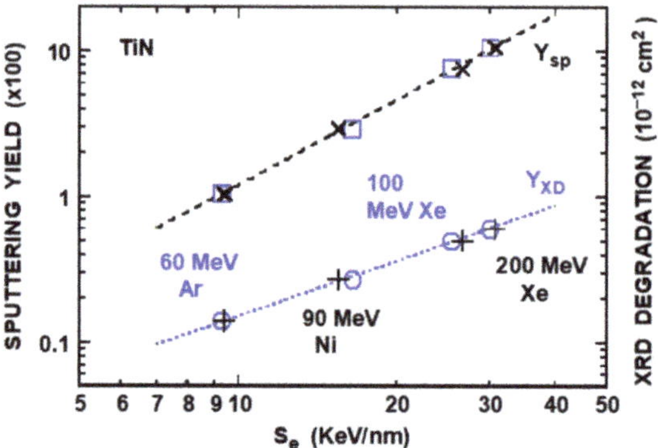

Figure 10. XRD intensity degradation Y_{XD} (10^{-12} cm^2) (o, +) and sputtering yields Ysp (Ti + N) ($\square$, x) vs. electronic stopping power S_e (keV/nm). S_e is calculated by TRIM1997 (o, $\square$) and by SRIM2013 (+, x). Power-law fits are indicated by dotted lines: $Y_{XD} = (0.0224S_e)^{1.26}$ and $Y_{sp} = (1.17S_e)^{1.95}$.

4. Discussion

4.1. Comparison of Lattice Disordering with Sputtering

The electronic stopping power (S_e) dependence of lattice disordering Y_{XD}, together with electronic sputtering, is summarized in Table 6, recognizing that most of the data have used TRIM1997. Results using SRIM2013 and TRIM1997 are compared in Section 3. Both exponents of the power-law fits are similar for SiO_2, ZnO, Fe_2O_3, TiN and WO_3 films, as well as for KBr and SiC. As mentioned in Section 3, it can be seen that the exponent of the lattice disordering N_{XD} is comparable with that of sputtering N_{sp}, except for Fe_2O_3, in which N_{sp} is exceptionally close to unity, as in the case of Cu_2O ($N_{sp} = 1.0$) [56] and CuO ($N_{sp} = 1.08$) [59]. The similarity of the exponent of lattice disordering and sputtering for SiO_2, ZnO, Fe_2O_3, TiN, WO_3, KBr and SiC imply that both phenomena originate from similar mechanisms, despite the fact that small displacements and annealing and/or the reduction in disordering via ion-induced defects are involved in the lattice disordering, whereas large displacements are involved in sputtering. The result of Fe_2O_3 indicates that the electronic excitation is more effective for lattice disordering. In the case of CuO, N_{XD} is nearly zero [59]. In Table 6, Y_{XD} (10^{-12} cm^2) at S_e = 10 keV/nm and Y_{XD}/Y_{sp} ($\times 10^{-15}$ cm^2) are listed. It is found that the ratio Y_{XD}/Y_{sp} is an order of 10^{-15} cm^2, except for ZnO, where the sputtering yields are exceptionally small. More data of lattice disordering would be desired for further discussion.

Table 6. Summary of electronic stopping power (S_e in keV/nm) dependence of lattice disordering $Y_{XD} = (B_{XD}S_e)^{N_{XD}}$ for the present results of SiO$_2$, ZnO, Fe$_2$O$_3$ and TiN films, and sputtering yields $Y_{sp} = (B_{sp}S_e)^{N_{sp}}$ of the present result for TiN. Lattice disordering and sputtering yields of WO$_3$ film from [58,72], those of KBr and SiC from [56] and sputtering yields of SiO$_2$, ZnO and Fe$_2$O$_3$ (see Section 3). Constant B_{XD} and B_{sp} and the exponent N_{XD} and N_{sp} are obtained using TRIM1997 and those using SRIM2013 are in parentheses. Y_{XD} at S_e = 10 keV and Y_{XD}/Y_{sp} (10^{-15} cm^2) are given.

Sample	B_{XD} (nm/keV)	N_{XD}	B_{sp} (nm/keV)	N_{sp}	Y_{XD} (10^{-12} cm^2) (S_e = 10 keV/nm)	Y_{XD}/Y_{sp} (10^{-15} cm^2)
SiO$_2$	0.055 (0.0545)	3.4 (2.9)	0.58 (0.62)	3.0 (3.0)	0.13	0.67
ZnO	0.057 (0.0585)	1.32 (1.16)	0.175	1.57	0.476	198
Fe$_2$O$_3$	0.029 (0.028)	2.54 (2.28)	1.16 (2.2)	1.25 (1.05)	0.043	2.0
TiN	0.0224	1.26	1.17	1.95	0.15	1.26
WO$_3$	0.07355	2.65	0.65	3.6	0.44	0.53
KBr	0.127	2.4	0.77	3.0	1.78	3.9
SiC	0.0377	1.97	1.86	1.53	0.15	1.7

4.2. Electron–Lattice Coupling

Three models have been suggested for atomic displacement induced by electronic excitation: Coulomb explosion (CE) [3,4], thermal spike (TS) [50] and exciton model [30,95–97]. The neutralization time of the ionized region along the ion path is generally too short, and the fraction of the charged sputtered ions is small, e.g., 100 MeV Xe ions on SiO$_2$ glass [48]. Hence, the CE model is unsound. However, a small atomic separation during the short time might be enough for electron–lattice coupling (a key for electronic excitation effects), which will be discussed later. A crude estimation of the evaporation yield for SiO$_2$ based on the TS model appears to be far smaller than the experimental sputtering yield [55] and thus the TS model is also unsound. Moreover, the electron–lattice coupling or transfer mechanism of electronic energy into the lattice is not clear in the model. In the exciton model, the non-radiative decay of self-trapped excitons (STX, i.e., localized excited-state of electronic system coupled with lattice) leads to atomic displacement. According to the exciton model (or bandgap scheme), it is anticipated that the energy of the atoms in motion from the non-radiative decay of STX is comparable with the bandgap, leading to a larger sputtering yield with a larger bandgap, discarding the argument for the efficiency of STX generation from the electron–hole pairs, which is inversely proportional to the bandgap. This bandgap scheme is examined below. The effective depth contributing to the electronic sputtering of WO$_3$ has been obtained to be 40 nm, which is nearly independent of S_e [98], which would shed light on understanding the electronic sputtering; therefore, more data are desired.

The electronic sputtering yield Y_{sp} super-linearly depends on the electronic stopping power (S_e), and Y_{sp} at S_e = 10 keV/nm is taken to be a representative value, which is plotted as a function of the bandgap (E_g) in Figure 11 from [56], including the present TiN result. The optical absorbance (defined as $\log_{10}(I_o/I)$, I_o and I being the incident and transmitted photon intensities) of TiN films are measured, and the direct bandgap E_g is obtained to be 4.5 eV for a film thickness of 25–50 nm, which decreases to 2.8 eV for a film thickness of ~180 nm by using the relation: (absorbance $\bullet$ photon energy)^2 is proportional to photon energy—E_g. The thickness dependence of E_g is under investigation by considering the influence of the reflectivity, film growth conditions and experimental problems, such as stray light, etc. A large variation has been reported for E_g, 4.0 eV (film thickness of 260 nm on Si substrate) (Popovic et al.) [99], 3.4 eV (thickness of ~100 nm on glass substrate) (Solovan et al.) [100] and 2.8–3.2 eV (film thickness of 460 nm on glass substrate) (Kavitha et al.) [101]. In this study, E_g is taken to be 4 eV and this choice is tolerable in the following discussion. It has been reported that the bandgap is reduced by 0.06 eV under a 400 KeV Xe ion implantation at 10^{16} cm^{-2} [99]. High-energy ion impact effects on optical properties are under way. It can be observed that the bandgap scheme seems to work for $E_g > 3$ eV [56]. A large deviation (two orders of magnitude) from the upper limit (dashed line indicated in Figure 11) is observed for ZrO_2, MgO, $MgAl_2O_4$ and Al_2O_3. The existence of STX is known for limited materials, rare gas solids, SiO_2 and alkali halides [30,95,96]. The STX does not exist for MgO and probably does not exist for Al_2O_3 [102]. The deviation for MgO and Al_2O_3 could be explained by the non-existence of STX. The numbers of electron–hole pairs leading to STX are inversely proportional to E_g, which could be a reason for the dependence of the sputtering yields for $E_g < 3$ eV. In any case, the single parameter of the band gap is insufficient for the explanation of the bandgap dependence of the sputtering yields.

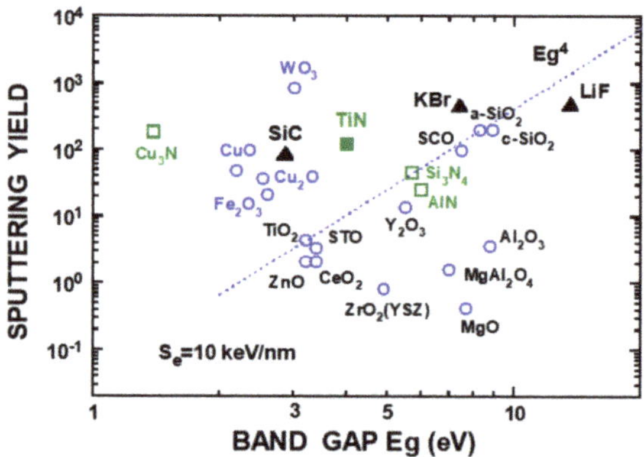

Figure 11. Sputtering yield at $S_e = 10$ keV/nm vs. bandgap. Data from [56], TiN (present result) and LiF data from [62]. Dotted line is a guide for eyes ($E_g > 3$ eV).

Martin et al. [102] argued that STX exists for materials with small elastic constants. Following this suggestion, sputtering yields are plotted as a function of the elastic constant (C_{11}) in Figure 12. Here, C_{11} (GPa) is taken to be 87 (SiO_2), 348 ($SrTiO_3$), 497 (Al_2O_3), 294 (MgO), 270 (TiO_2), 210 (ZnO), 299 ($MgAl_2O_4$), 242 (Fe_2O_3), 35 (KBr) and 114 (LiF) [85], and, for other materials, 403 (CeO_2) [103], 224 (Y_2O_3) [104], 400 (ZrO_2) [105], 13 (WO_3) [106], 126 (Cu_2O) [107], 135 (CuO) [108], 388 (Si_3N_4) as an average of the values [109,110], 345 (polycrystalline-AlN) [111], which is smaller by 16% than 410 (AlN single crystal) [112], 234 (Cu_3N) [113], 500 (SiC) [114] and 625 (TiN) [115]. It can be observed for oxides (the most abundant data are available at present) that Y_{sp} decreases exponentially with an increase in the elastic constant for $C_{11} < 300$ GPa, except for MgO and ZrO_2. Y_{sp} for nitrides and SiC is larger than that for oxides at a given C_{11}, and these are to be separately treated. It can be understood that the elastic constant represent the resistance of lattice deformation by electronic energy deposition. However, a single parameter, either the bandgap or elastic constant, is not adequate, and at least one more parameter is necessary. Furthermore, parameters other than those mentioned above are to be explored. More data for nitrides, alkali halides and especially carbides are desired.

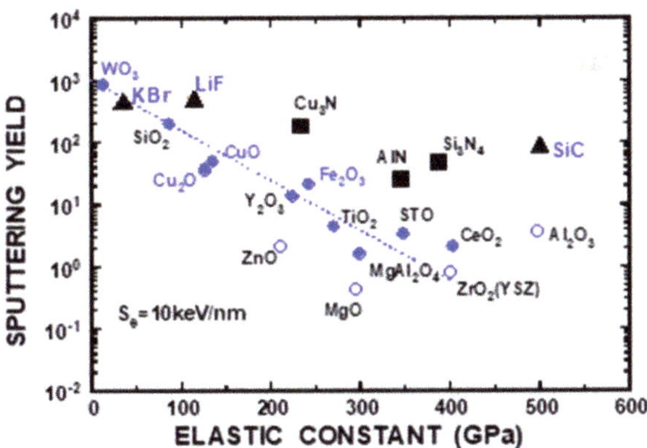

Figure 12. Sputtering yield at S_e = 10 keV/nm vs. elastic constant. Sputtering yield of TiN (present result), LiF from [62] and others from [56]. Dotted line is a guide for eyes for the most abundant available data of oxides (o, •).

Finally, a mechanism for the electron–lattice coupling is discussed. In an ionized region along the ion path, Coulomb repulsion leads to atomic motion, which is not adequate to cause sputtering because of its short neutralization time. Nevertheless, displacement comparable with the lattice vibration amplitude (one tenth of the average atomic separation, d_{av} of ~0.25 nm for a-SiO_2) is highly achievable during the neutralization time. As a first step, the time required for the Si^+–O^+ displacement of 0.025 nm (one tenth of d_{av}) from d_{av} is estimated to be ~15 fs using a formula [116]. Also, the time is estimated to be ~ 15 fs and ~ 12 fs for the Zn^+–O^+ displacement of 0.02 nm from d_{av} of 0.2 nm in ZnO and for the Ti^+–N^+ displacement of 0.02 nm from d_{av} of 0.2 nm in TiN, respectively. A similar situation has been reported for the Fe^+–O^+ displacement of 0.01 nm in Fe_2O_3 (~7 fs) [60], the K^+–Br^+ displacement of 0.01 nm in KBr (~9 fs) and the Si^+–C^+ displacement of 0.01 nm in SiC (~6 fs). These suggest a possibility that a small displacement comparable with the lattice vibration amplitude caused by Coulomb repulsion during the short neutralization time leads to the generation of a highly excited-state coupled with the lattice (h-ESCL), and h-ESCL is considered to be equivalent to STX or multi STX. The non-radiative decay of h-ESCL leads to atomic displacement (a larger displacement results in sputtering and smaller displacement results in phonon generation or lattice distortion).

5. Conclusions

We have measured the lattice disordering of polycrystalline SiO_2, ZnO, Fe_2O_3 and TiN films, as well as the sputtering yield of TiN, by high-energy ion impact. It is found that lattice disordering is caused by electronic excitation and the degradation of the XRD intensity fits to the power-law on the electronic stopping power. The exponent in the fit of the XRD degradation is comparable with that of the electronic sputtering yield for these films, as well as the published results of WO_3, KBr and SiC, implying that both lattice disordering and sputtering originate from similar mechanisms. In the case of Fe_2O_3, on the other hand, the exponent of the lattice disordering is larger by twice than that of the sputtering (the exponent for the sputtering is close to unity). The exciton mechanism seems to work for Eg > 3 eV, with some exceptions, and the elastic constant is examined as another scaling parameter for the electronic sputtering yields. A possibility of electron–lattice coupling is discussed based on a crude estimation that an atomic displacement comparable with the vibration amplitude due to Coulomb repulsion during the short neutralization time in the ionized region along the ion path can be attainable and, thus, the generation of a highly excited state coupled with the lattice is highly achievable, resulting in atomic displacement.

Author Contributions: Conceptualization and writing, N.M.; investigation, M.S., S.O. and B.T. All authors have read and agreed to the published version of the manuscript.

Funding: This research received no external funding.

Institutional Review Board Statement: Not applicable.

Informed Consent Statement: Not applicable.

Data Availability Statement: Not applicable.

Acknowledgments: One of the authors (N.M.) thanks for utilization of XRD equipments of RIGAKU ULTIMA IV and RAD IIc at Radioisotope Research Center, Nagoya University and of AN Van de Graaff accelerator, Nagoya University for Rutherford backscattering spectroscopy (RBS) measurement. High-energy ion irradiation has been performed using a TANDEM accelerator at Tokai Research Center, JAEA.

Conflicts of Interest: The authors declare no conflict of interest.

References

1. Young, D.A. Etching of Radiation Damage in Lithium Fluoride. *Nature* **1958**, *182*, 375–377. [CrossRef] [PubMed]
2. Silk, E.C.H.; Barnes, R.S. Examination of Fission Fragment Tracks with an Electron Microscope. *Phil. Mag.* **1959**, *4*, 970–972. [CrossRef]
3. Fleischer, R.L.; Price, P.B.; Walker, R.M. Ion Explosion Spike Mechanism for Formation of Charged-Particle Tracks in Solids. *J. Appl. Phys.* **1965**, *36*, 3645–3652. [CrossRef]
4. Fleischer, R.L. *Tracks to Innovation*; Springer: New York, NY, USA, 1998; pp. 1–193.
5. Meftah, A.; Brisard, F.; Costantini, J.M.; Dooryhee, E.; Hage-Ali, M.; Hervieu, M.; Stoquert, J.P.; Studer, F.; Toulemonde, M. Track formation in SiO_2 quartz and the thermal-spike mechanism. *Phys. Rev. B* **1994**, *49*, 12457–12463. [CrossRef]
6. Toulemonde, M.; Bouffard, S.; Studer, F. Swift heavy ions in insulating and conducting oxides: Tracks and physical properties. *Nucl. Instr. Meth. B* **1994**, *91*, 108–123. [CrossRef]
7. Ramos, S.M.M.; Bonardi, N.; Canut, B.; Bouffard, S.; Della-Negra, S. Damage creation in α-Al_2O_3 by MeV fullerene impacts. *Nucl. Instr. Meth. B* **1998**, *143*, 319–332. [CrossRef]
8. Skuratov, V.A.; Zinkle, S.J.; Efimov, A.E.; Havancsak, K. Surface defects in Al_2O_3 and MgO irradiated with high-energy heavy ions. *Surf. Coat. Technol.* **2005**, *196*, 56–62. [CrossRef]
9. Khalfaoui, N.; Stoquert, J.P.; Haas, F.; Trautmann, C.; Meftah, A.; Toulemonde, M. Damage creation threshold of Al_2O_3 under swift heavy ion irradiation. *Nucl. Instr. Meth. B* **2012**, *286*, 247–253. [CrossRef]
10. O'Connell, J.H.; Rymzhanov, R.A.; Skuratov, V.A.; Volkov, A.E.; Kirilkin, N.S. Latent tracks and associated strain in Al_2O_3 irradiated with swift heavy ions. *Nucl. Instr. Meth. B* **2016**, *374*, 97–101. [CrossRef]
11. Kluth, P.; Pakarinen, O.H.; Djurabekova, F.; Giulian, R.; Ridgway, M.C.; Byrne, A.P. Nanoscale density fluctuations in swift heavy ion irradiated amorphous SiO_2. *J. Appl. Phys.* **2011**, *110*, 123520. [CrossRef]
12. Benyagoub, A.; Toulemonde, M. Ion tracks in amorphous silica. *J. Mater. Res.* **2015**, *30*, 1529–1543. [CrossRef]
13. Zinkle, S.J.; Skuratov, V.A.; Hoelzer, D.T. On the conflicting roles of ionizing radiation in ceramics. *Nucl. Instr. Meth. B* **2002**, *191*, 758–766. [CrossRef]
14. Van Vuuren, A.J.; Ibrayeva, A.D.; Skuratov, V.A.; Zdorovets, M.V. Analysis of the microstructural evolution of silicon nitride irradiated with swift Xe ions. *Ceram. Int.* **2020**, *46*, 7155–7160. [CrossRef]
15. Kitayama, T.; Morita, Y.; Nakajima, K.; Narumi, K.; Saitoh, Y.; Matsuda, M.; Sataka, M.; Tsujimoto, M.; Isoda, S.; Toulemonde, M.; et al. Formation of ion tracks in amorphous silicon nitride films with MeV C_{60} ions. *Nucl. Instr. Meth. B* **2015**, *356–357*, 22–27. [CrossRef]
16. Mota-Santiago, P.; Vazquez, H.; Bierschenk, T.; Kremer, F.; Nadzri, A.; Schauries, D.; Djurabekova, F.; Nordlund, K.; Trautmann, C.; Mudie, S.; et al. Nanoscale density variations induced by high energy heavy ions in amorphous silicon nitride and silicon dioxide. *Nanotechnology* **2018**, *29*, 144004. [CrossRef]
17. Kucheyev, S.O.; Timmers, H.; Zou, J.; Williams, J.S.; Jagadish, C.; Li, G. Lattice damage produced in GaN by swift heavy ions. *J. Appl. Phys.* **2004**, *95*, 5360–5365. [CrossRef]
18. Mansouri, S.; Marie, P.; Dufour, C.; Nouet, G.; Monnet, I.; Lebius, H.; Benamara, Z.; Al-Douri, Y. Swift heavy ion effects in gallium nitride. *Int. J. Nanoelectron. Mater.* **2008**, *1*, 101–106.
19. Sall, M.; Monnet, I.; Moisy, F.; Grygiel, C.; Grygiel, C.; Jubolt-Leclerc, S.; Della-Negra, S.; Toulemonde, M.; Balanzat, E. Track formation in III-N semiconductors irradiated by swift heavy ions and fullerene and re-evaluation of the inelastic thermal spike model. *J. Mater Sci.* **2015**, *50*, 5214–5227. [CrossRef]
20. Moisy, F.; Sall, M.; Grygiel, C.; Ribet, A.; Balanzat, E.; Monnet, I. Optical bandgap and stress variations induced by the formation of latent tracks in GaN under swift heavy ion irradiation. *Nucl. Instr. Meth. B* **2018**, *431*, 12–18. [CrossRef]
21. Kamarou, A.; Wesch, W.; Wendler, E.; Undisz, A.; Rettenmayr, M. Swift heavy ion irradiation of InP: Thermal spike modeling of track formation. *Phys. Rev. B* **2006**, *73*, 184107. [CrossRef]

22. Ochedowski, O.; Osmani, O.; Schade, M.; Bussmann, B.K.; Ban-d'Etat, B.; Lebius, H.; Schleberger, M. Graphitic nanostripes in silicon carbide surfaces created by swift heavy ion irradiation. *Nat. Commun.* **2014**, *5*, 3913. [CrossRef] [PubMed]
23. Alencar, I.; Silva, M.R.; Leal, R.; Grande, P.L.; Papaleo, R.M. Impact Features Induced by Single Fast Ions of Different Charge-State on Muscovite Mica. *Atoms* **2021**, *9*, 17. [CrossRef]
24. Okubo, N.; Ishikawa, N.; Sataka, M.; Jitsukawa, S. Surface amorphization in Al_2O_3 induced by swift heavy ion irradiation. *Nucl. Instr. Meth. B* **2013**, *314*, 208–210. [CrossRef]
25. Grygiel, C.; Moisy, F.; Sall, M.; Lebius, H.; Balanzat, E.; Madi, T.; Been, T.; Marie, D.; Monnet, I. In-situ kinetics of modifications induced by swift heavy ions in Al_2O_3: Colour center formation, structural modification and amorphization. *Acta Mater.* **2017**, *140*, 157–167. [CrossRef]
26. Dhar, S.; Bolse, W.; Lieb, K.-P. Ion-beam induced amorphization and dynamic epitaxial recrystallization in α-quartz. *J. Appl. Phys.* **1999**, *85*, 3120–3123. [CrossRef]
27. Rymzhanov, R.A.; Medvedev, N.; O'Connell, J.H.; van Vuuren, A.J.; Skuratov, V.A.; Evolkov, A.E. Recrystallization as the governing mechanism of ion track formation. *Sci. Rep.* **2019**, *9*, 3837. [CrossRef]
28. Williams, J.S. Damage Formation, Amorphization and Crystallization in Semiconductors at Elevated Temperature. Chapter 6. In *Ion Beam Modification of Solids*; Wesch, W., Wendler, E., Eds.; Springer: Berlin, Germany, 2016; pp. 243–285.
29. Gupta, S.; Gehrke, H.G.; Krauser, J.; Trautmann, C.; Severin, D.; Bender, M.; Rothard, H.; Hofsass, H. Conducting ion tracks generated by charge-selected swift heavy ions. *Nucl. Instr. B* **2016**, *381*, 76–83. [CrossRef]
30. Itoh, N.; Duffy, D.M.; Khakshouri, S.; Stoneham, A.M. Making tracks: Electronic excitation roles in forming swift heavy ion tracks. *J. Phys. Condens. Matter* **2009**, *21*, 474205. [CrossRef]
31. Rogers, M.D. Mass Transport and Grain Growth induced by Fission Fragments in Thin Films of Uranium Dioxide. *J. Nucl. Mater.* **1965**, *16*, 298–305. [CrossRef]
32. Rogers, M.D. Ejection of Uranium Atoms from Sintered UO_2. *J. Nucl. Mater.* **1967**, *22*, 103–105. [CrossRef]
33. Nilsson, G. Ejection of Uranium Atoms from Sintered UO_2 by Fission Fragments in Different Gases and at Different Gas Pressures. *J. Nucl. Mater.* **1966**, *20*, 215–230. [CrossRef]
34. Meins, C.K.; Griffith, J.E.; Qiu, Y.; Mendenhall, M.H.; Seiberling, L.E.; Tombrello, T.A. Sputtering of UF_4 by High Energy Heavy Ions. *Rad. Eff.* **1983**, *71*, 13–33. [CrossRef]
35. Bouffard, S.; Duraud, J.P.; Mosbah, M.; Schlutig, S. Angular distribution of the sputtered atoms from UO_2 under high electronic stopping power irradiation. *Nucl. Instr. Meth. B* **1998**, *141*, 372–377. [CrossRef]
36. Schlutig, S. Contribution a L'etude de la Pulverisation et de L'endommagement du Dioxyde D'uranuim par les Ions Lourds Rapides. Ph.D. Thesis, University of Caen, Caen, France, March 2001.
37. Brown, W.L.; Lanzerotti, L.J.; Poate, J.M.; Augustyniak, W.M. "Sputtering" of Ice by MeV Light Ions. *Phys. Rev. Lett.* **1978**, *40*, 1027–1030. [CrossRef]
38. Brown, W.L.; Augustyniak, W.M.; Brody, E.; Cooper, B.; Lanzerotti, L.J.; Ramirez, A.; Evatt, R.; Johnson, R.E. Energy Dependence of the Erosion of H_2O Ice Films by H and He Ions. *Nucl. Instr. Meth.* **1980**, *170*, 321–325. [CrossRef]
39. Bottiger, J.; Davies, J.A.; L'Ecuyer, J.; Matsunami, N.; Ollerhead, R. Erosion of Frozen-Gas Films by MeV Ions. *Rad. Eff.* **1980**, *49*, 119–124. [CrossRef]
40. Baragiora, R.A.; Vidal, R.A.; Svendsen, W.; Schou, J.; Shi, M.; Bahr, D.A.; Atteberrry, C.L. Sputtering of water ice. *Nucl. Instr. Meth. B* **2003**, *209*, 294–303. [CrossRef]
41. Dartois, E.; Auge, B.; Boduch, P.; Brunetto, R.; Chabot, M.; Domaracka, A.; Ding, J.J.; Kamalou, O.; Lv, X.Y.; Rothard, H.; et al. Heavy ion irradiation of crystalline ice: Cosmic ray amorphisation cross-section and sputtering yield. *Astron. Astrophys.* **2015**, *576*, A125. [CrossRef]
42. Galli, A.; Vorburger, A.; Wurz, P.; Tulej, M. Sputtering of water ice films: A re-assessment with singly and doubly charged oxygen and argon ions, molecular oxygen, and electrons. *Icarus* **2017**, *291*, 36–45. [CrossRef]
43. Brown, W.L.; Lanzerotti, L.J.; Marcantonio, K.J.; Johnson, R.E.; Reimann, C.T. Sputtering of Ices by High Energy Particle Impact. *Nucl. Instr. Meth. B* **1986**, *14*, 392–402. [CrossRef]
44. Mejia, C.; Bender, M.; Severin, D.; Trautmann, C.; Boduch, P.H.; Bordalo, V.; Domaracka, A.; Lv, X.Y.; Martinez, R.; Rothard, H. Radiolysis and sputtering of carbon dioxide ice induced by swift Ti, Ni and Xe ions. *Nucl. Instr. Meth. B* **2015**, *365*, 477–481. [CrossRef]
45. Qiu, Y.; Griffith, J.E.; Meng, W.J.; Tombrello, T.A. Sputtering of Silicon and its Compounds in the Electronic Stopping Region. *Rad. Eff.* **1983**, *70*, 231–236. [CrossRef]
46. Sugden, S.; Sofield, C.J.; Murrell, M.P. Sputtering by MeV ions. *Nucl. Instr. Meth. B* **1992**, *67*, 569–573. [CrossRef]
47. Matsunami, N.; Sataka, M.; Iwase, A. Electronic sputtering of oxides by high energy ion impact. *Nucl. Instr. Meth. B* **2002**, *193*, 830–834. [CrossRef]
48. Matsunami, N.; Fukuoka, O.; Shimura, T.; Sataka, M.; Okayasu, S. A multi-exciton model for the electronic sputtering of oxides. *Nucl. Instr. Meth. B* **2005**, *230*, 507–511. [CrossRef]
49. Arnoldbik, W.M.; Tomozeiu, N.; Habraken, F.H.P.M. Electronic sputtering of thin SiO_2 films by MeV heavy ions. *Nucl. Instr. Meth. B* **2003**, *203*, 151–157. [CrossRef]
50. Toulemonde, M.; Assmann, W.; Trautmann, C.; Gruner, F. Jetlike Component in Sputtering of LiF Induced by Swift Heavy Ions. *Phys. Rev. Lett.* **2002**, *88*, 057602. [CrossRef]

51. Toulemonde, M.; Assmann, W.; Trautmann, C. Electronic sputtering of vitreous SiO_2: Experimental and modeling results. *Nucl. Instr. Meth B* **2016**, *379*, 2–8. [CrossRef]

52. Griffith, J.E.; Weller, R.A.; Seiberling, L.E.; Tombrello, T.A. Sputtering of Uranium Tetrafluoride in the Electronic Stopping Region. *Rad. Eff.* **1980**, *51*, 223–232. [CrossRef]

53. Toulemonde, M.; Assmann, W.; Muller, D.; Trautmann, C. Electronic sputtering of LiF, CaF_2, LaF_3 and UF_4 with 197 MeV Au ions: Is the stoichiometry of atom emission preserved? *Nucl. Instr. Meth. B* **2017**, *406*, 501–507. [CrossRef]

54. Matsunami, N.; Sataka, M.; Iwase, A.; Okayasu, S. Electronic excitation induced sputtering of insulating and semiconducting oxides by high energy heavy ions. *Nucl. Instr. Meth. B* **2003**, *209*, 288–293. [CrossRef]

55. Matsunami, N.; Sataka, M.; Okayasu, S.; Tazawa, M. Electronic sputtering of nitrides by high-energy ions. *Nucl. Instr. B* **2007**, *256*, 333–336. [CrossRef]

56. Matsunami, N.; Okayasu, S.; Sataka, M.; Tsuchiya, B. Electronic sputtering of SiC and KBr by high energy ions. *Nucl. Instr. Meth. B* **2020**, *478*, 80–84. [CrossRef]

57. Matsunami, N.; Sataka, M.; Okayasu, S.; Ishikawa, N.; Tazawa, M.; Kakiuchida, H. High-energy ion irradiation effects on atomic structures and optical properties of copper oxide and electronic sputtering. *Nucl. Instr. Meth. B* **2008**, *266*, 2986–2989. [CrossRef]

58. Matsunami, N.; Sataka, M.; Okayasu, S.; Kakiuchida, H. Ion irradiation effects on tungsten-oxide films and charge state effect on electronic erosion. *Nucl. Instr. Meth. B* **2010**, *268*, 3167–3170. [CrossRef]

59. Matsunami, N.; Sakuma, Y.; Sataka, M.; Okayasu, S.; Kakiuchida, H. Electronic sputtering of CuO films by high-energy ions. *Nucl. Instr. Meth. B* **2013**, *314*, 55–58. [CrossRef]

60. Matsunami, N.; Okayasu, S.; Sataka, M. Electronic excitation effects on Fe_2O_3 films by high-energy ions. *Nucl. Instr. Meth. B* **2018**, *435*, 142–145. [CrossRef]

61. Matsunami, N.; Kakiuchida, H.; Tazawa, M.; Sataka, M.; Sugai, H.; Okayasu, S. Electronic and atomic structure modifications of copper nitride films by ion impact and phase separation. *Nucl. Instr. Meth. B* **2009**, *267*, 2653–2656. [CrossRef]

62. Assmann, W.; d'Etat, B.B.; Bender, M.; Boduch, P.; Grande, P.L.; Lebius, H.; Lelievre, D.; Marmitt, G.G.; Rothard, H.; Seidl, T.; et al. Charge-state related effects in sputtering of LiF by swift heavy ions. *Nucl. Instr. Meth B* **2017**, *392*, 94–101. [CrossRef]

63. Toulemonde, M.; Assmann, W.; Ban-d'Etat, B.; Bender, M.; Bergmaier, A.; Boduch, P.; Della-Negra, S.; Duan, J.; El-Said, A.S.; Gruner, F.; et al. Sputtering of LiF and other halide crystals in the electronic energy loss regime. *Eur. Phys. J. D* **2020**, *74*, 144. [CrossRef]

64. Matsunami, N.; Sataka, M.; Okayasu, S.; Tsuchiya, B. Charge State Effect of High Energy Ions on Material Modification in the Electronic Stopping Region. *Atoms* **2021**, *9*, 36. [CrossRef]

65. Ziegler, J.F.; Biersack, J.P.; Littmark, U. *The Stopping and Range of Ions in Solids*; Pergamon Press: New York, NY, USA, 1985; pp. 1–321.

66. Ziegler, J.F.; Ziegler, M.D.; Biersack, J.P. SRIM-The stopping and range of ions in matter (2010). *Nucl. Instr. Meth. B* **2010**, *268*, 1818–1823. [CrossRef]

67. Grande, P.L.; Schiwietz, G. Convolution approximation for the energy loss, ionization probability and straggling of fast ions. *Nucl. Instr. Meth. B* **2009**, *267*, 859–865. [CrossRef]

68. Matsunami, N.; Sataka, M.; Iwase, A. Sputtering of high T_c superconductor $YBa_2Cu_3O_{7-\delta}$ by high energy heavy ions. *Nucl. Instr. Meth. B* **2001**, *175-177*, 56–61. [CrossRef]

69. Meftah, A.; Assmann, W.; Khalfaoui, N.; Stoquert, J.P.; Studer, F.; Toulemonde, M.; Trautmann, C.; Voss, K.-O. Electronic sputtering of $Gd_3Ga_5O_{12}$ and $Y_3Fe_5O_{12}$ garnets: Yield, stoichiometry and comparison to track formation. *Nucl. Instr. Meth. B* **2011**, *269*, 955–958. [CrossRef]

70. Shinde, N.S.; Matsuanmi, N.; Shimura, T.; Sataka, M.; Okayasu, S.; Kato, T.; Tazawa, M. Alignment of grain orientation in polycrystalline-SiO_2 films induced by high-energy heavy ions. *Nucl. Instr. Meth. B* **2006**, *245*, 231–234. [CrossRef]

71. Matsunami, N.; Itoh, M.; Kato, M.; Okayasu, S.; Sataka, M.; Kakiuchida, H. Ion induced modifications of Mn-doped ZnO films. *Nucl. Instr. Meht. B* **2015**, *365*, 191–195. [CrossRef]

72. Matsunami, N.; Kato, M.; Sataka, M.; Okayasu, S. Disordering of ultra thin WO_3 films by high-energy ions. *Nucl. Instr. Meth. B* **2017**, *409*, 272–276. [CrossRef]

73. Matsunami, N.; Sogawa, T.; Sakuma, Y.; Ohno, N.; Tokitani, M.; Masuzaki, S. Deuterium retention in Fe_2O_3 under low-energy deuterium-plasma exposure. *Phys. Scr. T* **2011**, *145*, 014042. [CrossRef]

74. Matsunami, N.; Ohno, N.; Tokitani, M.; Tsuchiya, B.; Sataka, M.; Okayasu, S. Modifications of WNOx films by keV D and H ions. *Surf. Coat. Technol.* **2020**, *394*, 125798. [CrossRef]

75. Matsunami, N.; Okayasu, S.; Sataka, M.; Tsuchiya, B. Ion irradiation effects on WN_xO_y thin films. *Nucl. Instr. Meth. B* **2018**, *435*, 146–151. [CrossRef]

76. Matsunami, N.; Teramoto, T.; Okayasu, S.; Sataka, M.; Tsuchiya, B. Modifications of WN_xO_y films by ion impact. *Surf. Coat. Technol.* **2018**, *355*, 84–89. [CrossRef]

77. Matsunami, N.; Itoh, M.; Takai, Y.; Tazawa, M.; Sataka, M. Ion beam modification of ZnO thin films on MgO. *Nucl. Instr. Meth. B* **2003**, *206*, 282–286. [CrossRef]

78. Matsunami, N.; Itoh, M.; Kato, M.; Okayasu, S.; Sataka, M.; Kakiuchida, H. Growth of Mn-doped ZnO thin films by rf-sputter deposition and lattice relaxation by energetic ion impact. *Appl. Surf. Sci.* **2015**, *350*, 31–37. [CrossRef]

79. Joint Committee on Powder Diffraction Standards (JCPDS). *JCPDS File 381420*; International Centre for Diffraction Data (ICDD): Newtown Square, PA, USA.

80. Storm, E.; Israel, H.I. Photon Cross Sections from 1 keV to 100 MeV for Elements $Z = 1$ to $Z = 100$. *Atom. Data Nucl. Data Tables* **1970**, *7*, 565–681. [CrossRef]

81. Shima, K.; Kuno, N.; Yamanouchi, M.; Tawara, H. Equilibrium Charge Fractions of Ions of $Z = 4$–92 Emerging from a Carbon Foil. *Atom. Data Nucl. Data Tables* **1992**, *51*, 173–241. [CrossRef]

82. Schiwietz, G.; Grande, P.L. Improved charge-state formulas. *Nucl. Instr. Meth B* **2001**, *175–177*, 125–131. [CrossRef]

83. DuBois, R.D.; Santos, A.C.F.; Olson, R.E. Scaling laws for electron loss from ion beams. *Nucl. Instr. Meth. A* **2005**, *544*, 497–501. [CrossRef]

84. DuBois, R.D.; Santos, A.C.F. Target-Scaling Properties for Electron Loss by Fast Heavy Ions (Chapter 8). In *Atomic Processes in Basic and Applied Physics*; Shevelko, V., Tawara, H., Eds.; Springer: Berlin, Germany, 2012; pp. 185–209.

85. Lide, D.R. (Ed.) *CRC Handbook of Chemistry and Physics*, 84th ed.; CRC Press: Boca Raton, FL, USA, 2003.

86. Rodrigues, G.C.; Indelicato, P.; Santos, J.P.; Patte, P.; Parente, F. Systematic calculation of total atomic energies of ground state configurations. *Atom. Data Nucl. Data Tables* **2004**, *86*, 117–233. [CrossRef]

87. Sigmund, P. Sputtering by Ion Bombardment: Theoretical Concepts. Chapter 2. In *Sputtering by Particle Bombardment*; Behrisch, R., Ed.; Springer: Berlin, Germany, 1981; Volume 1, pp. 9–71.

88. Betz, G.; Wehner, G.K. Sputtering of Multicomponent Materials. Chapter 2. In *Sputtering by Particle Bombardment*; Behrisch, R., Ed.; Springer: Berlin, Germany, 1983; Volume 2, pp. 11–90.

89. Paul, H. The Stopping Power of Matter for Positive Ions. Chapter 7. In *Modern Practices in Radiation Therapy*; Natanasabapathi, G., Ed.; InTech: Rijeka, Croatia, 2012; pp. 113–132.

90. Besenbacher, F.; Bottiger, J.; Graversen, O.; Hansen, J.L.; Sorensen, H. Stopping power of solid argon for helium ions. *Nucl. Instr. Meth.* **1981**, *188*, 657–667. [CrossRef]

91. Kittel, C. *Introduction to Solid State Physics*, 6th ed.; John Wiley & Sons Inc.: New York, NY, USA, 1986; pp. 1–646.

92. ICRU Report 49. *Stopping Powers and Ranges for Protons and Alpha Particles*; International Commission on Radiation Units and Measurements: Bethesda, MD, USA, 1993; pp. 1–286.

93. Arnau, A.; Bauer, P.; Kastner, F.; Salin, A.; Ponce, V.H.; Fainstein, P.D.; Echenique, P.M. Phase effect in the energy loss of hydrogen projectiles in zinc targets. *Phys. Rev. B* **1994**, *49*, 6470. [CrossRef]

94. Ranjan, R.; Allain, J.P.; Hendricks, M.R.; Ruzic, D.N. Absolute sputtering yield of Ti/TiN by Ar^+/N^+ at 400–700 eV. *J. Vac. Sci. Tech. A* **2001**, *19*, 1004–1007. [CrossRef]

95. Itoh, N. Creation of lattice defects by electronic excitation in alkali halides. *Adv. Phys.* **1982**, *31*, 491–551. [CrossRef]

96. Itoh, N.; Stoneham, A.M. *Materials Modification by Electronic Excitation*; Cambridge University Press: Cambridge, UK, 2001; pp. 1–520.

97. Duffy, D.M.; Daraszewicz, S.L.; Mulroue, J. Modeling the effects of electronic excitations in ionic-covalent materials. *Nucl. Instr. Meth. B* **2012**, *277*, 21–27. [CrossRef]

98. Matsunami, N.; Sataka, M.; Okayasu, S. Effective depth of electronic sputtering of WO_3 films by high-energy ions. *Nucl. Instr. Meth. B* **2019**, *460*, 185–188. [CrossRef]

99. Popovic, M.; Novakovic, M.; Mitric, M.; Zhang, K.; Rakocevic, Z.; Bibic, N. Xenon implantation effects on the structural and optical properties of reactively sputtered titanium nitride thin films. *Mat. Res. Bull.* **2017**, *91*, 36–41. [CrossRef]

100. Solovan, M.; Brus, V.V.; Maistruk, E.V.; Maryanchuk, P.D. Electrical and Optical Properties of TiN Films. *Inorgan. Mat.* **2014**, *50*, 40–45. [CrossRef]

101. Kavitha, A.; Kannan, R.; Reddy, P.S.; Rajashabala, S. The effect of annealing on the structural, optical and electrical properties of Titanium Nitride (TiN) thin films prepared by DC magnetron sputtering with supported discharge. *J. Mat. Sci. Mat. Electron.* **2016**, *27*, 10427–10434. [CrossRef]

102. Martin, P.; Guizard, S.; Daguzan, P.; Petite, G.; D'Oliveira, P.; Meynadier, P.; Perdrix, M. Subpicosecond study of carrier trapping dynamics in wide-band-gap crystals. *Phys. Rev. B* **1997**, *9*, 5799–5810. [CrossRef]

103. Nakajima, A.; Yoshihara, A.; Ishigame, M. Defect-induced Raman spectra in doped CeO_2. *Phys. Rev. B* **1994**, *50*, 13297–13307. [CrossRef]

104. Palko, J.W.; Kriven, W.M.; Sinogeikin, S.V.; Bass, J.D.; Sayir, A. Elastic constnts of yttria (Y_2O_3) monocrystals to high temperatures. *J. Appl. Phys.* **2001**, *89*, 7791–7796. [CrossRef]

105. Kandil, H.M.; Greiner, J.D.; Smith, J.F. Single-Crystal Elastic Constnats of Yttria-Stabilized Ziroconia in the Range 20° to 700 °C. *J. Am. Ceram. Soc.* **1984**, *67*, 341–346. [CrossRef]

106. Arab, M.; Dirany, N.; David, M. BAW resonator as elastic characterization tools of WO_3 thin films. *Mater. Today Proc.* **2016**, *3*, 152–156. [CrossRef]

107. Beg, M.M.; Shapiro, S.M. Study of phonon dispersion relations in cuprous oxide by inelastic neutron scattering. *Phys. Rev. B* **1976**, *13*, 1728–1734. [CrossRef]

108. Yao, B.; Zhou, X.; Liu, M.; Yu, J.; Cao, J.; Wang, L. First-principles calculations on phase transformation and elastic properties of CuO under pressure. *J. Comp. Electron.* **2018**, *17*, 1450–1456. [CrossRef]

109. Hay, J.C.; Sun, E.Y.; Pharr, G.M.; Becher, P.F.; Alexander, K.B. Elastic Anisotropy of β–Silicon Nitride Whiskers. *J. Am. Ceram. Soc.* **1998**, *81*, 2661–2669. [CrossRef]

110. Vogelgesang, R.; Grimsditch, M.; Wallace, J.S. The elastic constants of single crystal β–Si_3N_4. *Appl. Phys. Lett.* **2000**, *76*, 982–984. [CrossRef]
111. Carlotti, G.; Gubbiotti, G.; Hickernell, F.S.; Liaw, H.M.; Socino, G. Comparative study of the elastic properties of polycrystalline aluminum nitride films on silicon by Brillouin light scattering. *Thin Solid Films* **1997**, *310*, 34–38. [CrossRef]
112. McNeil, L.; Grimsditch, M.; French, R.H. Vibrational Spectroscopy of Aluminum Nitride. *J. Am. Ceram. Soc.* **1993**, *76*, 1132–1136. [CrossRef]
113. Hou, Z.F. Effects of Cu, N and Li interaction on the structural stability and electronic structure of cubic Cu_3N. *Solid State Sci.* **2008**, *10*, 1651–1657. [CrossRef]
114. Kamitani, K.; Grimsditch, M.; Nipko, J.C.; Loong, C.K.; Okada, M.; Kimura, I. The elastic constants of silicon carbide: A Brillouin-scattering study of 4H and 6H SiC single crystals. *J. Appl. Phys.* **1997**, *82*, 3152–3154. [CrossRef]
115. Kim, J.O.; Achenbach, J.D.; Mirkarimi, P.B.; Shinn, M.; Barnett, S.A. Elastic constants of single-crystal transition-metal nitride films measured by line-focus acoustic microscopy. *J. Appl. Phys.* **1992**, *72*, 1805–1811. [CrossRef]
116. Gemmell, D.S. Determining the Stereochemical Structures of Molecular Ions by "Coulomb-Explosion" Techniques with Fast (MeV) Molecular Ion Beams. *Chem. Rev.* **1980**, *80*, 301–311. [CrossRef]

quantum beam science

MDPI

Article

Phase Transformation by 100 keV Electron Irradiation in Partially Stabilized Zirconia

Yasuki Okuno [1,*] and Nariaki Okubo [2]

[1] KINKEN, Tohoku University, 2-1 Katahira, Aoba-ku, Sendai City 980-8577, Miyagi, Japan
[2] Japan Atomic Energy Agency, 2-4 Shirakata, Tokai-Mura 319-1195, Ibaraki-Ken, Japan; okubo.nariaki@jaea.go.jp
* Correspondence: okuno.yasuki@imr.tohoku.ac.jp; Tel.: +81-022-215-2068

Abstract: Partially stabilized zirconia (PSZ) is considered for use as an oxygen-sensor material in liquid lead-bismuth eutectic (LBE) alloys in the radiation environment of an acceleration-driven system (ADS). To predict its lifetime for operating in an ADS, the effects of radiation on the PSZ were clarified in this study. A tetragonal PSZ was irradiated with 100 keV electrons and analyzed by X-ray diffraction (XRD). The results indicate that the phase transition in the PSZ, from the tetragonal to the monoclinic phase, was caused after the irradiation. The deposition energy of the lattice and the deposition energy for the displacement damage of a 100 keV electron in the PSZ are estimated using the particle and heavy ion transport code system and the non-ionizing energy loss, respectively. The results suggest that conventional radiation effects, such as stopping power, are not the main mechanism behind the phase transition. The phase transition is known to be caused by the low-temperature degradation of the PSZ and is attributed to the shift of oxygen ions to oxygen sites. When the electron beam is incident to the material, the kinetic energy deposition and excitation-related processes are caused, and it is suggested to be a factor of the phase transition.

Keywords: phase transition; electron irradiation; partially stabilized zirconia; XRD; radiation simulation

Citation: Okuno, Y.; Okubo, N. Phase Transformation by 100 keV Electron Irradiation in Partially Stabilized Zirconia. *Quantum Beam Sci.* **2021**, *5*, 20. https://doi.org/10.3390/qubs5030020

Academic Editors: Akihiro Iwase and Rozaliya Barabash

Received: 5 April 2021
Accepted: 11 June 2021
Published: 25 June 2021

1. Introduction

An acceleration-driven system (ADS) is a powerful tool for effectively transmuting minor actinides in the double-strata fuel cycle strategy for separation/conversion technology [1,2]. As a neutron source and coolant, such systems use a liquid lead–bismuth eutectic (LBE) alloy as a nuclear spallation target [3]. The LBE alloy is designed to flow through piping at 2 m/s. It causes erosion and corrosion in the pipes [4,5], which are suppressed by adjusting the oxygen concentration by about 10^{-7} wt.% in the LBE alloy [6]. Therefore, an oxygen sensor is required to control the oxygen concentration in the LBE alloy.

Oxygen sensors based on yttria-stabilized zirconia (YSZ) and a Pt/air electrode are used worldwide to monitor the oxygen concentration in LBE alloys [7,8]. Such oxygen sensors must have high fracture toughness because a high load is applied by the flowing LBE alloy. The fracture toughness of YSZ depends on the yttrium concentration [9]. YSZ with 3 mol% yttrium is called partially stabilized zirconia (3Y–PSZ).

Under high stress, 3Y–PSZ undergoes a phase transition from a tetragonal (t) phase to a monoclinic (m) phase [10–12]. Because the m phase contributes to arresting crack propagation, it has been shown to play an important role in endowing this material with high fracture toughness. Therefore, if the fracture toughness can be maintained in the ADS operating environment, the PSZ may become an important material for use in ADS oxygen sensors.

As the LBE alloy flows through the entire ADS cooling system, it becomes radioactive through the spallation reaction. The LBE alloy is predicted to expose the oxygen sensor to a high radiation field of over 1 kGy/h [6], which is expected to lead to radiation damage in the oxygen-sensor material.

In a previous study of pure zirconia irradiated with 340 keV Xe and 800 keV Bi ion beams [13–15], an m-to-t phase transition was observed. The phase transition between m and t is considered to be caused by the cumulative transfer of energy laid down in the displacement cascades. On the other hand, the t phase of 3Y-PSZ is a metastable phase and is related to its strong mechanical properties. However, the t-to-m transition is triggered by mechanical stress and thermal annealing, and may be affected by the deposition energy of the radiation because of the metastable state phase.

To use the PSZ device in a radiation environment, the behavior of the PSZ in the metastable t phase must be investigated. In this study, the effect of PSZ on a 100 keV electron beam was investigated, using a crystal structure and radiation simulation analysis.

2. Materials and Methods

The PSZ specimens consisted of zirconia doped with 3 mol% yttria (3Y–PSZ, Tsukuba Ceramic Works) and were cut into 4.0 mm (length) × 10.0 mm (width) × 3.0 mm (depth) specimens. They were sintered at 1500 °C. The top and bottom surfaces were mirror polished. To remove the phase transition during the shape processing, the PSZ specimens were processed with a quench heat treatment at 900 °C and air cooled.

The specimens were irradiated with a 100 keV electron beam generated by a Cockcroft–Walton electron accelerator at Osaka Prefecture University (OPU). In order to avoid generating heat in a small area with the micro-focus beam, the electron beam was defocused to about 2 cmΦ. The area of irradiation was extended to 6 cm × 10 cm much more than the surface of the sample by the scanning coil. The required electron fluence was determined from the number of charges flowing to the ground collected in a Faraday cup and the flux was 5×10^{12} cm^2/s during the irradiation. Before and after irradiation, the specimens were subjected to X-ray diffraction (XRD), and the crystal structure was analyzed using the $\theta-2\theta$ method. The XRD measurements were carried out using Cu Kα radiation.

3. Results

Figure 1 shows the XRD pattern of 3Y-PSZ before and after irradiation with a 100 keV electron beam. In Figure 1a, the main XRD peaks are assigned to all measurement regions. This result indicates that the t (111) phase dominates the 3Y-PSZ specimen at approximately $2\theta = 30°$, before and after irradiation [16]. The peaks of the m (11 − 1) and (111) phases at approximately $2\theta = 28°$ and 31.5° do not appear because the signal intensity is much lower than that of the t phase. A peak is known to exist in the m (11 − 1) phase.

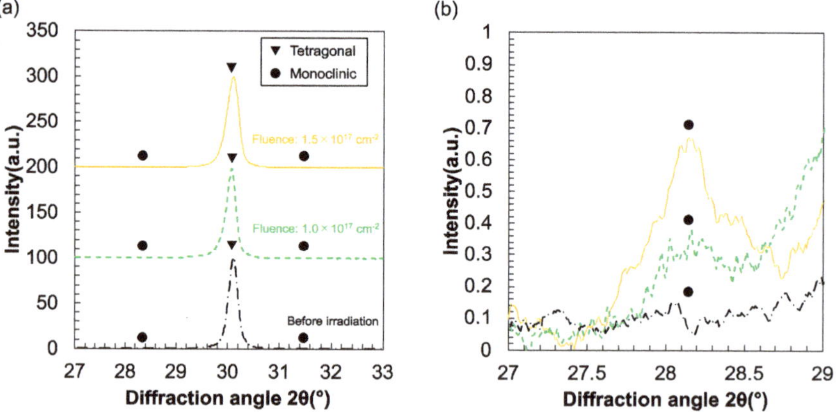

Figure 1. XRD patterns of a PSZ specimen before and after irradiation with 100 keV electron beams. The black line is before irradiation. (**a**) is the spectrum of the measured range shifted up by 100; (**b**) is the spectrum focused on m (11 − 1) between 27 and 29°.

In Figure 1b, to focus on the m (11 − 1) peak, the diffraction angle was set between 27° and 29°. The results show that the intensity of the peak monotonically increases after irradiation, depending on the fluence of 100 keV electrons.

The fraction C_M of the m phase to the t phase can be evaluated from the peak areas of the most intense signals of t (101), m (11 − 1), and t (111), using the following equation [13]:

$$C_M = 0.82 \times \frac{I_m(11-1) + I_m(111)}{I_t(101)} \times 100\% \tag{1}$$

where I_m and I_t are the peak intensities of m and t, respectively. Figure 2 shows the fractional content of the m phase as a function of the electron fluence. This result indicates that the fractional content of m increases as the absorbed dose increases. Therefore, the phase transformation in 3Y–PSZ is caused by electron irradiation.

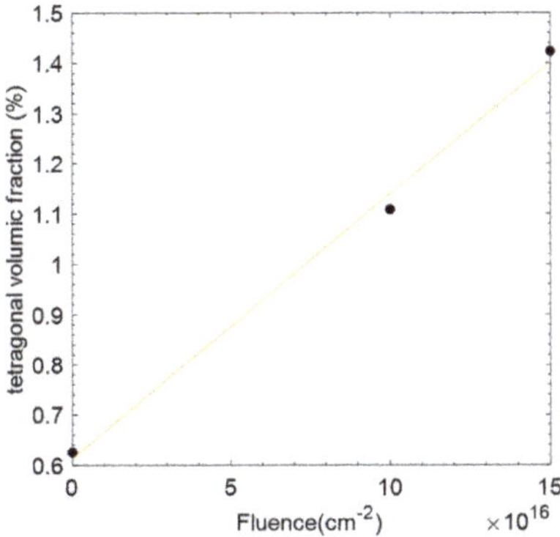

Figure 2. Fraction of monoclinic zirconia as a function of the fluence of 100 keV electrons.

Because the phase transition is detected by XRD, the absorbed dose should be calculated in the shallow surface region that is probed by XRD. This region was determined by the X-ray penetration depth. The fraction $G(x)$ of the total diffracted intensity, due to a surface layer of depth xx, can be expressed using the following [17,18]:

$$G(x) = 1 - \exp\left(\frac{-2\mu x}{\sin\theta}\right) \tag{2}$$

where μ is the absorption coefficient and $-2\mu x/\sin\theta$ is the effective path length for X-rays to penetrate to a depth x at a given Bragg angle θ. For the PSZ, the path length of Cu Kα X-rays incident at $\theta = 28°$ is approximately 9 µm, as shown in Equation (2), with $G(x) = 0.99$. The electrons are known to have a short flight distance in the materials. To investigate the range of the flight distance, the Monte Carlo calculation for radiation behavior was conducted, using the particle and heavy ion transport code system (PHITS) [19].

Figure 3 shows the depth distribution of the dose from the surface of the PSZ, in the case of irradiation with one of the 100 keV electrons. This result indicates that the value of the dose rate peaks at approximately 5 µm, and continues to 30 µm. Therefore, the flight range of a 100 keV electron was considered to be longer than the observation range of the XRD analysis.

When charged particles, e.g., electrons and ions, pass through materials, the radiation effect is expressed as a loss of energy, due to an interaction called total stopping power (S). One type of S is collision stopping power (S_{col}), e.g., electronic stopping power (S_e), nuclear stopping power (S_n), and radiative stopping power (S_{rad}). S_n is the main factor that causes the displacement and migration of atoms.

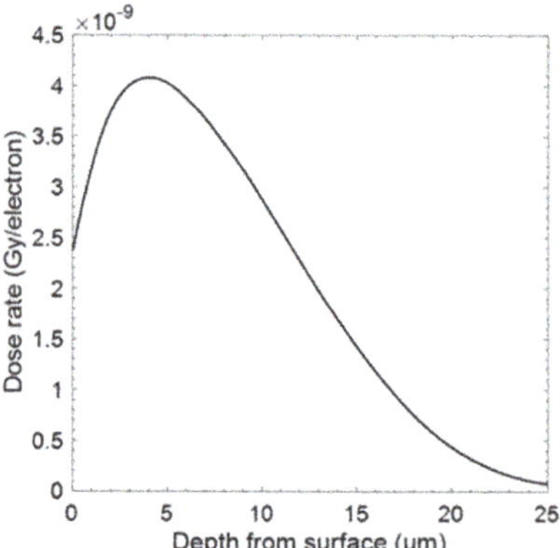

Figure 3. Calculation result of dose rate as a function of the depth from surface.

In an effort to describe the fraction of energy that moves into displacements such as the S_n, due to radiation, the non-ionizing energy loss (NIEL) was developed and is expressed as follows [20–22]:

$$\text{NIEL}\,(E) = \frac{N_A}{A} \int_{\theta_{min}}^{\pi} L[T(\theta, E)]\,T(\theta, E)\,\frac{\mathrm{d}\sigma(\theta, E)}{\mathrm{d}\Omega}\,d\Omega \tag{3}$$

where N_A is Avogadro's number, A is the atomic mass, E is the energy of the incident particle, θ is the scattering angle, σ is the scattering cross-section, and Ω is the solid angle of scattering. The equation also requires information regarding the differential cross-section for atomic displacements $(\mathrm{d}\sigma(\theta, E)/\mathrm{d}\Omega)$, the average recoil energy of the target atoms $(T(\theta, E))$, and the Lindhard partition factor $(L[T(\theta, E)])$, which partitions the energy into ionizing and non-ionizing events.

Figure 4 shows the NIEL vs. energy for Zr, Y, and O atoms for electrons. The result shows that the NIEL value for each element rapidly increases above a certain energy threshold (E_{th}). An increase in the NIEL value means that the kinetic energy given by the electron beam to the atom exceeds the displacement threshold energy (E_d) and causes the production of a primary knock-on atom (PKA).

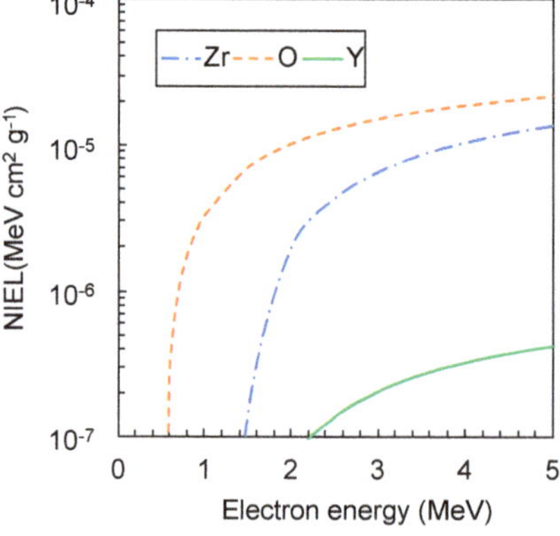

Figure 4. Non-ionizing energy loss (NIEL) vs. electron energy calculated for Zr, Y, and O atoms. The values of E_d for Zr, Y, and O are 80, 80, and 120 eV, respectively.

However, the E_{th} values of Y, Zr, and O are higher than the energy of the incident electrons by approximately 2200, 1250, and 600 keV, respectively. Therefore, the phase transition is considered to be caused by the transfer of kinetic energy lower than the recoil effect, similar to PKA.

From the calculation result in Figure 3, the deposition energy such as the S_e is also estimated to be 1.9×10^{-14} eV when an electron passes the lattice of t-PSZ having a volume of 137 Å^3, which might be much lower than the energy that causes the phase transition.

The phase transition from t to m in the PSZ is also known to be a thermal effect which is between 150 and 300 °C, as shown in Figure 5 [23]. The electron beam that heats the sample surface is considered to be about 80 mW because the condition of electron beam is a defocus-beam, rather than a micro-focus beam. The region such as 25 μm from the surface, which is main energy deposition region indicated from Figure 3, is 11 K/s. The sample temperature between the electron irradiation was less than 40 °C. For the above reasons, in the irradiation, the sample temperature is lower than the annealing temperature which causes a phase transition from the t-to-m phase. Therefore, the observed phase transition caused by the irradiation with 100 keV electrons is not considered to be a radiation effect such as the Se, the Sn and the thermal effect.

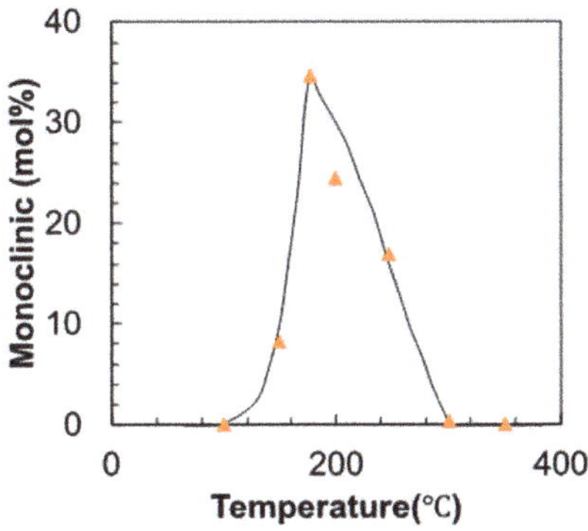

Figure 5. Relationship between the amount of the monoclinic phase and annealing temperature in PSZ fabricated above 1500 °C [23]. The annealing time is 50 h.

4. Discussion

The crystal structures of the t and m phases of PSZ are shown in Figure 6 [24]. This phase transformation is considered to be a martensitic transition. Generally, an athermal diffusionless martensitic transition occurs quickly, with the motion of the phase boundary as high as the speed of sound [25]. The overall transition proceeds in two major stages [26]. First, the transition of the lattice structure from tetragonal to monoclinic occurs by the shearing displacement of zirconium ions. The second stage involves the sliding of oxygen ions to the oxygen sites in the monoclinic lattice. The displacement of oxygen ions from the ideal fluorite positions along the c-axis was investigated by X-ray diffraction (XRD) [27]. Therefore, the 100 keV electron is considered to cause the shearing displacement of zirconium ions and the sliding of oxygen ions. In [9], the reported activation energies for the phase transition are close to 100 kJ/mol (~1 eV), which is similar to the activation energy for the sliding of oxygen ions to oxygen sites. In order to find the mechanism of the phase transition, the factor of the sliding of oxygen ions to oxygen sites should be considered.

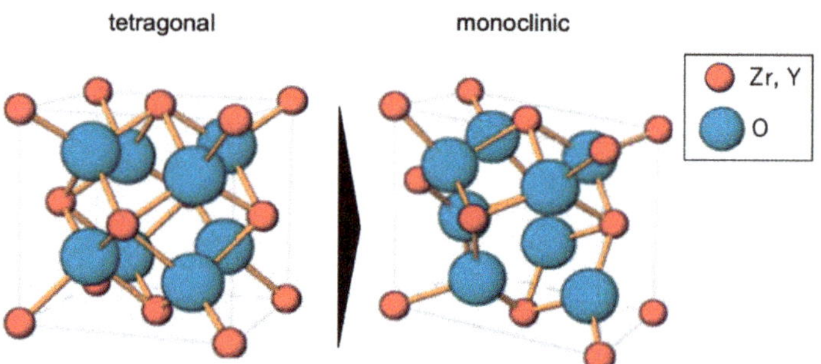

Figure 6. Crystal structure of monoclinic and tetragonal partially stabilized zirconia [24].

One of the factors is the kinetic energy transfer due to the elastic collision between the electron and target atoms. The kinetic energy of the target atom from the incident electron is expressed as follows:

$$E_p = \frac{2M_e}{M_\mathrm{T}} \cdot \frac{1}{M_e c^2} \left(E + 2M_e c^2 \right) E \ \sin^2 \frac{\theta}{2} \tag{4}$$

where E, θ, c, M_T, and M_e denote the energy of the incident electron, scattering angle, speed of light, mass of the target atom, and electron mass, respectively. Therefore, in the case of $\theta = 0°$, E_p takes the maximum value ($E_{p,max}$), which is represented by the following equation:

$$E_{p,\ max} = \frac{2E\left(E + 2M_e c^2 \right)}{M_\mathrm{T} c^2} \tag{5}$$

Figure 7 shows the $E_{p,max}$ for Zr, Y, and O atoms, as a function of the energy of the incident electron. The result shows that the value of $E_{p,max}$ for the O atom is 1 eV at the energy of the 7.2 keV incident electron. This suggests that a 100 keV electron can cause the migration of oxygen ions to oxygen sites, by elastic scattering, to trigger the phase transition.

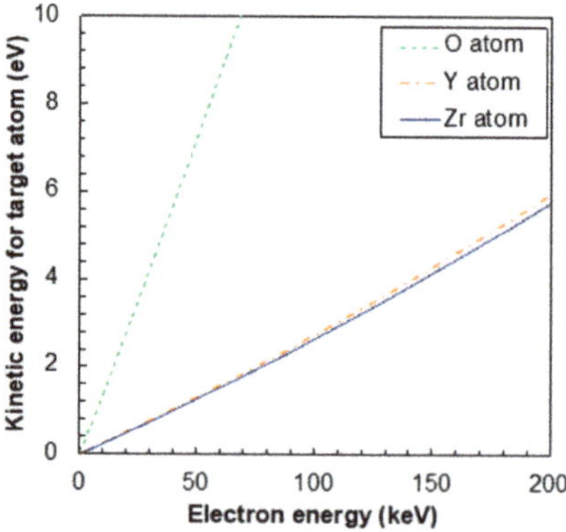

Figure 7. Kinetic energy transfer for Zr, Y, and O atoms due to elastic scattering as a function of the energy of the incident electron.

Another factor to consider are the excitation-related processes [28]. The amorphous-to-crystalline phase transition via the excitation-related processes is often observed in amorphous ceramics such as Al_2O_3, $ScPO_4$ and $LaPO_4$ [29]. The behavior of phase transition via excitation-related processes is known to have an approximate relationship with the Bethe formula [30], which is as the following:

$$-\frac{dE}{dx} = 2\pi e^4 N_A \frac{\rho Z}{AE} \ln\left(\sqrt{\frac{E}{2}}\frac{E}{I}\right)$$

(6)

$$I = 9.76\overline{Z} + \frac{58.5}{\overline{Z}^{0.19}},$$

(7)

where e is the elementary charge, N_A is the Avogadro number, ρ is the mass density, Z is the atomic number, A is the atomic weight, E is the electron energy, is the base of natural logarithm, I is the mean excitation energy, and $\overline{Z}$ is the mean atomic number.

Figure 8 shows the calculation result of the stopping power for electron in ZrO_2 as a function of the energy of the incident electron. The result indicates that the stopping power increases with the decreasing incident electron energy. This tendency is in contrast to the kinetic energy of the elastic collision as shown in Figure 7. The possible factors that cause the phase transformation are the transfer of kinetic energy or the electronic excitation effect. In further research, in order to investigate the mechanism of the m→t phase transition in PSZ, it is important to determine the energy dependence of the phase transformation by low-energy electron beams. If the phase transformation rate increases with decreasing incident electron energy, the electronic excitation effect is the main factor. On the other hand, when the tendency between the energy and the phase transformation rate is opposite, the influence of kinetic energy application is considered to be large.

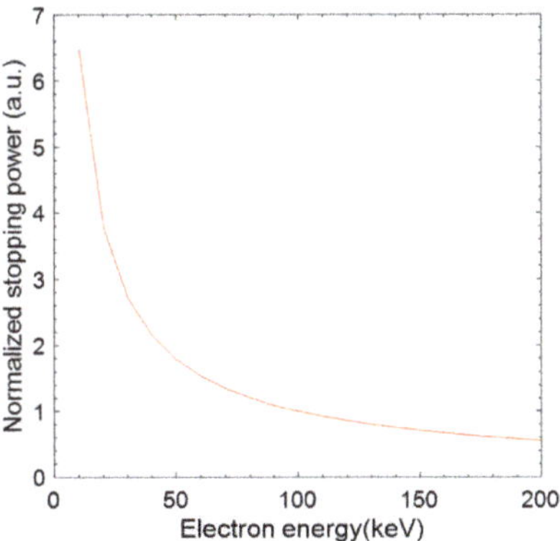

Figure 8. Stopping power for incident electrons normalized at 100 keV electron energy in ZrO_2.

5. Conclusions

It was observed that the monoclinic phase in 3Y-PSZ increased with increasing electron-beam fluence. The calculation results using PHITS suggested that the energy deposited by a 100 keV electron to the PSZ lattice was not the main reason for the phase transition. The NIEL calculation indicated that the atoms in the PSZ only recoiled under electron irradiation at energies over 0.56 MeV; 100 keV electrons cannot cause displacement damage.

However, the kinetic energy or the excitation-related processes from incident electrons in the PSZ might be higher than the energy needed to shift oxygen ions to oxygen sites, and is suggested to be the cause of the phase transition.

Author Contributions: Conceptualization, Y.O. and Y.O.; methodology, Y.O.; software, Y.O.; validation, Y.O.; formal analysis, Y.O.; investigation, Y.O.; resources, N.O.; data curation, Y.O.; writing—original draft preparation, Y.O.; writing—review and editing, Y.O.; visualization, Y.O.; supervision, N.O.; project administration, N.O.; funding acquisition, N.O. All authors have read and agreed to the published version of the manuscript.

Funding: This research received no external funding.

Acknowledgments: We would like to thank Takashi Oka and Shuichi Okuda (OPU) for supporting our electron irradiation tests.

Conflicts of Interest: The authors declare no conflict of interest.

References

1. Tsujimoto, K.; Oigawa, H.; Ouchi, N.; Kikuchi, K.; Kurata, Y.; Mizumoto, M.; Sasa, T.; Saito, S.; Nishihara, K.; Umeno, M.; et al. Research and Development Program on Accelerator Driven Subcritical System in JAEA. *J. Nucl. Sci. Technol.* **2007**, *44*, 483. [CrossRef]
2. Oigawa, H.; Tsujimoto, K.; Nishihara, K.; Sugawara, T.; Kurata, Y.; Takei, H.; Saito, S.; Sasa, T.; Obayashi, H. Role of ADS in the back-end of the fuel cycle strategies and associated design activities: The case of Japan. *J. Nucl. Mater.* **2011**, *415*, 229. [CrossRef]
3. Kurata, Y. Corrosion experiments and materials developed for the Japanese HLM systems. *J. Nucl. Mater.* **2011**, *415*, 254. [CrossRef]
4. Yeliseyeva, O.; Tsisar, V.; Benamati, G. Influence of temperature on the interaction mode of T91 and AISI 316L steels with Pb–Bi melt saturated by oxygen. *Corros. Sci.* **2008**, *50*, 1672. [CrossRef]
5. Schroer, C.; Skrypnik, A.; Wedemeyer, O.; Konys, J. Oxidation and dissolution of iron in flowing lead–bismuth eutectic at 450 °C. *Corros. Sci.* **2012**, *61*, 63. [CrossRef]
6. Deloffre, P.; Terlain, A.; Barbier, F. Corrosion and deposition of ferrous alloys in molten lead–bismuth. *J. Nucl. Mater.* **2002**, *301*, 35–39. [CrossRef]
7. Fernandez, J.A.; Abella, J.; Barcelo, J.; Victori, L. Development of an oxygen sensor for molter 44.5% lead–55.5% bismuth alloy. *J. Nucl. Mater.* **2002**, *301*, 47–52. [CrossRef]
8. Schroer, C.; Konys, J.; Verdaguer, A.; Abellà, J.; Gessi, A.; Kobzova, A.; Babayan, S.; Courouau, J.-L. Design and testing of electrochemical oxygen sensors for service in liquid lead alloys. *J. Nucl. Mater.* **2011**, *415*, 338–347. [CrossRef]
9. Chevalier, J.; Cremillard, L.; Virkar, A.V.; Clarke, D.R. The Tetragonal-Monoclinic Transformation in Zirconia: Lessons Learned and Future Trends. *J. Amer. Ceram. Soc.* **2009**, *92*, 1901–1920. [CrossRef]
10. Ramos, C.M.; Tabata, A.S.; Cesar, P.F.; Rubo, J.H.; Fraisconi, P.A.S.; Borges, A.F.S. Application of Micro-Raman Spectroscopy to the Study of Yttria-Stabilized Tetragonal Zirconia Polycrystal (Y-TZP) Phase Transformation. *Appl. Spectrosc.* **2015**, *69*, 810–814. [CrossRef]
11. Paul, A.; Vaidhyanathan, B.; Binner, J. Micro-Raman spectroscopy of indentation induced phase transformation in nanozirconia ceramics. *Adv. Appl. Ceram.* **2011**, *110*, 114–119. [CrossRef]
12. Melk, L.; Turon-Vinas, M.; Roa, J.J.; Antti, M.L.; Anglada, M. The influence of unshielded small cracks in the fracture toughness of yttria and of ceria stabilised zirconia. *J. Eur. Ceram. Soc.* **2016**, *36*, 147. [CrossRef]
13. Simeone, D.; Bechade, J.L.; Gosset, D.; Chevarier, A.; Daniel, P.; Pilliaire, H.; Baldinozzi, G. Investigation on the zirconia phase transition under irradiation. *J. Nucl. Mater.* **2000**, *281*, 171–181. [CrossRef]
14. Simeone, D.; Gossset, D.; Bechade, J.L.; Chevarier, A. Analysis of the monoclinic–tetragonal phase transition of zirconia under irradiation. *J. Nucl. Mater.* **2002**, *300*, 27–38. [CrossRef]
15. Simeone, D.; Baldinozzi, G.; Gosset, D.; Dutheil, M.; Bulou, A.; Hansen, T. Monoclinic to tetragonal semireconstructive phase transition of zirconia. *Phys. Rev. B* **2003**, *67*, 1–8. [CrossRef]
16. Leib, E.W.; Vainio, U.; Pasquarelli, R.M.; Kus, J.; Czaschke, C.; Walter, N.; Janssen, R.; Müller, M.; Schreyer, A.; Weller, H.; et al. Synthesis and thermal stability of zirconia and yttria-stabilized zirconia microspheres. *J. Colloid Interface Sci.* **2015**, *448*, 582. [CrossRef]
17. Ishikawa, N.; Sonoda, T.; Okamoto, Y.; Sawabe, T.; Takegahara, K.; Kosugi, S.; Iwase, A. X-ray Study of Radiation Damage in UO2 Irradiated with High-Energy Heavy Ions. *J. Nucl. Mater.* **2011**, *419*, 392–396. [CrossRef]
18. Blanton, T.N.; Barnes, C.L.; Lelental, M. The effect of X-ray penetration depth on structural characterization of multiphase Bi-Sr-Ca-Cu-O thin films by X-ray diffraction techniques. *Phys. C* **1991**, *173*, 3–4. [CrossRef]
19. Sato, T.; Iwamoto, Y.; Hashimoto, S.; Ogawa, T.; Furuta, T.; Abe, S.; Kai, T.; Tsai, P.; Matsuda, N.; Iwase, H.; et al. Features of Particle and Heavy Ion Transport code System (PHITS) version 3.02. *J. Nucl. Sci. Technol.* **2018**, *55*, 684. [CrossRef]
20. Messenger, S.R.; Burke, E.A.; Xapsos, M.A.; Walters, R.J.; Jackson, E.M.; Weaver, B.D. Nonionizing Energy Loss (NIEL) for Heavy Ions. *IEEE Trans. Nucl. Sci.* **1999**, *46*, 1595. [CrossRef]
21. Jun, I.; Xapsos, M.A.; Messenger, S.R.; Burke, E.A.; Walters, R.J.; Summers, G.P.; Jordan, T. Proton nonionizing energy loss (NIEL) for device applications. *IEEE Trans. Nucl. Sci.* **2003**, *50*, 1924.
22. Summers, G.P.; Burke, E.A.; Xapsos, M.A. Displacement damage analogs to ionizing radiation effects. *Radiat. Meas.* **1995**, *24*, 1–8. [CrossRef]
23. Sato, T.; Ohtaki, S.; Shimada, M. Transformation of yttria partially stabilized zirconia by low temperature annealing in air. *J. Mater. Sci.* **1985**, *20*, 1466. [CrossRef]
24. Brog, J.P.; Chanez, A.C.; Crochet, A.; Fromm, K.M. Polymorphism, what it is and how to identify it: A systematic review. *RSC Adv.* **2013**, *3*, 16905. [CrossRef]

25. Cohen, M.; Wayman, C.M. Fundamentals of martensitic reactions. In *Metallurgical Treatises*; Tien, J.K., Elliott, J.F., Eds.; TMS-AIME: Warrendale, PA, USA, 1983; pp. 445–468.
26. Ge, Q.L.; Lei, T.C.; Mao, J.F.; Zhou, Y. In situ transmission electron microscopy observations of the tetragonal-to-monoclinic phase transformation of zirconia in AI2Oa-ZrO_2 (2 mol % Y_2O_3) composite. *J. Mater. Sci. Lett.* **1993**, *12*, 819–822. [CrossRef]
27. Lamas, D.G.; Walsöe de Reca, N.E. Oxygen ion displacement in compositionally homogeneous, nanocrystalline ZrO_2–12 mol% Y_2O_3 powders. *Mater. Lett.* **1999**, *41*, 204–208. [CrossRef]
28. Nakamura, R.; Ishimaru, M.; Yasuda, H.; Nakajima, H. Atomic rearrangements in amorphous Al2O3 under electron-beam irradiation. *J. Appl. Phys.* **2013**, *113*, 064312. [CrossRef]
29. Meldrum, A.; Boatner, L.A.; Ewing, R.C. Electron-irradiation-induced nucleation and growth in amorphous $LaPO_4$, $ScPO_4$, and zircon. *J. Mater. Res.* **1997**, *12*, 1816. [CrossRef]
30. Nguyen-Truong, H.T. Modified Bethe formula for low-energy electron stopping power without fitting parameters. *Ultramicroscopy* **2015**, *149*, 26–33. [CrossRef] [PubMed]

quantum beam science

Article

Comprehensive Understanding of Hillocks and Ion Tracks in Ceramics Irradiated with Swift Heavy Ions

Norito Ishikawa [1],* , Tomitsugu Taguchi [2] and Hiroaki Ogawa [1]

[1] Nuclear Science and Engineering Center, Japan Atomic Energy Agency (JAEA), Tokai, Ibaraki 319-1195, Japan; ogawa.hiroakixx@jaea.go.jp
[2] Quantum Beam Science Research Directorate, National Institutes for Quantum and Radiological Science and Technology (QST), Tokai, Ibaraki 319-1106, Japan; taguchi.tomitsugu@qst.go.jp
* Correspondence: ishikawa.norito@jaea.go.jp; Tel.: +81-29-282-6089; Fax: +81-29-282-6716

Received: 13 November 2020; Accepted: 8 December 2020; Published: 9 December 2020

Abstract: Amorphizable ceramics ($LiNbO_3$, $ZrSiO_4$, and $Gd_3Ga_5O_{12}$) were irradiated with 200 MeV Au ions at an oblique incidence angle, and the as-irradiated samples were observed by transmission electron microscopy (TEM). Ion tracks in amorphizable ceramics are confirmed to be homogenous along the ion paths. Magnified TEM images show the formation of bell-shaped hillocks. The ion track diameter and hillock diameter are similar for all the amorphizable ceramics, while there is a tendency for the hillocks to be slightly bigger than the ion tracks. For $SrTiO_3$ (STO) and 0.5 wt% niobium-doped STO (Nb-STO), whose hillock formation has not been fully explored, 200 MeV Au ion irradiation and TEM observation were also performed. The ion track diameters in these materials are found to be markedly smaller than the hillock diameters. The ion tracks in these materials exhibit inhomogeneity, which is similar to that reported for non-amorphizable ceramics. On the other hand, the hillocks appear to be amorphous, and the amorphous feature is in contrast to the crystalline feature of hillocks observed in non-amorphizable ceramics. No marked difference is recognized between the nanostructures in STO and those in Nb-STO. The material dependence of the nanostructure formation is explained in terms of the intricate recrystallization process.

Keywords: swift heavy ion; hillocks; ion tracks; ion irradiation; TEM

1. Introduction

Long nanometer-sized damage trails are created in ceramics continuously along the trajectories of swift heavy ions (SHIs), if the energy transfer from a SHI to an electron system of ceramics is sufficiently high [1–3]. Such characteristic damage is called an ion track. The mechanism of ion track formation has been extensively studied so far, and it has been one of the central topics in the research on ion-solid interactions. Ceramics can be categorized into two groups; amorphizable and non-amorphizable ceramics. If amorphous ion tracks are created, the ceramics are called amorphizable ceramics. The electronic stopping power dependence of ion track sizes in many amorphizable ceramics has been successfully predicted using the thermal spike model [4–8]. According to the model, the amorphous ion track formation is attributable to a local temperature rise sufficient to cause local melting along an ion path. On the other hand, there are non-amorphizable ceramics in which a crystal structure within an ion track region is not amorphized [9–14]. It has been found that in non-amorphizable ceramics, ion track size is markedly smaller than the size of the melt predicted by the thermal spike models [13,14]. Recent molecular dynamics (MD) simulation has revealed that the small ion tracks in non-amorphizable ceramics are attributable to fast recrystallization after transient melting [15–18]. It has been pointed out that the ionic nature of atomic bonding in non-amorphizable ceramics may be responsible for such fast recrystallization in non-amorphizable ceramics [9,19]. However, there is currently no consensus on which material property makes the distinction between amorphizable and non-amorphizable ceramics. Therefore, it is important to examine the material dependence of ion track formation in terms of amorphization/recrystallization.

Ion track formation caused by SHIs is often accompanied by the formation of hillocks (so-called surface ion tracks) [20–35]. Our recent studies revealed that both ion tracks and hillocks are amorphous in the case of amorphizable ceramics ($Y_3Fe_5O_{12}$ (YIG)) [36,37], whereas crystalline hillocks are found in the case of non-amorphizable ceramics (CaF_2, SrF_2, BaF_2, and CeO_2) [36,38]. Since the surface protrusion is a direct consequence of local melting along the ion path (melting of the ion track region), the observation of crystalline hillocks in non-amorphizable ceramics also provides a strong evidence of recrystallization after melting of the ion track region. It was also found that the hillock diameter always coincides with the diameter of a melt predicted by the thermal spike model for both amorphizable and non-amorphizable ceramics. This means that a hillock diameter value is affected by a melting process, but it remains unchanged after the subsequent recrystallization process. It is likely that a hillock size reflects melting, whereas an ion track size reflects both melting and subsequent recrystallization. Therefore, a comparative analysis of ion tracks and hillocks allows us to elucidate the whole processes of melting and amorphization/recrystallization. Moreover, there must be a variety of hillock and ion track morphologies due to an intricate recrystallization process, which can be the origin of the material dependence of nanostructure formation.

We recently proposed a method for precise measurement of a hillock size by transmission electron microscopy (TEM) [36–38]. The method is useful for the direct observation of a hillock side-view. It allows the accurate measurement of hillock dimensions and the identification of hillock crystal structure. In our previous study, we have studied hillocks and ion tracks in non-amorphizable ceramics such as CaF_2, SrF_2, and BaF_2, whereas only one amorphizable ceramic (YIG) has been studied [36,37]. In the present study, we have further studied the relationship between the hillock diameter and ion track diameter for various types of ceramics. First, we show results of the TEM observations for SHI-irradiated amorphizable ceramics such as $LiNbO_3$, $ZrSiO_4$, and $Gd_3Ga_5O_{12}$ (GGG) and then for $SrTiO_3$ (STO) and 0.5 wt% niobium-doped STO (Nb-STO), whose hillock formation has not been fully explored. The electrical resistivity of STO increases by more than nine orders of magnitude by doping with 0.5 wt% Nb, whereas its crystal structure remains unaffected by the doping [39]. The effect of Nb doping on its hillock morphology is also investigated in this study. Based on the TEM investigations of both amorphizable and non-amorphizable ceramics, we discuss how the SHI-induced nanostructure formation depends on the intricate recrystallization process.

2. Experiments

Thin samples for TEM observations were prepared before irradiation by the following procedure. The original materials were $ZrSiO_4$ (98%) powder (Kojundo Chemical Laboratory Co., Ltd., Saitama, Japan), $LiNbO_3$ single crystals (Onizawa Fine Product Co., Ltd., Ibaraki, Japan), $Gd_3Ga_5O_{12}$ (GGG) single crystals (Rare Metallic Co., Ltd., Tokyo, Japan), $SrTiO_3$ (STO) single crystals (Shinkosha Co., Ltd., Kanagawa, Japan), and 0.5 wt% niobium-doped STO (Nb-STO) single crystals (Shinkosha Co., Ltd., Kanagawa, Japan). The original materials were finely ground using an agate mortar and pestle. For the preparation of $LiNbO_3$ specimens, first, 3 mm diameter nickel grids with 2000 lines/inch were coated using 0.5% Neoprene W in toluene (Nisshin EM Co., Ltd., Tokyo, Japan), which works as an adhesive agent, and then the ground $LiNbO_3$ powder was randomly dispersed on the grids. For the preparation of materials other than $LiNbO_3$, the ground samples were dispersed in ethanol by an ultrasonic bath for a few minutes, and then the ethanol was dropped on 3 mm diameter 200 mesh copper grids covered with porous carbon films. The grids were then air-dried at room temperature. The samples on the grids were subsequently irradiated with 200 MeV Au^{32+} ions at an oblique incidence angle (45° relative to normal direction) at room temperature in a tandem accelerator at JAEA-Tokai (Japan Atomic Energy Agency, Tokai Research and Development Center, Tokai, Japan). The charge state (32+) was chosen to ensure the charge of the incident ions to have the average value of the equilibrium charge. The samples were irradiated with ions at 1×10^{11} ions/cm^2. The as-irradiated samples were examined using a transmission electron microscope (TEM, Model 2100F, JEOL Ltd., Tokyo, Japan) operated at 200 kV. The ion track size was measured using the TEM images taken at a low magnification, wherein clear line-like contrasts were expected to be imaged. On the other hand, the hillock size was measured using the TEM images taken at a high magnification, wherein a clear contour of hillocks was expected to be obtained. The electronic stopping power (S_e) was estimated using SRIM-2008 [40,41].

3. Results and Discussions

3.1. Dimensions of Hillocks and Ion Tracks in LiNbO$_3$, ZrSiO$_4$, and GGG

We have observed nanostructures created by SHI irradiation in amorphizable ceramics using TEM. Figure 1a–c show the bright field images of ion tracks in LiNbO$_3$, ZrSiO$_4$, and Gd$_3$Ga$_5$O$_{12}$ irradiated with 200 MeV Au at an oblique incidence angle. In the figures, the ion tracks are imaged as line-like contrasts. The diameter of the ion tracks can be estimated by measuring the width of the line-like contrasts. However, for most of the ion track images, it is difficult to distinguish the ion track contrasts from hillock contrasts. However, if the hillocks are created at the edge of the samples, it is possible to infer the hillock side-view. It can be concluded from the images that hillocks have similar diameters to those of ion tracks. Size distribution of ion tracks is shown in Figure 2. The average sizes of the hillocks and ion tracks are summarized in Table 1.

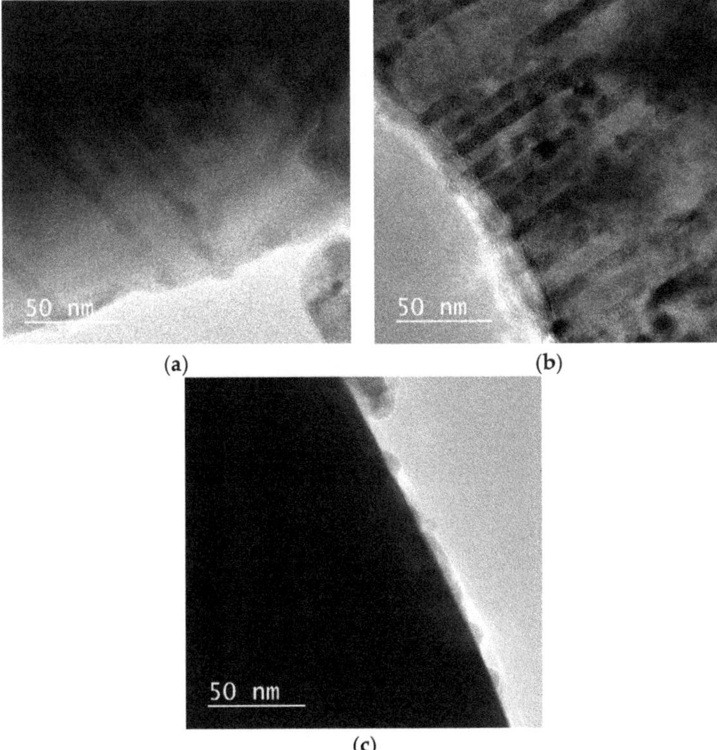

Figure 1. Bright field images of ion tracks induced in (**a**) LiNbO$_3$, (**b**) ZrSiO$_4$, and (**c**) Gd$_3$Ga$_5$O$_{12}$ irradiated with 200 MeV Au^{32+} at an oblique incidence angle. The images were taken at relatively low magnification.

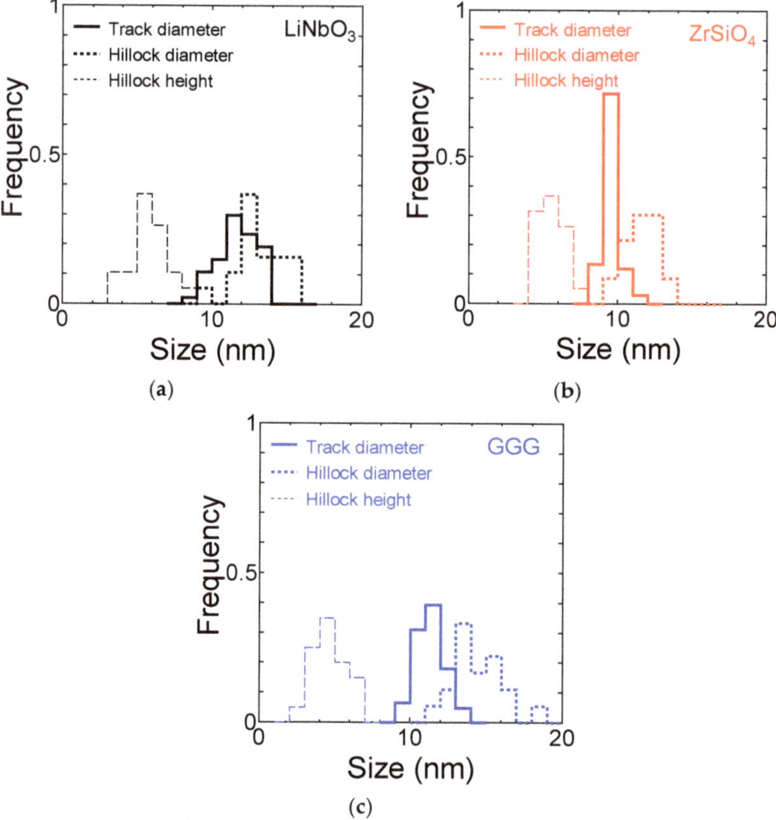

Figure 2. Size distribution of the track diameter, the hillock diameter, and the hillock height in (**a**) LiNbO$_3$, (**b**) ZrSiO$_4$, and (**c**) Gd$_3$Ga$_5$O$_{12}$ irradiated with 200 MeV Au^{32+} at an oblique incidence angle.

Table 1. Average diameter of ion tracks (D$_{track}$), average diameter of hillocks (D$_{hillock}$), average height of hillocks (H$_{hillock}$) are listed with standard deviations for LiNbO$_3$, ZrSiO$_4$, and GGG irradiated with 200 MeV Au^{32+}. The corresponding S$_e$ values are also listed.

	D$_{track}$ (nm)	D$_{hillock}$ (nm)	H$_{hillock}$ (nm)	Se (keV/nm)
LiNbO$_3$	11.7 ± 1.3	13.2 ± 1.5	5.8 ± 1.2	28.1
ZrSiO$_4$	9.5 ± 0.5	11.6 ± 1.1	5.3 ± 0.8	29.6
GGG	11.3 ± 0.9	14.7 ± 2.0	4.7 ± 1.0	34.2

Previous literatures have reported ion track sizes in LiNbO$_3$ [42–45], ZrSiO$_4$ [46,47], and GGG [45,48]. They have confirmed that the ion tracks in these ceramics are amorphous. A good summary of the ion track data in LiNbO$_3$ is given in Ref. [45], in which the dependence of ion track size on the electronic stopping power (S$_e$) is shown. The S$_e$-dependence demonstrates that an ion track size depends primarily on the electronic stopping power, whereas low ion velocity acts as a secondary factor that contributes to a bigger ion track owing to the velocity effect [49]. The literature also demonstrates that ion track sizes are in accordance with the values predicted by the thermal spike model. The present study shows that the ion track diameter in LiNbO$_3$ irradiated with 200 MeV Au (S$_e$ = 28.1 keV/nm) is 11.7 ± 1.3 nm, whereas the previous study showed that the ion track diameter in LiNbO$_3$ irradiated with 201 MeV U ions (S_e = 28.1 keV/nm) was around 13 nm [44,45], indicating that the present study is in accordance with the previous results within the experimental error.

Previous studies reported that the ion track diameter for $ZrSiO_4$ irradiated with 10 GeV Pb ions (S_e = 20.0 keV/nm) was 5.2 ± 0.2 nm [46] and that for $ZrSiO_4$ irradiated with 2.9 GeV Pb ions (S_e = 33.6 keV/nm) [47] was 8 nm. The present study shows that the ion track diameter for $ZrSiO_4$ irradiated with 200 MeV Au ions is 9.5 ± 0.5 nm, exhibiting large ion tracks. Since the large ion tracks observed in the present study could be due to the relatively high electronic stopping power (S_e = 29.6 keV/nm) and the velocity effect [49], there is no contradiction between the present and previous results.

A good summary of the previous ion track data in GGG is given in Ref. [45]. The present study shows that the ion track diameter for GGG irradiated with 200 MeV Au (S_e = 34.0 keV/nm) is 11.3 ± 0.9 nm, whereas a previous study shows a similar ion track diameter (12.4 ± 1.6 nm for GGG irradiated with 250 MeV Pb ions (S_e = 34.0 keV/nm)) [45]. Therefore, the present result is consistent with the previous results. The present results are also consistent with the prediction made by the thermal spike model.

Figure 3a–c show magnified images of hillocks in $LiNbO_3$, $ZrSiO_4$, and GGG, respectively, irradiated with 200 MeV Au at an oblique incidence angle. As shown in the figures, hillocks are successfully observed in those materials. It is also found that the hillocks are clearly amorphous, which confirms the amorphizable nature of these ceramics. The side-view of the hillocks allows measuring both hillock diameter and height. The distribution of the hillock sizes is shown in Figure 2 together with that of the ion track sizes. The average sizes of the hillocks and track sizes are summarized in Table 1. As demonstrated in the table, the hillock diameter is always similar to the ion track size, although the former tends to be slightly larger than the latter. A hillock height seems nearly half of the hillock diameter.

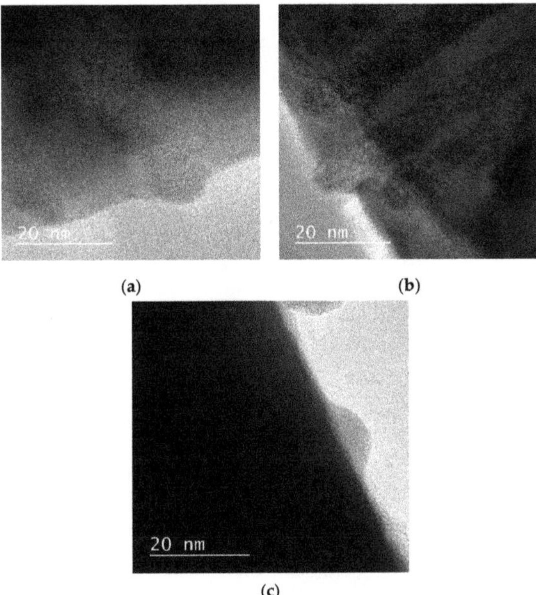

(a)   (b)

(c)

Figure 3. Bright field images of hillocks induced in (**a**) $LiNbO_3$, (**b**) $ZrSiO_4$, and (**c**) $Gd_3Ga_5O_{12}$ irradiated with 200 MeV Au^{32+} at an oblique incidence angle. The images were taken at relatively high magnification.

Although the hillock shape seems to be semi-spherical in the low magnification images, the magnified images demonstrate the formation of bell-shaped hillocks. Bell-shaped hillocks can also be found in $Y_3Fe_5O_{12}$ [36,37] and $Y_3Al_5O_{12}$ [16], that are also categorized as amorphizable ceramics.

3.2. Formation Process of Hillocks and Ion Tracks in LiNbO₃, ZrSiO₄, and GGG

The present study shows that both hillocks and ion tracks exhibit amorphous features, and they have similar sizes in amorphizable ceramics (LiNbO₃, ZrSiO₄, and GGG). The present results agree with the prediction of the thermal spike model. According to the thermal spike model, SHIs cause transient melting along the ion path, when the electronic stopping power exceeds a certain threshold value. During such transient melting, thermal pressure and volume change caused by solid-liquid transition lead to surface protrusion of the melt. In the amorphizable ceramics, rapid cooling after melting results in the formation of amorphous ion tracks and hillocks.

It is important to note that, although the hillock diameter appears to be similar to the ion track diameter, the magnified TEM images shows that the former is always slightly larger than the latter. This can be ascribed to the spreading tendency of a liquid on a solid surface. It is likely that the hillock shape is determined by the balance of the adhesive (the liquid wanting to maintain contact with the solid) and cohesive forces within the liquid (both internal cohesive force and surface tension) during melting. If the adhesive force dominates, the protruded part of the melt can spread over the surface, leading to the formation of bell-shaped hillocks. Conversely, it can turn into a spherical shape, if the cohesive force dominates. Such spherical hillocks have been reported in some ceramics irradiated with high-energy fullerene ions having a very high S_e [50,51]. Large volume of surface protrusion induced by high S_e may be closely related to the formation of spherical hillocks.

3.3. Hillocks and Ion Tracks in SrTiO₃ and Nb-Doped SrTiO₃

Figures 4 and 5 show the bright field image of the ion tracks in STO and Nb-STO irradiated with 200 MeV Au at an oblique incidence angle, respectively. The ion track has a bright core surrounded by a dark fringe in the underfocus condition (Figures 4a and 5a), whereas it has a dark core surrounded by a bright fringe in the overfocus condition (Figures 4b and 5b). The Fresnel fringe is an indication of the formation of ion tracks with a lower density than that of the surrounding matrix [52,53]. Such focus-dependent Fresnel contrast is not found in amorphizable ceramics (e.g., YIG, LiNbO₃, ZrSiO₄, and GGG).

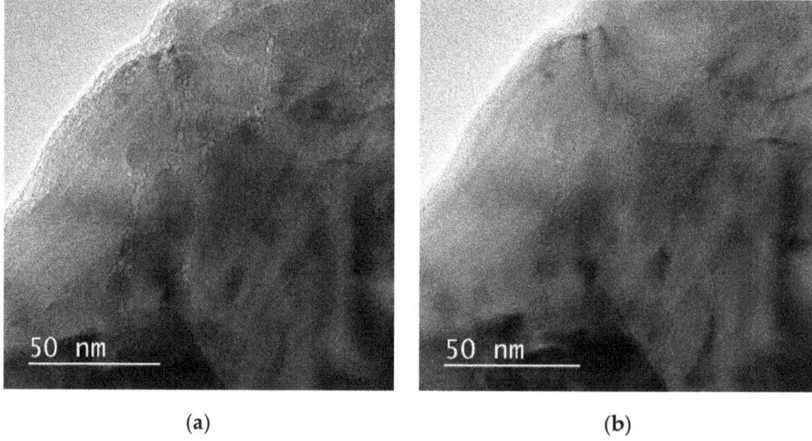

Figure 4. Bright field images of ion tracks induced in $SrTiO_3$ irradiated with 200 MeV Au^{32+} at an oblique incidence angle. The images were taken in (**a**) underfocus and (**b**) overfocus conditions.

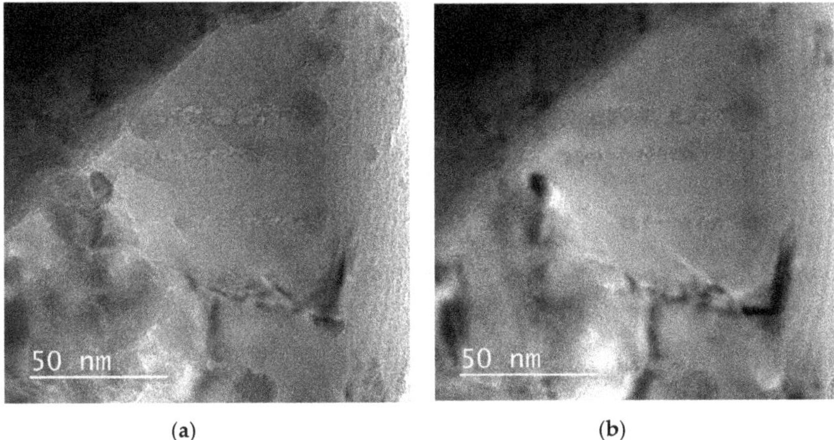

Figure 5. Bright field images of ion tracks induced in Nb-doped $SrTiO_3$ irradiated with 200 MeV Au^{32+} at an oblique incidence angle. The images were taken in (**a**) underfocus condition and (**b**) overfocus condition.

The ion tracks in STO and Nb-STO have inhomogeneous morphology in contrast to the homogenous morphology of the ion tracks in amorphizable ceramics. The inhomogeneity of the ion tracks is a common feature observed in non-amorphizable ceramics such as CeO_2 [38], CaF_2, SrF_2, and BaF_2 [36], in which partial recrystallization after transient melting is the likely cause of the inhomogeneous ion track formation. Black contrasts corresponding to hillocks are found at the end of the ion tracks. Figures 6 and 7 show magnified images of a hillock created at the edge of the thin TEM samples of STO and Nb-STO, respectively. Bell-shaped hillocks are observed in STO and Nb-STO. In both materials, hillocks are found to be amorphous. It is interesting to find that amorphous hillocks are created, although the ion track region is recrystallized.

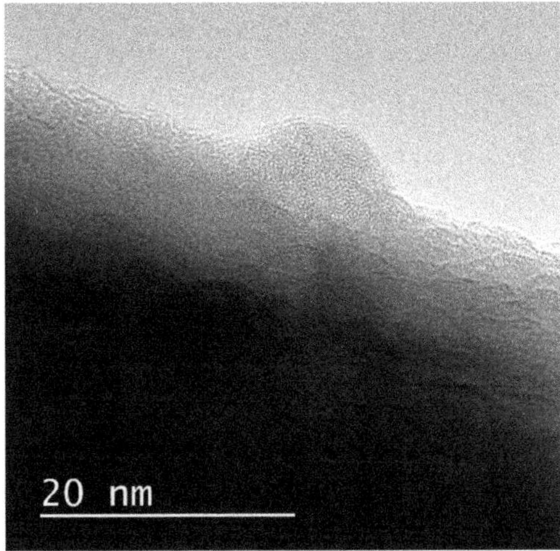

Figure 6. Bright field images of hillocks induced in $SrTiO_3$ irradiated with 200 MeV Au^{32+} at an oblique incidence angle.

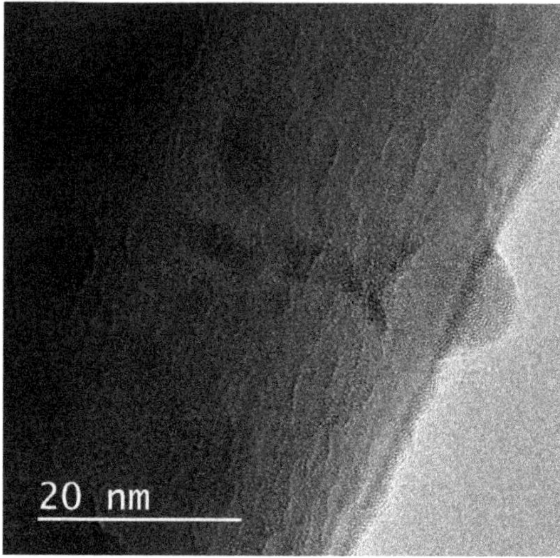

Figure 7. Bright field images of hillocks induced in Nb-doped $SrTiO_3$ irradiated with 200 MeV Au^{32+} at an oblique incidence angle.

Since it is difficult to find inhomogeneous ion tracks in TEM, we could not find sufficient numbers of ion tracks and hillocks to perform a statistical analysis. Therefore, in this study, we only show the size range of ion tracks and hillocks observed in STO and Nb-STO in Table 2. As demonstrated in the table, there is a marked difference in the diameters of ion tracks and hillocks in both STO and Nb-STO. The large hillock diameter compared with the ion track diameter is also found in typical non-amorphizable ceramics (CaF_2, SrF_2, and BaF_2). The present result can be a strong evidence that supports the recrystallization of the ion track regions in STO and Nb-STO. Conversely, the amorphous hillocks are signs of the failure of recrystallization. It seems that STO and Nb-STO are intermediate ceramics between amorphizable and non-amorphizable ceramics.

Table 2. Approximate values of ion track diameter (D_{track}), hillock diameter ($D_{hillock}$), and hillock height ($H_{hillock}$) in $SrTiO_3$ (STO) and 0.5 wt% niobium-doped STO (Nb-STO). The corresponding S_e values are also listed.

	D_{track} (nm)	$D_{hillock}$ (nm)	$H_{hillock}$ (nm)	Se (keV/nm)
STO	3~5	12~15	4~5	28.6
Nb-STO	3~4	11~13	4~5	28.5

Here, it is important to discuss whether the ion tracks are actually crystalline or not, since some of the previous studies claimed that they are amorphous in STO. According to Ref. [54], analysis of X-ray diffraction peaks in ion-irradiated STO supports creation of amorphous ion tracks due to single impacts. In the same literature, although crystalline tracks containing defects are observed by TEM, the authors claimed that the crystalline tracks are created owing to 200 keV electron beam exposure which can cause amorphous–crystalline transition of ion track areas. Conversely, our TEM results of STO and Nb-STO support creation of crystalline ion tracks rather than amorphous ion tracks. For example, an inhomogenous feature of ion tracks in STO and Nb-STO is similar to that observed in partially recrystallized ion tracks in non-amorphizable ceramics. Moreover, the markedly smaller ion track diameter than the hillock diameter demonstrates that the ion track region is partially recrystallized after transient melting.

The clear difference in the morphology between the hillocks in STO and Nb-STO was not observed in this study. Therefore, the influence of Nb-doping on the hillock formation is not found in this study. A previous study using TEM and small angle X-ray scattering (SAXS) reported that 1 wt% Nb-doping does not affect track formation [55]. Track formation is easy in insulators since the energy of hot electrons is transferred to the lattice in insulators before being cooled down by free electrons. In contrast, track formation is difficult in metals since high electronic conductivity allows rapid energy diffusion in an electron subsystem. Although the electrical conductivity of Nb-STO is more than nine orders of magnitude higher than that of STO, the electron density may still be too low to influence the formation of hillocks and ion tracks. It should be noted that the electrical conductivity of metals is still much higher than that of Nb-STO. The difference of bonding character (metallic bonding in metals vs. ionic/covalent bonding in STO) should be also contributing to the less sensitivity of track formation in metals [55].

3.4. Factors That Determine the Recrystallization Process

The presence of recrystallization process separates non-amorphizable ceramics from amorphizable ceramics. Two factors that determine the recrystallization process have been proposed; (1) simplicity of lattice structure and (2) the strength of ionic bonding. The previous literature [15] proposed that a simple lattice structure with the smallest number of the peaks in the pair correlation function is in correlation with the recrystallization efficiency. Actually, amorphous ion tracks and hillocks are observed in YIG and GGG that have complicated crystal structure, whereas crystalline ion tracks and hillocks are observed in rather simple structured compounds (CaF_2, SrF_2, BaF_2, and CeO_2 with fluorite structure). STO has a relatively simple cubic structure with space group of $Pm\bar{3}m$, where there are only 5 atoms in the unit cell. The relatively simple crystal structure of STO can be the origin of the recrystallization of the ion track region. However, creation of amorphous hillocks in STO demonstrates that STO and Nb-STO are intermediate ceramics between amorphizable and non-amorphizable ceramics. Here, it is worth noting that STO has a perovskite structure which is composed of TiO_2 planes and SrO planes, which make the structure somehow complicated. This unique structure of STO seems to be responsible for the intermediate behavior of ion-induced nanostructure formation. Therefore, we believe that the correlation of recrystallization effectiveness with the simplicity of the crystal structure is one of the promising hypotheses, and its validity should be further examined.

Another possible factor is the strength of ionic bonding, i.e., materials with a higher degree of ionicity recrystallize easily [19,56]. Non-amorphizable ceramics include many ionic crystals. It is conceivable that long-range ionic forces rather than short-range covalent interaction contribute to rapid recrystallization [19,57]. STO has mixed ionic-covalent bonding properties. While a hybridization of the O-2p states with the Ti-3d states within the TiO_6 octahedra leads to a pronounced covalent bonding, Sr^{2+} and O^{2-} ions exhibit an ionic bonding character. Although STO is one of the complex oxides, the ionic bonding character in STO may be also responsible for the partial recrystallization of ion tracks. Therefore, the two factors mentioned above is important to explain why STO and Nb-STO are intermediate ceramics between amorphizable and non-amorphizable ceramics.

It is worth discussing why the hillock region fail to recrystallize, whereas the ion track region recrystallizes in STO. The previous MD simulation demonstrated that oxygen atoms settle in their original sites faster than metal atoms in SHI-irradiated MgO and Al_2O_3, and metallic atoms then adjust to a layer of oxygen that has already been built [15]. The study concluded that that the ability of oxygen atoms to reach their equilibrium sites governs the recrystallization process after melting. It is conceivable that in STO, oxygen loss in the hillock region effectively hinders recovery of the lattice structure. The hillock region tends to be more oxygen deficient than the ion track region, since oxygen atoms tend to escape from the irradiated surface. An important role of oxygen deficiencies should be further investigated in the future.

3.5. Summary of Formation Processes of Hillocks and Ion Tracks

To understand the difference between nanostructure formation in amorphizable and non-amorphizable ceramics, it is important to describe the process in terms of melting and successive recrystallization.

It is convenient to start the discussion with CeO_2, which is one of the non-amorphizable ceramics. According to Ref. [38], hillocks are found to be spherical in CeO_2 irradiated with 200 MeV Au. It was found that hillocks (spherical objects) have crystalline features, where the lattice orientation of hillocks was aligned with that of the matrix. This points out to the process consisting of the following three steps as explained in Figure 8a. (1) A molten region is created along the ion path. A part of the molten region protrudes above the surface because of the thermal pressure and additional pressure due to volume change caused by solid-liquid transition. (2) During cooling, the molten region embedded in the matrix begins to recrystallize. The partial recrystallization results in ion tracks smaller than those expected from the size of the melt. The shape change (spheroidization) of the protruded part strongly suggests that the protruded part remains liquid for a long period of time which is long enough for the molten protrusion to change its shape. The protruded part of the molten region can spheroidize if the surface tension is strong enough. (3) A droplet at the surface starts to recrystallize epitaxially using the matrix as a template lattice, so that a crystalline nanosphere having the same crystal orientation as that of the matrix is formed. This result strongly suggests that the molten region embedded in the matrix is solidified before the protruded part of the melt is solidified. It is reasonable to assume that this sequence of the solidification process applies to all SHI-irradiated ceramics. Such a sequence of solidification process is supported by the MD simulation of the nanostructure formation in CaF_2 [16].

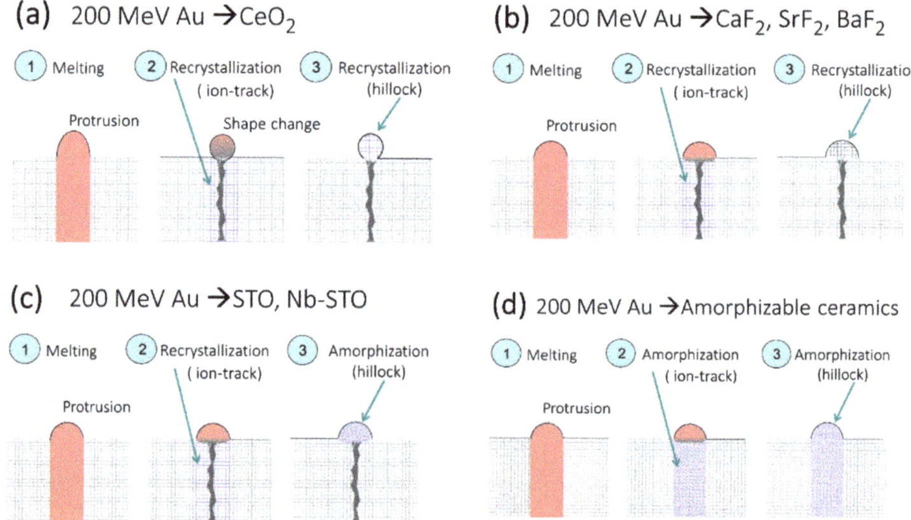

Figure 8. Schematic formation processes of ion tracks and hillocks in (**a**) CeO_2, (**b**) fluorides (CaF_2, SrF_2, and BaF_2), (**c**) STO and Nb-STO, and (**d**) amorphizable ceramics (YIG, $LiNbO_3$, $ZrSiO_4$, and GGG).

The likely process of nanostructure formation in fluorides (CaF_2, SrF_2, and BaF_2) irradiated with 200 MeV Au is presented in Figure 8b. According to our previous study [36], most of the crystalline hillocks in these fluorides are nearly semispherical, although some of the hillocks have nearly spherical shape. Since recrystallization plays an important role in both fluorides and CeO_2, the formation process in the fluorides should be similar to that of CeO_2 (Figure 8a). The difference between the spherical shape of hillocks and non-spherical shape can be explained by the difference in volume of protrusion.

If the volume of protrusion is large, the hillock shape tentatively has an unstable shape because of its high aspect ratio of height to width. It is conceivable that the unstable shape can rapidly turn into a stable spherical shape. Similar spherical hillocks have been already reported in some ceramics ($Gd_2Zr_2O_7$ [50] and YIG [51]) irradiated with high-energy fullerene ions with very high S_e, supporting the proposition that larger volume of the protrusion is the key to the formation of spherical hillocks.

The likely process of nanostructure formation in STO and Nb-STO is presented in Figure 8c. Although the first and second steps of the process in the figure are the same as those in Figure 8b, the third step of the process is different. Even though the molten region embedded in the matrix recrystallizes partially, the protruded part of the melt failed to recrystallize, resulting in amorphization. The failure of recrystallization only in the hillock region can be ascribed to oxygen deficiencies as discussed above.

The likely process in amorphizable ceramics (YIG, $LiNbO_3$, $ZrSiO_4$, and GGG) is presented in Figure 8d. The process consists of the following three steps. (1) A molten region is created along the ion path. A part of the molten region protrudes above the surface. (2) During cooling, the molten region embedded in the matrix starts to solidify, resulting in the amorphization of the molten region. Herein, the size of the molten region corresponds to that of the ion tracks. (3) The protruded part of the melt also starts to amorphize. This means that the hillock diameter is always similar to the ion track diameter. A subtle change of hillock shape leading to a slightly larger hillock diameter than the ion track diameter is not specified in the schematic, since this effect produces only a small difference between the hillock diameter and ion track diameter.

4. Conclusions

Amorphizable ceramics such as $LiNbO_3$, $ZrSiO_4$, and $Gd_3Ga_5O_{12}$ were irradiated with 200 MeV Au ions at an oblique incidence angle. Line-like homogeneous ion tracks and bell-shaped hillocks are observed by TEM. The ion track and hillock diameters are similar for all the amorphizable ceramics, although the hillock diameter is found to be slightly larger than the ion track diameter. The TEM images of ion tracks in STO and Nb-STO irradiated with 200 MeV Au show similar features to those observed in the non-amorphizable ceramics. For example, the ion track diameter is markedly smaller than the hillock diameter, and they exhibit inhomogeneity. However, the hillocks in these ceramics are found to be amorphous which is in contrast to the crystalline feature of hillocks observed in the non-amorphizable ceramics. Therefore, it can be concluded that STO and Nb-STO are intermediate ceramics between amorphizable and non-amorphizable ceramics. No marked difference is observed between hillock formation in STO and that in Nb-STO. The material dependence of nanostructure formation can be ascribed to the intricate recrystallization process. The present results support that (1) simplicity of lattice structure and (2) the strength of ionic bonding can be the factors that determine the recrystallization effectiveness.

Author Contributions: Conceptualization, N.I.; TEM observation, N.I. and T.T.; Sample preparation, N.I.; Ion irradiation, N.I. and H.O.; Manuscript writing, N.I. All authors have read and agreed to the published version of the manuscript.

Funding: Part of the present work was financially supported by JSPS KAKENHI Grant Number 20K05389.

Acknowledgments: The authors are grateful to the technical staff of the tandem accelerator at JAEA-Tokai for supplying high-quality ion beams. One of the authors (N.I.) thanks A. Iwase and A. Kitamura for their constant support during the research.

Conflicts of Interest: The authors declare no conflict of interest.

References

1. Toulemonde, M.; Assmann, W.; Dufour, C.; Meftah, A.; Trautmann, C. Nanometric transformation of the matter by short and intense electronic excitation: Experimental data versus inelastic thermal spike model. *Nucl. Instrum. Methods Phys. Res. Sect. B Beam Interact. Mater. Atoms* **2012**, *277*, 28–39. [CrossRef]

2. Szenes, G. Comparison of two thermal spike models for ion–solid interaction. *Nucl. Instrum. Methods Phys. Res. Sect. B Beam Interact. Mater. Atoms* **2011**, *269*, 174–179. [CrossRef]

3. Itoh, N.; Duffy, D.M.; Khakshouri, S.; Stoneham, A.M. Making tracks: Electronic excitation roles in forming swift heavy ion tracks. *J. Phys. Condens. Matter* **2009**, *21*, 474205. [CrossRef] [PubMed]

4. Toulemonde, M.; Studer, F. Comparison of the radii of latent tracks induced by high-energy heavy ions in $Y_3Fe_5O_{12}$ by HREM, channelling Rutherford backscattering and Mössbauer spectrometry. *Philos. Mag. A* **1988**, *58*, 799–808. [CrossRef]

5. Toulemonde, M. Nanometric phase transformation of oxide materials under GeV energy heavy ion irradiation. *Nucl. Instrum. Methods Phys. Res. Sect. B Beam Interact. Mater. Atoms* **1999**, *156*, 1–11. [CrossRef]

6. Szenes, G. Monoatomic and cluster ion irradiation induced amorphous tracks in yttrium iron garnet. *Nucl. Instrum. Methods Phys. Res. Sect. B Beam Interact. Mater. Atoms* **1998**, *146*, 420–425. [CrossRef]

7. Szenes, G. Ion-velocity-dependent track formation in yttrium iron garnet: A thermal-spike analysis. *Phys. Rev. B* **1995**, *52*, 6154–6157. [CrossRef]

8. Szenes, G. Information provided by a thermal spike analysis on the microscopic processes of track formation. *Nucl. Instrum. Methods Phys. Res. Sect. B Beam Interact. Mater. Atoms* **2002**, *191*, 54–58. [CrossRef]

9. Toulemonde, M.; Dufour, C.; Meftah, A.; Paumier, E. Transient thermal processes in heavy ion irradiation of crystalline inorganic insulators. *Nucl. Instrum. Methods Phys. Res. Sect. B Beam Interact. Mater. Atoms* **2000**, *166*, 903–912. [CrossRef]

10. Schwartz, K.; Trautmann, C.; Steckenreiter, T.; Geiß, O.; Krämer, M. Damage and track morphology in LiF crystals irradiated with GeV ions. *Phys. Rev. B* **1998**, *58*, 11232–11240. [CrossRef]

11. Berthelot, A.; Hémon, S.; Gourbilleau, F.; Dufour, C.; Domengès, B. and Paumier, E. Behaviour of a nanometric SnO_2 powder under swift heavy-ion irradiation: From sputtering to splitting. *Phil. Mag. A* **2000**, *80*, 2257. [CrossRef]

12. Takaki, S.; Yasuda, K.; Yamamoto, T.; Matsumura, S.; Ishikawa, N. Structure of ion tracks in ceria irradiated with high energy xenon ions. *Prog. Nucl. Energy* **2016**, *92*, 306–312. [CrossRef]

13. Toulemonde, M.; Benyagoub, A.; Trautmann, C.; Khalfaoui, N.; Boccanfuso, M.; Dufour, C.; Gourbilleau, F.; Grob, J.J.; Stoquert, J.P.; Costantini, J.M.; et al. Dense and nanometric electronic excitations induced by swift heavy ions in an ionic CaF_2 crystal: Evidence for two thresholds of damage creation. *Phys. Rev. B* **2012**, *85*, 054112. [CrossRef]

14. Toulemonde, M.; Assmann, W.; Dufour, C.; Meftah, A.; Studer, F.; Trautmann, C. Experimental Phenomena and Thermal Spike Model Description of Ion Tracks in Amorphisable Inorganic Insulators. *Mat. Fys. Medd.* **2006**, *52*, 263–292.

15. Rymzhanov, R.A.; Medvedev, N.; O'Connell, J.H.; Van Vuuren, A.J.; Skuratov, V.A.; Volkov, A. Recrystallization as the governing mechanism of ion track formation. *Sci. Rep.* **2019**, *9*, 3837. [CrossRef] [PubMed]

16. Rymzhanov, R.A.; O'Connell, J.; Van Vuuren, A.J.; Skuratov, V.A.; Medvedev, N.; Volkov, A. Insight into picosecond kinetics of insulator surface under ionizing radiation. *J. Appl. Phys.* **2020**, *127*, 015901. [CrossRef]

17. Jiang, W.; Devanathan, R.; Sundgren, C.; Ishimaru, M.; Sato, K.; Varga, T.; Manandhar, S.; Benyagoub, A. Ion tracks and microstructures in barium titanate irradiated with swift heavy ions: A combined experimental and computational study. *Acta Mater.* **2013**, *61*, 7904–7916. [CrossRef]

18. Zhang, J.; Lang, M.; Ewing, R.C.; Devanathan, R.; Weber, W.J.; Toulemonde, M. Nanoscale phase transitions under extreme conditions within an ion track. *J. Mater. Res.* **2010**, *25*, 1344–1351. [CrossRef]

19. Trachenko, K. Understanding resistance to amorphization by radiation damage. *J. Phys. Condens. Matter* **2004**, *16*, R1491–R1515. [CrossRef]

20. Aumayr, F.; Facsko, S.; El-Said, A.S.; Trautmann, C.; Schleberger, M. Single ion induced surface nanostructures: A comparison between slow highly charged and swift heavy ions. *J. Phys. Condens. Matter* **2011**, *23*, 393001. [CrossRef]

21. El-Said, A.S.; Aumayr, F.; Della-Negra, S.; Neumann, R.; Schwartz, K.; Toulemonde, M.; Trautmann, C.; Voss, K.-O. Scanning force microscopy of surface damage created by fast C60 cluster ions in CaF2 and LaF3 single crystals. *Nucl. Instrum. Methods Phys. Res. Sect. B Beam Interact. Mater. Atoms* **2007**, *256*, 313–318. [CrossRef]

22. Muller, C.; Voss, K.O.; Lang, M.; Neumann, R. Correction of systematic errors in scanning force microscopy images with application to ion track micrographs. *Nucl. Instrum. Methods Phys. Res. Sect. B Beam Interact. Mater. Atoms* **2003**, *212*, 318–325. [CrossRef]

23. Popok, V.N.; Jensen, J.; Vuckovic, S.; Macková, A.; Trautmann, C. Formation of surface nanostructures on rutile (TiO_2): Comparative study of low-energy cluster ion and high-energy monoatomic ion impact. *J. Phys. D Appl. Phys.* **2009**, *42*, 205303. [CrossRef]

24. Meftah, A.; Benhacine, H.; Benyagoub, A.; Grob, J.; Izerrouken, M.; Kadid, S.; Khalfaoui, N.; Stoquert, J.; Toulemonde, M.; Trautmann, C. Data consistencies of swift heavy ion induced damage creation in yttrium iron garnet analyzed by different techniques. *Nucl. Instrum. Methods Phys. Res. Sect. B Beam Interact. Mater. Atoms* **2016**, *366*, 155–160. [CrossRef]

25. Khalfaoui, N.; Rotaru, C.; Bouffard, S.; Toulemonde, M.; Stoquert, J.; Haas, F.; Trautmann, C.; Jensen, J.; Dunlop, A. Characterization of swift heavy ion tracks in CaF_2 by scanning force and transmission electron microscopy. *Nucl. Instrum. Methods Phys. Res. Sect. B Beam Interact. Mater. Atoms* **2005**, *240*, 819–828. [CrossRef]

26. Khalfaoui, N.; Rotaru, C.; Bouffard, S.; Jacquet, E.; Lebius, H.; Toulemonde, M. Study of swift heavy ion tracks on crystalline quartz surfaces. *Nucl. Instrum. Methods Phys. Res. Sect. B Beam Interact. Mater. Atoms* **2003**, *209*, 165–169. [CrossRef]

27. Awazu, K.; Wang, X.; Fujimaki, M.; Komatsubara, T.; Ikeda, T.; Ohki, Y. Structure of latent tracks in rutile single crystal of titanium dioxide induced by swift heavy ions. *J. Appl. Phys.* **2006**, *100*, 044308. [CrossRef]

28. Akcöltekin, E.; Akcöltekin, S.; Osmani, O.; Duvenbeck, A.; Lebius, H.; Schleberger, M. Swift heavy ion irradiation of $SrTiO_3$ under grazing incidence. *New J. Phys.* **2008**, *10*, 53007. [CrossRef]

29. Muller, C.; Cranney, M.; El-Said, A.; Ishikawa, N.; Iwase, A.; Lang, M.; Neumann, R. Ion tracks on LiF and CaF_2 single crystals characterized by scanning force microscopy. *Nucl. Instrum. Methods Phys. Res. Sect. B Beam Interact. Mater. Atoms* **2002**, *191*, 246–250. [CrossRef]

30. Skuratov, V.; Zagorski, D.; Efimov, A.E.; Kluev, V.; Toporov, Y.; Mchedlishvili, B. Swift heavy ion irradiation effect on the surface of sapphire single crystals. *Radiat. Meas.* **2001**, *34*, 571–576. [CrossRef]

31. Schwen, D.; Bringa, E.M.; Krauser, J.; Weidinger, A.; Trautmann, C.; Hofsäss, H. Nano-hillock formation in diamond-like carbon induced by swift heavy projectiles in the electronic stopping regime: Experiments and atomistic simulations. *Appl. Phys. Lett.* **2012**, *101*, 113115. [CrossRef]

32. Müller, A.; Müller, C.; Neumann, R.; Ohnesorge, F. Scanning force microscopy of heavy-ion induced damage in lithium fluoride single-crystals. *Nucl. Instrum. Methods Phys. Res. Sect. B Beam Interact. Mater. Atoms* **2000**, *166*, 581–585. [CrossRef]

33. Ramos, S.; Bonardi, N.; Canut, B.; Bouffard, S.; Della-Negra, S. Damage creation in α-Al_2O_3 by MeV fullerene impacts. *Nucl. Instrum. Methods Phys. Res. Sect. B Beam Interact. Mater. Atoms* **1998**, *143*, 319–332. [CrossRef]

34. Daya, D.B.; Hallén, A.; Eriksson, J.; Kopniczky, J.; Papaleo, R.; Reimann, C.; Håkansson, P.; Sundqvist, B.; Brunelle, A.; Della-Negra, S.; et al. Radiation damage features on mica and L-valine probed by scanning force microscopy. *Nucl. Instrum. Methods Phys. Res. Sect. B Beam Interact. Mater. Atoms* **1995**, *106*, 38–42. [CrossRef]

35. Daya, D.B.; Hallén, A.; Håkansson, P.; Sundqvist, B.; Reimann, C. Scanning force microscopy study of surface tracks induced in mica by 78.2-MeV ^{127}I ions. *Nucl. Instrum. Methods Phys. Res. Sect. B Beam Interact. Mater. Atoms* **1995**, *103*, 454–465. [CrossRef]

36. Ishikawa, N.; Taguchi, T.; Okubo, N. Hillocks created for amorphizable and non-amorphizable ceramics irradiated with swift heavy ions: TEM study. *Nanotechnology* **2017**, *28*, 445708. [CrossRef]

37. Ishikawa, N.; Taguchi, T.; Kitamura, A.; Szenes, G.; Toimil-Molares, M.E.; Trautmann, C. TEM analysis of ion tracks and hillocks produced by swift heavy ions of different velocities in $Y_3Fe_5O_{12}$. *J. Appl. Phys.* **2020**, *127*, 055902. [CrossRef]

38. Ishikawa, N.; Okubo, N.; Taguchi, T. Experimental evidence of crystalline hillocks created by irradiation of CeO_2 with swift heavy ions: TEM study. *Nanotechnology* **2015**, *26*, 355701. [CrossRef]

39. Karczewski, J.; Riegel, B.; Gazda, M.; Jasinski, P.; Kusz, B. Electrical and structural properties of Nb-doped $SrTiO_3$ ceramics. *J. Electroceramics* **2009**, *24*, 326–330. [CrossRef]

40. Ziegler, J.F.; Biersack, J.P.; Littmark, U. *The Stopping and Range of Ions in Solids*; Pergamon: New York, NY, USA, 1985.

41. Ziegler, J.F. SRIM-2003. *Nucl. Instrum. Methods B* **2004**, *219–220*, 1027–1036. [CrossRef]

42. Sachan, R.; Pakarinen, O.H.; Liu, P.; Patel, M.K.; Chisholm, M.F.; Zhang, Y.; Wang, X.L.; Weber, W.J. Structure and band gap determination of irradiation-induced amorphous nano-channels in $LiNbO_3$. *J. Appl. Phys.* **2015**, *117*, 135902. [CrossRef]

43. Canut, B.; Ramos, S.; Bonardi, N.; Chaumont, J.; Bernas, H.; Cottereau, E. Defect creation by MeV clusters in LiNbO$_3$. *Nucl. Instrum. Methods Phys. Res. Sect. B Beam Interact. Mater. Atoms* **1997**, *122*, 335–338. [CrossRef]

44. Canut, B.; Ramos, S.; Brenier, R.; Thevenard, P.; Loubet, J.; Toulemonde, M. Surface modifications of LiNbO$_3$ single crystals induced by swift heavy ions. *Nucl. Instrum. Methods Phys. Res. Sect. B Beam Interact. Mater. Atoms* **1996**, *107*, 194–198. [CrossRef]

45. Meftah, A.; Costantini, J.M.; Khalfaoui, N.; Boudjadar, S.; Stoquert, J.P.; Studer, F.; Toulemonde, M. Experimental determination of track cross-section in Gd$_3$Ga$_5$O$_{12}$ and comparison to the inelastic thermal spike model applied to several materials. *Nucl. Instrum. Methods B* **2005**, *237*, 563–574. [CrossRef]

46. Lang, M.; Lian, J.; Zhang, F.; Hendriks, B.W.; Trautmann, C.; Neumann, R.; Ewing, R.C. Fission tracks simulated by swift heavy ions at crustal pressures and temperatures. *Earth Planet. Sci. Lett.* **2008**, *274*, 355–358. [CrossRef]

47. Bursill, L.A.; Braunshausen, G. Heavy-ion irradiation tracks in zircon. *Philos. Mag. A* **1990**, *62*, 395–420. [CrossRef]

48. Costantini, J.; Miro, S.; Lelong, G.; Guillaumet, M.; Toulemonde, M. Damage induced in garnets by heavy ion irradiations: A study by optical spectroscopies. *Philos. Mag.* **2017**, *98*, 312–328. [CrossRef]

49. Meftah, A.; Brisard, F.; Costantini, J.M.; Hage-Ali, M.; Stoquert, J.P.; Studer, F.; Toulemonde, M. Swift heavy ions in magnetic insulators: A damage-cross-section velocity effect. *Phys. Rev. B* **1993**, *48*, 920–925. [CrossRef]

50. Zhang, J.; Lang, M.; Lian, J.; Liu, J.; Trautmann, C.; Della-Negra, S.; Toulemonde, M.; Ewing, R.C. Liquid-like phase formation in Gd$_2$Zr$_2$O$_7$ by extremely ionizing irradiation. *J. Appl. Phys.* **2009**, *105*, 113510. [CrossRef]

51. Jensen, J.; Dunlop, A.; Della-Negra, S.; Pascard, H. Tracks in YIG induced by MeV C$_{60}$ ions. *Nucl. Instrum. Methods Phys. Res. Sect. B Beam Interact. Mater. Atoms* **1998**, *135*, 295–301. [CrossRef]

52. Li, W.; Wang, L.; Sun, K.; Lang, M.; Trautmann, C.; Ewing, R.C. Porous fission fragment tracks in fluorapatite. *Phys. Rev. B* **2010**, *82*, 144109. [CrossRef]

53. Karlušić, M.; Ghica, C.; Negrea, R.; Siketić, Z.; Jakšić, M.; Schleberger, M.; Fazinić, S. On the threshold for ion track formation in CaF$_2$. *New J. Phys.* **2017**, *19*, 023023. [CrossRef]

54. Grygiel, C.; Lebius, H.; Bouffard, S.; Quentin, A.; Ramillon, J.M.; Madi, T.; Guillous, S.; Been, T.; Guinement, P.; Lelièvre, D.; et al. Online in situ x-ray diffraction setup for structural modification studies during swift heavy ion irradiation. *Rev. Sci. Instrum.* **2012**, *83*, 13902. [CrossRef] [PubMed]

55. Li, W.; Rodriguez, M.D.; Kluth, P.; Lang, M.; Medvedev, N.; Sorokin, M.; Zhang, J.; Afra, B.; Bender, M.; Severin, D.; et al. Effect of doping on the radiation response of conductive Nb–SrTiO$_3$. *Nucl. Instrum. Methods Phys. Res. Sect. B Beam Interact. Mater. Atoms* **2013**, *302*, 40–47. [CrossRef]

56. Lang, M.; Djurabekova, F.; Medvedev, N.; Toulemonde, M.; Trautmann, C. Fundamental Phenomena and Applications of Swift Heavy Ion Irradiations Comprehensive Nuclear Materials (Second Edition). *arXiv* **2020**, arXiv:2001.03711.

57. Sattonnay, G.; Thome, L.; Sellami, N.; Monnet, I.; Grygiel, C.; Legros, C.; Tétot, R. Experimental approach and atomistic simulations to investigate the radiation tolerance of complex oxides: Application to the amorphization of pyrochlores. *Nucl. Instrum. Methods Phys. Res. Sect. B Beam Interact. Mater. Atoms* **2014**, *326*, 228–233. [CrossRef]

Publisher's Note: MDPI stays neutral with regard to jurisdictional claims in published maps and institutional affiliations.

quantum beam science

Review

Control and Modification of Nanostructured Materials by Electron Beam Irradiation

Shun-Ichiro Tanaka

Micro System Integration Center (μSIC), Tohoku University, Aramaki, Sendai 980-0845, Japan; shunichiro.tanaka.d2@tohoku.ac.jp

Abstract: I have proposed a bottom-up technology utilising irradiation with active beams, such as electrons and ions, to achieve nanostructures with a size of 3–40 nm. This can be used as a nanotechnology that provides the desired structures, materials, and phases at desired positions. Electron beam irradiation of metastable θ-Al_2O_3, more than 10^{19} e/cm^2s in a transmission electron microscope (TEM), enables the production of oxide-free Al nanoparticles, which can be manipulated to undergo migration, bonding, rotation, revolution, and embedding. The manipulations are facilitated by momentum transfer from electrons to nanoparticles, which takes advantage of the spiral trajectory of the electron beam in the magnetic field of the TEM pole piece. Furthermore, onion-like fullerenes and intercalated structures on amorphous carbon films are induced through catalytic reactions. δ-, θ-Al_2O_3 ball/wire hybrid nanostructures were obtained in a short time using an electron irradiation flashing mode that switches between 10^{19} and 10^{22} e/cm^2s. Various α-Al_2O_3 nanostructures, such as encapsulated nanoballs or nanorods, are also produced. In addition, the preparation or control of Pt, W, and Cu nanoparticles can be achieved by electron beam irradiation with a higher intensity.

Keywords: electron irradiation; excited reaction field; transmission electron microscope; nanomaterials; manipulation; nanostructure; Al; Al_2O_3

Citation: Tanaka, S.-I. Control and Modification of Nanostructured Materials by Electron Beam Irradiation. *Quantum Beam Sci.* **2021**, *5*, 23. https://doi.org/10.3390/qubs5030023

Academic Editor: Akihiro Iwase

Received: 20 April 2021
Accepted: 12 July 2021
Published: 21 July 2021

Publisher's Note: MDPI stays neutral with regard to jurisdictional claims in published maps and institutional affiliations.

1. Introduction

The size range of several tens of nanometres represents a transition region from "top-down" to "bottom-up" processes in nanotechnology. The author has proposed a bottom-up technology utilising active beam irradiation to achieve nanostructures with a size of 3–40 nm. Recent developments in focusing and scanning technologies for active beams, such as electrons or ions, in tabletop apparatuses enable the evolution and control of various types of nanostructures, which can provide desirable hybrid structures, materials, and phases at the desired positions on the nanometre scale.

The source of the electron beam used by the author's group is a transmission electron microscope (TEM) equipped with either an LaB_6 filament or a field emission gun. Although TEM has been widely used as an analysis tool for studying nanostructure and element distributions, we consider a specimen stage of 3 mm in diameter as a reaction field, and focused electrons with an intensity more than 50 times higher than that used under normal observation conditions. The electron irradiation intensity ranged from 5×10^{19} to 4×10^{23} e/cm^2s (8×10^4–6×10^8 A/m^2) in our experiment, and we succeeded in producing oxide-free nanoparticles via electron irradiation of the oxide and facilitated their subsequent manipulation.

In this paper, three topics on the manipulation and control of ceramics and metals by electron beam irradiation are discussed: (1) the preparation of Al nanoparticles and their nanostructure evolution starting from metastable Al_2O_3, (2) their manipulation by electrons, and (3) the effects of electron irradiation on other nanoparticles, such as Cu, Pt, and W, to prepare nanofilms. This review is based on papers published between 1995 and 2005.

2. Electron Excited Reaction Field

2.1. Overview

In the "Exploratory Research for Advanced Technology" (ERATO) "Tanaka Solid Junction Project," JST [1], held in 1993–1998, I commenced an innovative challenge to fabricate nano-/microstructures by irradiation with an energy beam such as electrons and ions, based on the proposed concept of an "excited reaction field." One of the characteristics of such field is that the beams are obtained on the specimen stage of a TEM for electrons, and on the milling/thinning stage for ions. In other words, observation or specimen preparation apparatuses are utilised for their beam source and excited reaction fields. Irradiation with these energy beams has the following merits: first, it facilitates the selection of the site/energy/reaction. Second, it can induce nonequilibrium/catalytic reactions, making it possible to manipulate atomic clusters, synthesise nanomaterials, control the nanostructure and phase, and modify the nanospace.

2.2. Effects of Electron Beam Irradiation on θ-Al_2O_3

The normal electron beam intensity for observation by a TEM equipped with an LaB_6 filament is on the order of 10^{18} e/cm^2s (1600 A/m^2) for bright-field imaging and 5×10^{16} e/cm^2s s (80 A/m^2) for selected area diffraction measured by a fluorescent plate. We increased the electron beam density by increasing the current in the condenser lens to between 10^{19} and 10^{22} e/cm^2s to enable electron irradiation of the nanomaterials.

When an electron beam irradiates metastable θ-Al_2O_3 particles at an intensity of 10^{20} e/cm^2s in a TEM, successive reactions occur: ① Al nanoparticle inducement, ② rotation, revolution, and migration to coalesce/embed, and ③ formation of onion-like fullerenes and Al intercalation, as shown in Figure 1. Furthermore, ④ θ- or δ-Al_2O_3 nanostructures are obtained under flashing-mode irradiation (rapid switching between 10^{19} and 10^{22} e/cm^2s).

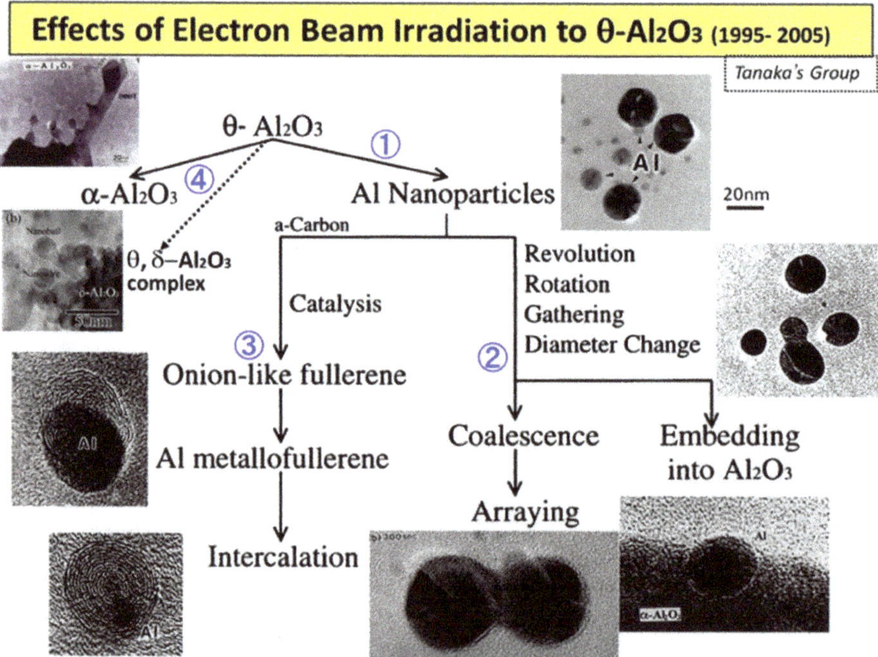

Figure 1. Evolution of the reactions and nanostructures induced by electron beam irradiation of metastable θ-Al_2O_3 in TEM. Numbers ①–④ denote the reaction routes explained [1].

3. Nanostructure Evolution

3.1. Al Nanoparticles

The starting material consists mainly of a metastable θ-Al_2O_3 monoclinic structure accompanied by orthorhombic δ-Al_2O_3 (Al_2O_3 allotrope), and it is different from stable α-Al_2O_3 with a trigonal structure. The θ-Al_2O_3 powder was synthesised using the vaporised metal combustion method (Admatechs Company Limited) [2].

Electron irradiation (2.1×10^{20} e/cm^2s for 50 s) of one θ-Al_2O_3 particle with a diameter of 100 nm on the φ3 mm specimen stage of the TEM led to the formation of Al nanoparticles with a diameter of 2–20 nm and a stable α-Al_2O_3 particle, as shown in Figure 2 [3]. The TEM used was a JEOL JEM-2010 equipped with an LaB_6 filament and a specimen chamber under a vacuum of 10^{-7} Pa, obtained by a direct coupling sputter ion pump. Under these conditions, the electrons cut the Al-O bond in θ-Al_2O_3, which was followed by the loss of oxygen to the vacuum and Al atom recombination to form nanoparticles, as schematically shown in Figure 3. In the electron-irradiated area, transformation or rearrangement from θ-Al_2O_3 to stable α-Al_2O_3 also occurred.

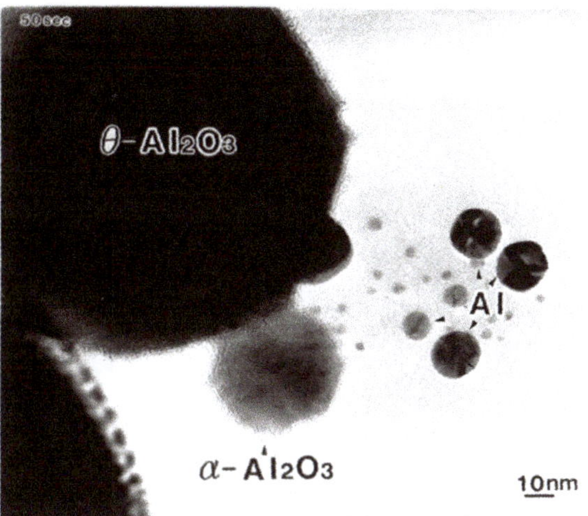

Figure 2. Al nanodecahedra of 10–17 nm in diameter and α-Al_2O_3 nanoparticles obtained by electron irradiation of metastable θ-Al_2O_3 with an intensity of 2.1×10^{20} e/cm^2s in TEM [3].

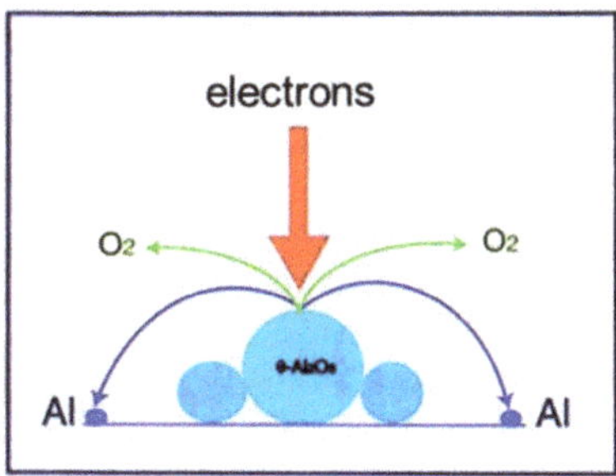

Figure 3. Schematic of Al nanodecahedra formation by electron irradiation in a TEM. Electrons cut the Al-O bond of metastable θ-Al_2O_3, which decomposes into Al and O atoms, and finally the Al atoms recombine to form twinned decahedra surrounded by {111} surfaces.

θ- and δ-Al_2O_3 appear as low-temperature phases in the allotropic transformation from γ-Al_2O_3 to α-Al_2O_3 upon heating. The starting powder used in this experiment was obtained by melting Al metal powder, vaporisation, collision of droplets, and quenching into the metastable phase [2]. The powder particles had a spherical shape with an average diameter of approximately 10 μm. No stable α-Al$_2$O$_3$ structure was observed by X-ray diffraction, and an equilibrium thermodynamic consideration was invalid for metastable θ-Al$_2$O$_3$, where the binding energy for Al-O was lower than that for α-Al$_2$O$_3$. The reaction was expected to proceed via a nonequilibrium route. Electron irradiation promoted the decomposition of metastable θ-Al$_2$O$_3$, recombination of Al atoms, loss of a part of oxygen atoms into vacuum, and transformation to stable α-Al$_2$O$_3$, as shown in Figure 3, which resembles one stile of the electron-stimulated desorption.

Al nanoparticles have a twinned decahedron structure surrounded by {111} surfaces, which has been reported in typical face-centred cubic noble metals, such as Au, Pt, or Ag, whereas no report has been published on Al because of its easily oxidised surface. Decahedra appeared with diameters in the range of 10–20 nm, and further electron irradiation of a set of nanodecahedra enabled their manipulation, as discussed in the next section.

3.2. Manipulation

In the ERATO project, TEM has been used not only as an observation apparatus but also as a tool for the manipulation of nanostructures by electrons on the specimen stage. The 200 keV JEM-2010 TEM has a pole piece with a magnetic field of 10^4 Gauss from top to bottom, where the specimen stage is located slightly above the centre. Based on the kinetic features analysed by Horiuchi et al. [4], the momentum transfer from electrons to nanoparticles is the source of their movement. The spiral trajectory of the electrons causes Al nanoparticles on the specimen stage to experience both a tangential force to rotate and revolve and a centripetal force to migrate, bond, and embed. The features around the specimen stage and the driving forces are illustrated in Figure 4.

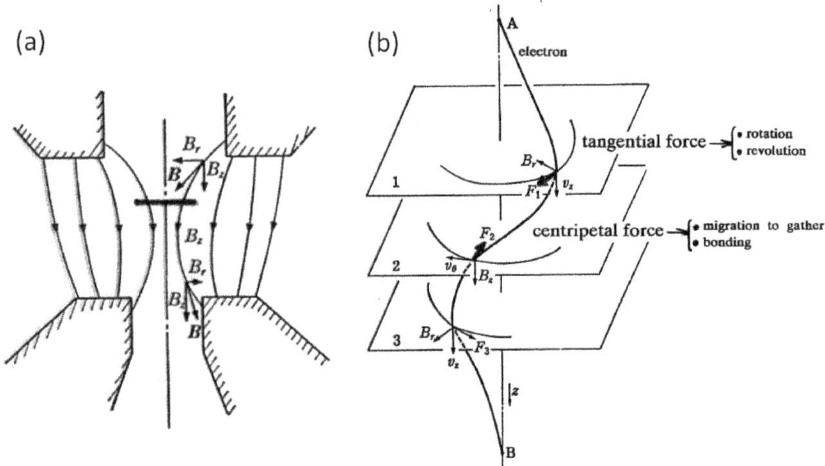

Figure 4. Model of the interaction between electrons and particles on the specimen stage in TEM. The magnetic field inside a pole piece is 10^4 Gauss in a 200 keV TEM. (**a**) Electrons follow a spiral trajectory in the magnetic field to transfer forces or momentum such that (**b**) the tangential force results in rotation and revolution and the centripetal force induces migration to gather, bond, and embed the particles.

3.2.1. Migration and Bonding

A set of Al nanodecahedra migrated and bonded to the irradiation centre of the electron beam, as shown in Figure 5. Analysis by superposition before and after irradiation, as shown in Figure 5c, revealed that migration, diameter decrease, revolution, and bonding occurred in the course of irradiation over 1200 s at 2.1×10^{20} e/cm^{2}s s, which was mainly due to the centripetal force, as shown in Figure 4. The bonding step of the two nanoparticles is shown in Figure 6, where the (111) planes are aligned parallel by rotation before necking (a) and then necking with Σ3 twinning (b) to eventual coalescence (c) [5–7].

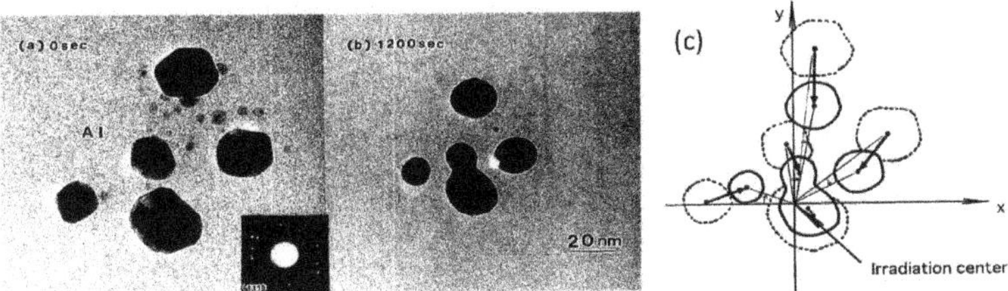

Figure 5. Effects of electron irradiation of a set of Al nanodecahedra inducing migration to the irradiation centre, as well as revolution and rotation. The intensity was 2.1×10^{20} e/cm^{2}s for (**a**) 0 s and (**b**) 1200 s. A schematic view of the effect is shown in (**c**) [6].

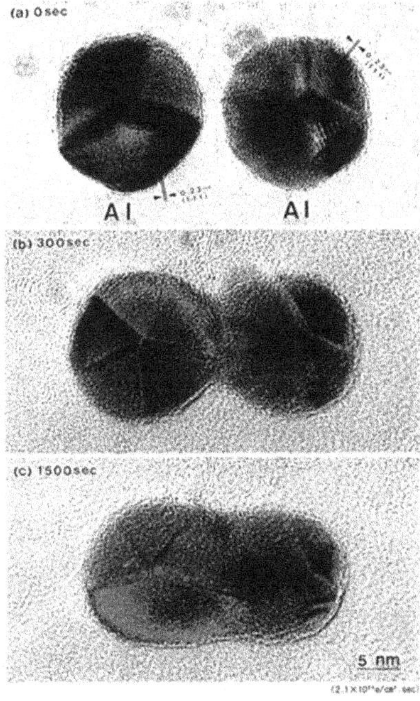

Figure 6. Al nanodecahedra (**a**) bonded by electron irradiation with an intensity of 2.1×10^{20} e/cm^{2}s after 300 s (**b**) and coalesced after 1500 s (**c**). The (111) planes in two Al nanodecahedra were aligned parallel by rotation and exhibited a Σ3 coincidence site lattice (CSL) boundary around the neck at 300 s [5,7].

3.2.2. Rotation and Revolution

The force acting on the nanoparticle in the magnetic field was also tangential, as well as centripetal, as shown in Figure 4, inducing a clockwise rotation of the nanoparticle on the specimen stage. The speed of rotation measured by the change in angle increased as the irradiation intensity increased for irradiation times less than 1000 s, as shown in Figure 7. The saturation of the rotation with longer exposure is considered to stem from a decrease in the diameter of the particle [6,8].

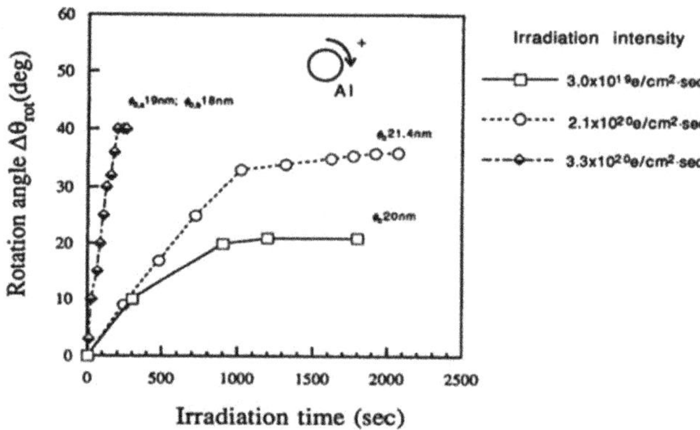

Figure 7. Rotation of Al nanodecahedra and its dependence on the irradiation intensity ranging from 3.0×10^{19} to 3.3×10^{20} e/cm^{2}s. The diameter of the nanoparticles is 20 nm, denoted on the line as φ_0 [6].

To confirm the effects of the tangential force on the nanoparticle, the magnetic direction in the pole piece was changed from bottom to top by reforming the TEM. Figure 8 shows the revolution behaviour of nanoparticles in both directions of the magnetic field. The clockwise and counterclockwise revolutions of the Al nanoparticles clearly depended on the direction of the magnetic field, and their speed increased as the irradiation intensity increased. The revolution of the nanoparticles was accompanied by their migration, as shown in Figure 5c.

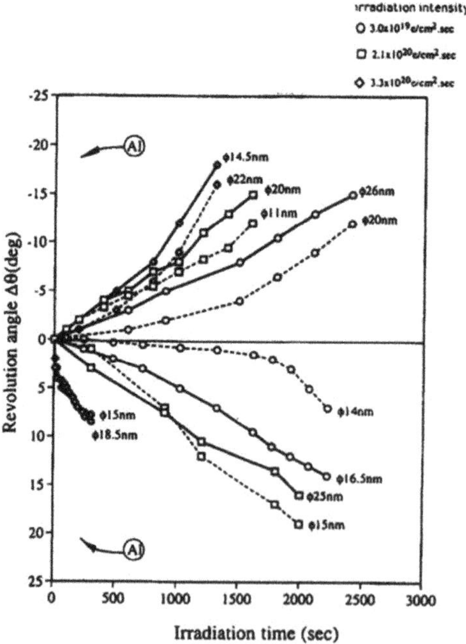

Figure 8. Clockwise (+) and counterclockwise (−) revolution of Al nanoparticles during electron irradiation controlled by changing the magnetic field direction. The case shown in Figure 5 induced the clockwise revolution, whereas reversal of the magnetic field (bottom to top) induced the opposite direction of revolution. Here, φ indicates the diameter of the initial Al particle [8].

3.2.3. Embedding

The forces discussed in Sections 3.2.1 and 3.2.2 manipulate the nanoparticles to cause migration and embedding into the substrate. Figure 9 shows an example of an Al nanoparticle embedded in the α-Al$_2$O$_3$ matrix following electron irradiation at 10^{20} e/cm^{2}s, which is attributed to an epitaxial relationship between the two substances. This technique can be utilised for the implantation of catalysts, such as Pt nanoparticles, at the desired position in a matrix [9].

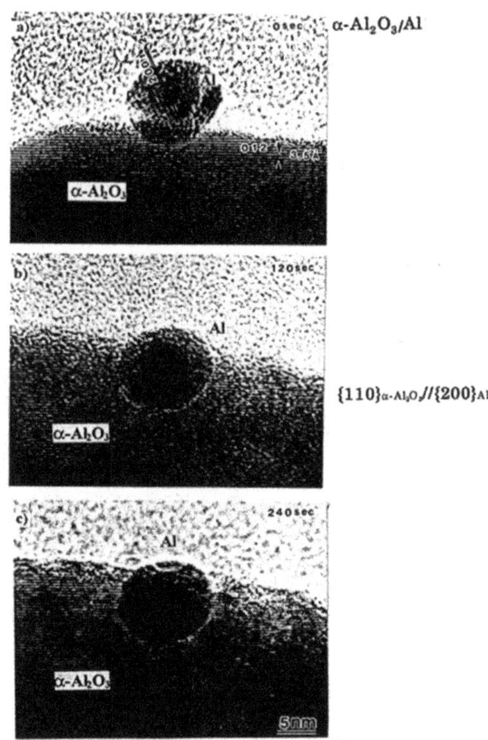

Figure 9. Al nanoparticle migrated and embedded into the α-Al$_2$O$_3$ matrix by electron irradiation of the order of 10^{20} e/cm^{2}s for (**a**) 0 s (**b**) 120 s and (**c**) 240 s. The α-Al$_2$O$_3$/Al interface structure exhibits a $\{11\text{–}20\}\alpha$-Al$_2$O$_3$//$\{200\}$Al epitaxial relationship even after electron irradiation for 240 s [9].

3.3. Fullerene and Intercalation

A series of nanostructures were formed on the amorphous carbon nanofilm of the specimen mesh, which was suspended on a Cu grid. Electron irradiation induced the formation of an onion-like fullerene nanostructure under the Al nanoparticles by the catalysis effect and promoted the intercalation of Al atoms between the graphite shells.

3.3.1. Onion-Like Fullerene

Giant onion-like fullerenes were induced from the amorphous carbon nanofilms under the Al nanodecahedra by electron beam irradiation, as shown in Figure 10. A catalytic reaction nucleated a graphitic flake along the edge of the Al nanoparticle, and prolonged electron irradiation induced the growth of the fullerene, shrinking of the nanoparticles, and intercalation of Al atoms between the shells, as shown in Figures 11 and 12 [5,10].

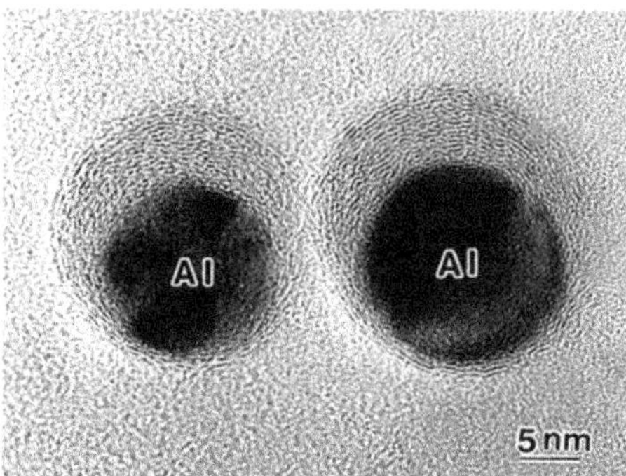

Figure 10. Giant onion-like fullerenes induced under Al nanoparticles by electron irradiation at an intensity of 3.0×10^{19} e/cm^{2}s for 2050 s. The Al nanoparticles are surrounded by onion-like fullerene shells [10].

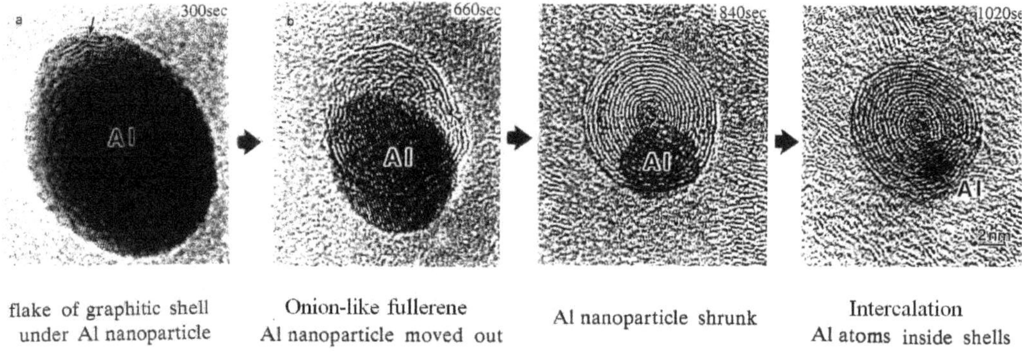

Figure 11. Series of reactions between Al nanoparticles and the amorphous carbon nanofilm (used as a specimen holder for TEM) induced by electron irradiation with an intensity of 10^{20} e/cm^{2}s. Reactions proceed from the catalytic formation of (**a**) a graphitic shell and (**b**) an onion-like fullerene under the Al nanoparticle, followed by (**c**) shrinking of the nanoparticle and (**d**) intercalation of Al atoms inside the shell [10].

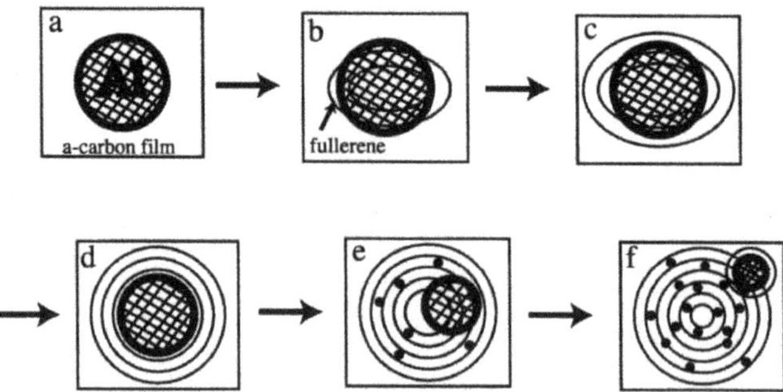

Figure 12. Schematic of a series of interaction behaviours between an Al nanoparticle and an amorphous carbon film under electron irradiation (a:initial). The following steps occurred: a nucleus of the giant onion-like fullerene was first induced under the Al particles (**b**). Al nanoparticles were encapsulated in the giant onion-like fullerene (**c,d**). Al nanoparticles moved outside of the giant onion-like fullerene (**e**), which also induced a new giant onion-like fullerene (**f**). Finally, intercalation progressed as Al atoms migrated inside the giant onion-like fullerene shells (**e,f**) [10].

3.3.2. Intercalation

To verify Al atom intercalation into the graphitic shells, an analysis using energy dispersive spectroscopy (EDS), for which the JEM-2010 TEM was equipped, was conducted on the electron irradiated specimen. Figure 13 shows the presence of Al atoms in the circled area of 7 nm in diameter, accompanied by C and Cu in the grid. Moreover, the expansion of the graphite (002) lattice spacing confirmed the presence of Al atoms between the layers, as shown in Figure 14. However, the growth saturated below the composition of Al_2C_6 owing to the blocking effect of the coexisting Cu atoms or the constraint of passing through multiple carbon layers. Electron energy loss spectra (EELS) also suggested the partial replacement of carbon atoms by Al or Cu atoms through σ bond, sp^2, decrease, as shown in Figure 15 [10–12].

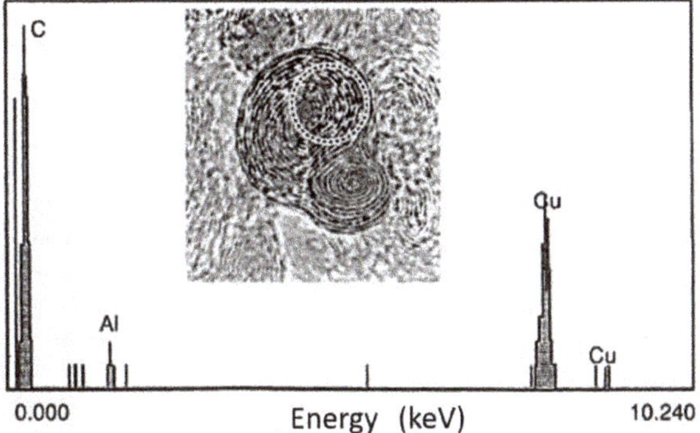

Figure 13. EDS analysis of the area inside the dotted circle in the nanostructure shown in Figure 11. The electron probe size was 7 nm in diameter. An Al peak was detected, indicating the possibility of intercalation, whereas the Cu peaks came from the Cu grid of the membrane [10].

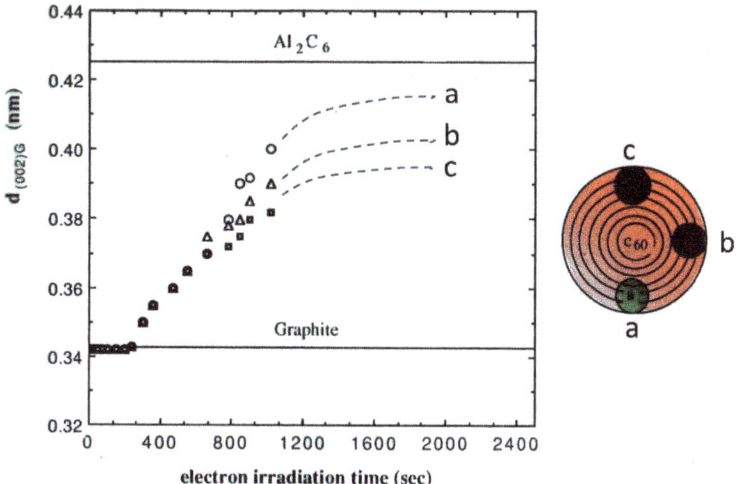

Figure 14. Expansion of the onion-like graphite lattice spacing $d_{(002)G}$ due to Al intercalation, under electron beam irradiation of 1.0×10^{20} e/cm^{2}s. The spacing increased with electron irradiation time and seemed to saturate at 0.425 nm, which coincides with the spacing of the compound Al_2C_6. The limit of lattice expansion stems either from the blocking effect due to the coexistence of Cu atoms or is constrained by multiple carbon layers [11].

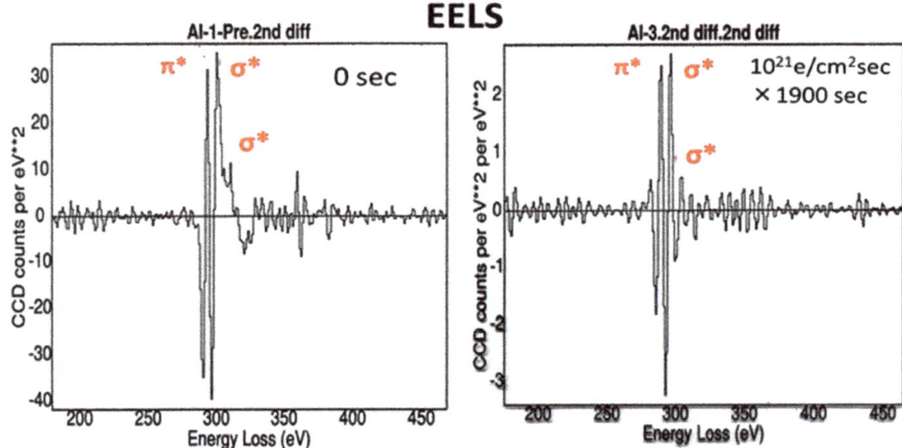

Figure 15. Second derivative of the EELS of the intercalated structure formed under electron irradiation of 1.0×10^{21} e/cm²s for 1900 s. No change is observed in the π-bond between layers after irradiation, whereas the σ-bond, sp² in-plane, decreased, which suggests the Al and Cu intercalated atoms partially replaced the carbon atoms [12].

4. Al_2O_3 Derivatives

Continuous electron irradiation of metastable θ-Al_2O_3 at 2.1×10^{20} e/cm²s produced Al nanoparticles and stable α-Al_2O_3 particles, as shown in Figure 2. When the intensity and area of electron irradiation changed abruptly (i.e., flashing of the beam), other-shaped nanosized aluminium oxides were obtained. In this section, I present the oxide derivatives of Al_2O_3 oxides with different shapes and phases.

4.1. α-Al_2O_3 Nanorods

Although α-Al_2O_3 with a polygonal shape was deposited beside θ-Al_2O_3 by the recombination of Al and O atoms, the α-Al_2O_3 nanorods shown in Figure 16 grew from the surface of the parent α-Al_2O_3. The rods had a faceted structure surrounded by {100} surfaces and grew in the {110} direction. Figure 16 was obtained in situ, in which the nanorods grew without an amorphous oxide film.

4.2. δ-, θ-Al_2O_3 Nanoballs/Nanowires

When metastable Al_2O_3 was irradiated by electrons with a higher density over a short time, the structure was retained, but a different nanostructure was also obtained. Kameyama and Tanaka abruptly switched between intensities of 5.5×10^{22} and 5×10^{19} e/cm²s in a so-called "flashing mode," maintaining each state for 0.1 s, which facilitated the formation of θ-, δ-Al_2O_3 nanorods/nanoballs, as shown in Figure 17, respectively [13].

Both θ-Al_2O_3 nanowires and nanoballs grown using flashing-mode electron beam irradiation are shown in Figure 17. They connected and grew toward the irradiation centre from the original θ-Al_2O_3 particle surface. The nanowire and nanoball had a (111)//(101) epitaxial relationship, as shown in Figure 18. The θ-Al_2O_3 nanoball was a sphere of 30 nm in diameter covered with an amorphous Al-O layer, suggesting that the impact of a higher irradiance electron beam resulted in a temperature rise of approximately 400 °C, reaction, and rapid cooling without phase transformation. The δ-Al_2O_3 nanowires and nanoballs shown in Figure 19 were obtained by applying the same flashing electron beam to δ-Al_2O_3 particles. The diameter of the δ-Al_2O_3 particles was 20 nm, which was slightly smaller than that of the θ-Al_2O_3 particles, and no explicit epitaxy relation was observed.

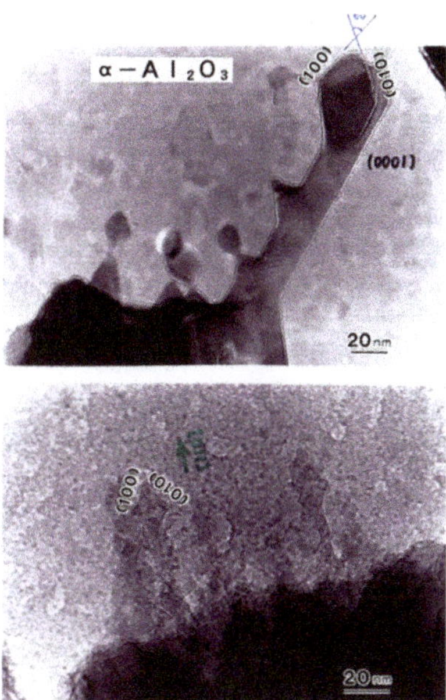

Figure 16. α-Al$_2$O$_3$ rods epitaxially grown from the θ-Al$_2$O$_3$ surface by electron beam irradiation at 2.1 × 10^{20} e/cm^{2}s, as shown in Figure 2. The rods grew in the {110} direction and were faceted by {100} surfaces [3].

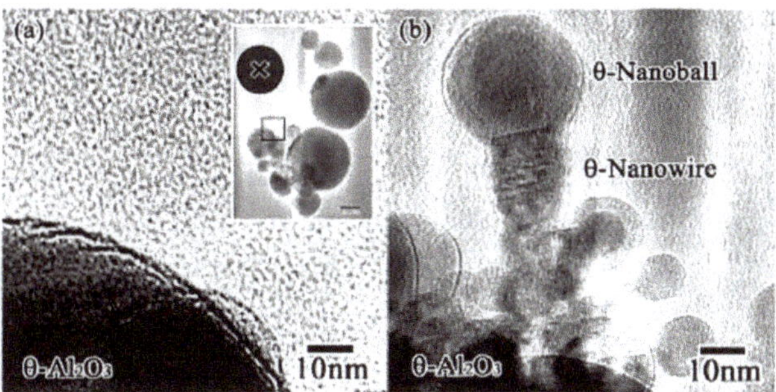

Figure 17. θ-Al$_2$O$_3$ nanowire and nanoball grown under flashing mode electron beam irradiation (i.e., rapid switching between intensities of 5.5 × 10^{22} and 5 × 10^{19} e/cm^{2}s). They connected and grew toward the irradiation centre from the original θ-Al$_2$O$_3$ particle surface. The nanowire grew epitaxially maintaining the same plane as the parent θ-Al$_2$O$_3$ particle. (**a**) θ-Al$_2$O$_3$ particles before irradiation, with X indicating the irradiation centre. (**b**) θ-Al$_2$O$_3$ nanowire and nanoball grown after electron beam irradiation [13].

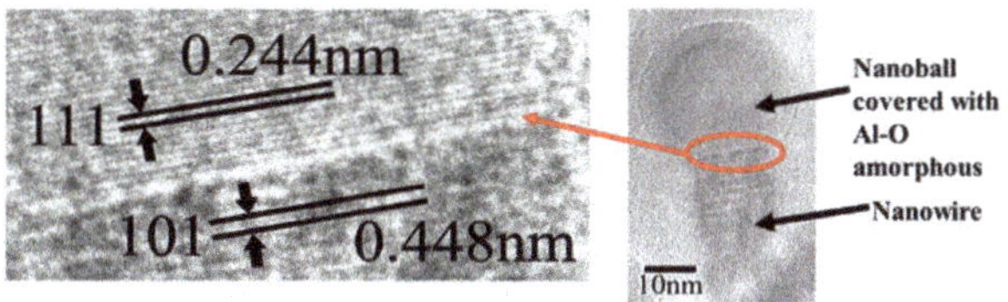

Figure 18. The interface of the θ-Al$_2$O$_3$ nanowire and nanoball, shown in Figure 17, exhibits a (111)//(101) epitaxial relationship. The nanoball is covered with an amorphous Al-O layer [14].

4.3. Al$_2$O$_3$ Nanoparticle Encapsulation

The surface of the Al$_2$O$_3$ particles was easily covered by an amorphous hydrocarbon layer through a reaction with residual gases in the TEM, where CO, H$_2$O, and H$_2$ remained even in a highly evacuated atmosphere. When the particles were irradiated by electrons through the outer layer, their inner volume was pulverised into smaller nanoparticles. An example is shown in Figure 20, for the case in which δ-Al$_2$O$_3$ particles with 200 nm in diameter were transformed into δ- and α-Al$_2$O$_3$ nanoballs with 2–20 nm in diameter, encapsulated by a hydrocarbon skin. This structure was generated by irradiation with an intensity of 1.6×10^{20} e/cm^{2}s, which was as high as that used for nanoparticle preparation and manipulation. Nanoparticle encapsulation technology may be applicable to drug delivery systems in medicine [15].

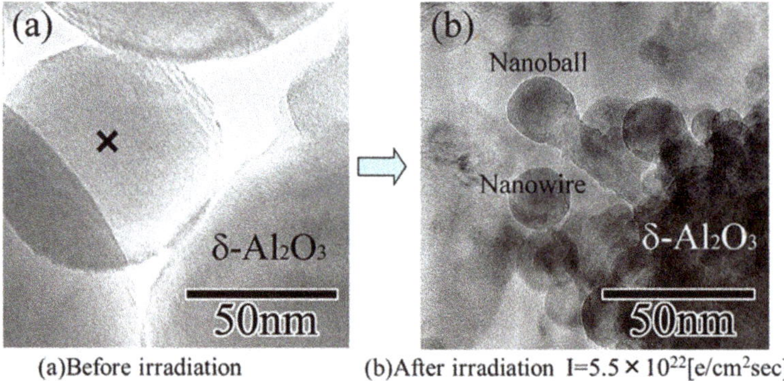

(a)Before irradiation   (b)After irradiation I=5.5 × 10^{22}[e/cm^{2}sec]

Figure 19. δ-Al$_2$O$_3$ nanowire and nanoball grown under flashing mode electron beam irradiation with intensities switching between 5.5×10^{22} and 5×10^{19} e/cm^{2}s. They connected and grew toward the irradiation centre from the original δ-Al$_2$O$_3$ particle surface. (**a**) δ-Al$_2$O$_3$ particles before irradiation, with X indicating the irradiation centre. (**b**) δ-Al$_2$O$_3$ nanowire and nanoball grown after electron beam irradiation [13].

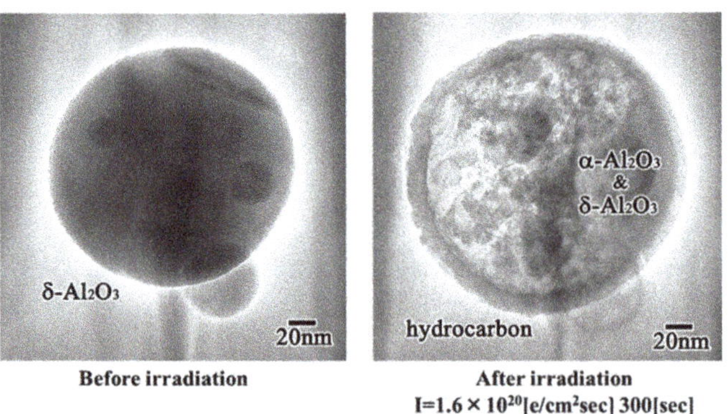

Before irradiation   After irradiation
I=1.6 × 10^{20}[e/cm^{2}sec] 300[sec]

Figure 20. δ- and α-Al$_2$O$_3$ nanoball-encapsulated structures obtained from δ-Al$_2$O$_3$ particles by electron beam irradiation at 1.6×10^{20} e/cm^{2}s for 300 s. An outer amorphous hydrocarbon layer is formed from residual gas contaminants in TEM [15].

5. Other Nanoparticles

5.1. W Nanoparticles and Manipulation

In Section 2, a novel method for preparing oxide-free Al nanoparticles from metastable oxides using electron beam irradiation was discussed. This method can be extended to other nanoparticles of easy oxide-forming elements, such as W. W has a heavier specific weight of 19.3 g/cm^3, and W-O has a larger bonding enthalpy such that a higher electron irradiation intensity is required to obtain W nanoparticles from WO_3. Although the fundamental electron optics in the TEM were the same as those used for Al nanoparticles, electrons from the field emission source provided a higher intensity of 10^{23} e/cm^{2}s than the 10^{20} e/cm^{2}s obtained from the LaB$_6$ filament. Using a Hitachi HF-2000 TEM equipped with a field emission gun with an intensity of 4×10^{23} e/cm^{2}s, Tamou and Tanaka reported the formation of W nanoparticles with an average diameter of 4.3 nm, as shown in Figure 21. The EELS spectra showed no oxygen atoms on the W surface, as shown in Figure 22 [16].

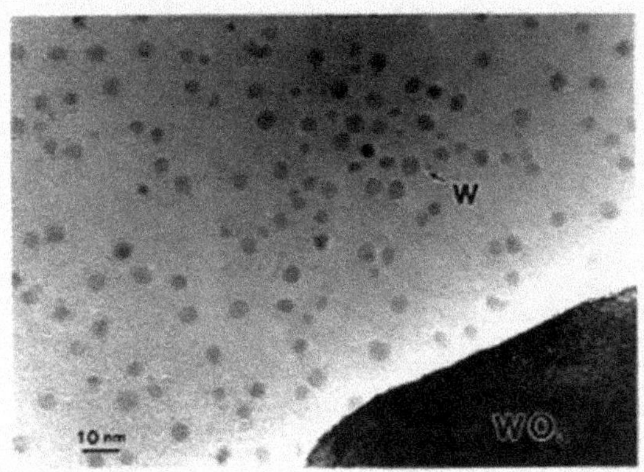

Figure 21. W nanoparticles deposited by electron beam irradiation of a WO_3 particle at 4×10^{23} e/cm^{2}s (6×10^8 A/m^2) for a few seconds. The diameter of the W particles ranged between 2 and 6 nm with an average of 4.3 nm [16].

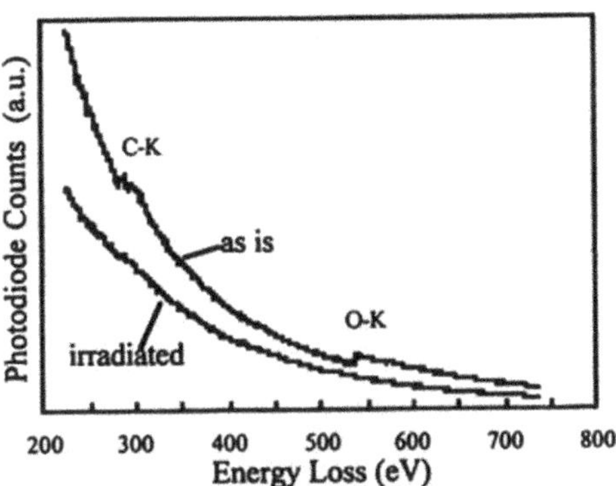

Figure 22. EELS spectra before and after electron irradiation of WO_3. Electron irradiation induced the disappearance of the O-Kα peak to form pure W nanoparticles [16].

5.2. W Migration to Bond and Fullerene Formation

Further electron irradiation of two W nanoparticles, obtained as in Figure 21 at 1.9×10^{21} e/cm^{2}s, which is an irradiation 10 times higher than that used to form Al as shown in Figure 6, induced migration, bonding, and coalescence, as shown in Figure 23 [16]. Graphitic shells also nucleated beneath the W nanoparticles from the amorphous carbon film and grew to onion-like fullerene, which is the same phenomenon observed with Al nanoparticles shown in Figures 10–12 [16].

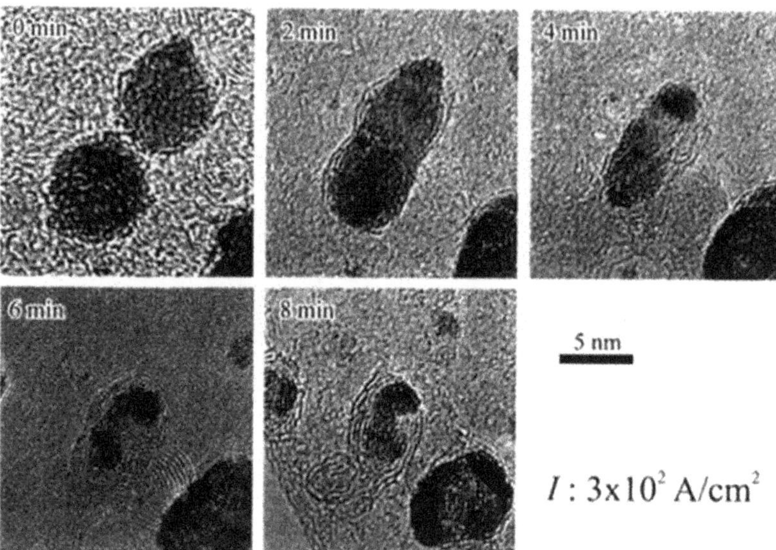

Figure 23. Effects of electron beam irradiation of W nanoparticles on an amorphous carbon film at 1.9×10^{21} e/cm^{2}s (300 A/cm^2). W nanoparticles migrated together and coalesced, followed by fullerene formation between the W particles and the carbon film. The general features of nanostructure evolution were the same as those observed with irradiation of Al nanoparticles shown in Figure 10 [16].

5.3. Bonding of Pt and Cu Nanoparticles

Electron irradiation of a group of Pt and Cu nanoparticles induced bonding to form nanofilms, as shown in Figures 24–26. In these cases, nanoparticles were prepared by Ar ion sputtering with a diameter of 10 nm for Pt and 50 nm for Cu. The irradiation intensity for Pt was the same as that for Al, whereas it was 100 times higher for Cu. Bonded Pt/Pt mainly showed three stable Σ3 twin boundaries. The Cu particles migrated to the irradiation centre and bonded, as shown in Figures 26 and 27. The driving force was also the momentum transfer from electrons in the pole piece of the TEM, as shown in Figures 4 and 5 [17–20].

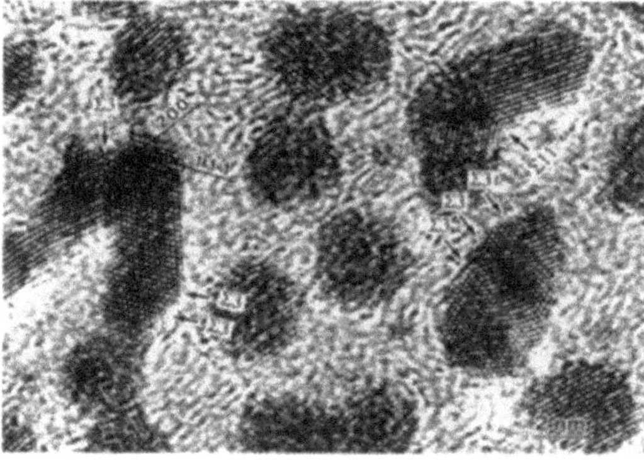

Figure 24. Pt nanoparticles bonded by electron irradiation with an intensity of 2.1×10^{20} e/cm^{2}s for 700 s. The bonded Pt/Pt nanoparticles had tilt boundaries of Σ3 and Σ11 [19].

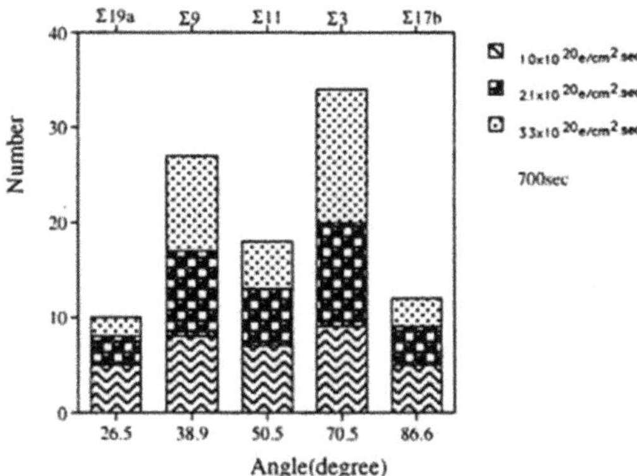

Figure 25. Histogram of tilt boundaries under electron irradiation at three intensities from 1.0 to 3.3×10^{20} e/cm^{2}s for 700 s. $\Sigma3$ CSL boundaries were predominant as a low energy structure [19].

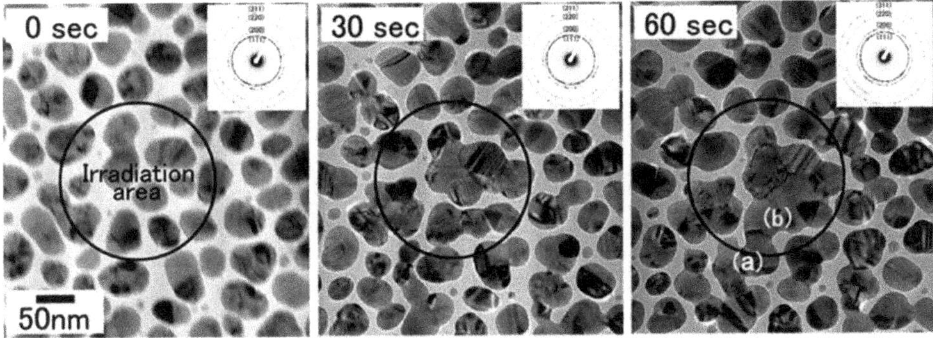

Figure 26. Change of the bright-field images and electron diffraction patterns of Cu nanoparticles irradiated with electrons at an intensity of 5.5×10^{22} e/cm^{2}s for 60 s. Cu nanoparticles migrated to the irradiation centre and bonded with each other in the marked irradiation area. The electron diffraction patterns, typical Cu Debye rings, did not change during irradiation. The nanostructures of the bonded interface (i.e., CSL) boundary, were obtained in regions (**a**) and (**b**) after 60 s of irradiation, as shown in Figure 27 [20].

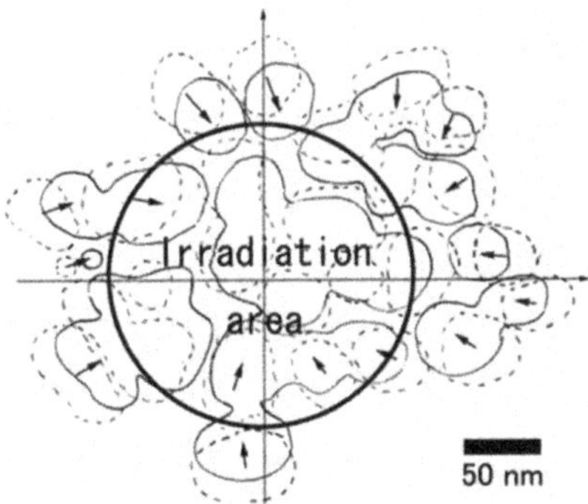

Figure 27. Superposed view of Cu nanoparticle migration and bonding during the irradiation times of 0 s (dotted line) and 30 s (solid line). The circle indicates the electron beam irradiation area of 200 nm diameter. Nanoparticles migrated toward the irradiation centre and finally bonded together. Unbonded small nanoparticles seemed to revolve clockwise around the irradiation centre [20].

6. Nature of Nanoparticle Manipulation and Nanostructure Modification by Electron Beam Irradiation

6.1. Temperature Rise in Al Nanoparticle Manipulation

In Section 3.2, various types of manipulation of Al nanoparticles on the TEM specimen stage were explained. These were migration, bonding, rotation, revolution, and embedding of the nanoparticles, and the driving forces were explained as tangential and centripetal forces, as shown in Figure 4. Another possibility of manipulation is the temperature rise caused by electron irradiation to induce their movement. Xu and Tanaka [10] estimated the temperature rise at the stage as 10° C at most, based on Equation (1) using Fisher's theory [21]:

$$Tm - Tg = r^2 I_0 \Delta E[a_0 + \ln(R/r)^2]/4kz, \tag{1}$$

where Tm is the maximum temperature of the carbon film, Tg is the temperature of the Cu support grid, namely Tm − Tg is the temperature rise by electron beam irradiation, r is the radius of the irradiation beam, I_0 is the intensity of the irradiation beam (10^{20} e/cm^{2}s), ΔE is the energy loss of the incident electron in the carbon film, when it is <1000 nm thick, a_0 is Euler's constant (0.5772), R is the distance between the irradiation beam centre and the Cu grid bar, k is the thermal conductivity of carbon, and z is the thickness of the carbon film (20 nm).

The heating effect in the localised area under the irradiation condition of 10^{20} e/cm^{2}s can be a minor effect, and the Lorenz force or the momentum transfer from electrons and ionised atoms is the major effect of the manipulation. This effect is clearly supported by the counterclockwise revolution caused by the magnetic field change, as shown in Figure 8 [8].

6.2. Temperature Rise in Al$_2$O$_3$ Nanocomplex and W Nanoparticles

When θ-Al$_2$O$_3$ was irradiated at a density of 10^{19}–10^{20} e/cm^{2}s, Al and α-Al$_2$O$_3$ were formed, as shown in Figures 2 and 16, where the reaction proceeded with a small temperature increase of the order of 10 °C, as shown in Section 6.1. On the contrary, the flashing mode of electron irradiation by rapid switching between intensities of 5.5 × 10^{22} and 5 × 10^{19} e/cm^{2}s was applied to θ- and δ-Al$_2$O$_3$ to induce Al$_2$O$_3$ nanoball/nanowire complexes, as shown in Figures 17 and 19 [13]. The higher electron beam intensity increased the temperature by more than 300 °C, as calculated through I_0 in Equation (1) by maintaining 5.5 × 10^{22} e/cm^{2}s even in a short time of less than 0.1 s. This temperature increase was also predicted by Yokota et al. [22]. Rapid and concentrated heat input at the localised resulted in an Al–O recombined nanoball/nanowire complex with epitaxy at the interface, as shown in Figure 18 [14].

Heavy atoms such as W required a higher irradiation intensity of 4 × 10^{23} e/cm^{2}s to obtain W nanoparticles, as shown in Figure 21 [16]. Although the binding energy of W–O in the starting material WO$_3$ was smaller than that of Al–O, the W atom is ten times heavier than the Al atom, and required a higher energy for sputtering. A temperature rise was also expected in this irradiation condition, but no melting was observed because of its higher melting point, 3680 K.

6.3. Lorentz Force in Nanoparticle Manipulation

To discuss the mechanism of nanoparticle manipulation in a TEM, the interaction between electrons and nanoparticles on the specimen stage in the magnetic field was analysed. The TEM used in this study was 200 keV JEM-2010, which has a pole piece with a magnetic field of 10^4 Gauss from top to bottom, where the specimen stage is located slightly above the centre plane. In Figure 4, the electron trajectory is illustrated schematically by Horiuchi et al. [4], and the Lorentz forces F_1 and F_2 arise from the magnetic components Br and Bz, respectively, with spirally running electrons. The specimen stage is located between planes 1 and 2, and Al, W, Pt, and Cu nanoparticles experience both a tangential force to rotate and revolve and a centripetal force to migrate, bond, and embed. The momentum transfer from electrons to nanoparticles is the source of this movement.

The Lorentz force exerted on one electron, F_e, was roughly estimated by Equation (2)

$$F_e = m_e v_e^2 / r, \tag{2}$$

where m_e is the mass of one static electron as 9.1094×10^{-31} kg, v_e is the velocity of electrons, considering the relativistic effects at 200 kV as $v_e = v_{200}\, 1.3914 = 2.900 \times 10^8$ m/s, and r is the distance between the nanoparticle and the irradiation centre. When assuming the experimental case of Al nanoparticles shown in Figure 5, r = 60 nm, the Lorentz force from one electron F_e was 1.277×10^{-6} N. The total Lorentz force, F, to the Al nanoparticle of 20 nm in diameter with an irradiation time of 1200 s at 10^{20} e/cm^{2}s was estimated to be 9.63×10^5 N. Although this is the maximum value, which occurs when electrons travel from plane 1 to 2 and the driving force of nanoparticle manipulation changes the direction from tangential to centripetal, it is too high for nanoparticle movement. The author proposes the following reasons: although the electron density was measured on the fluorescent plate beneath the stage, accelerated electron velocity decreased while travelling inside a TEM, and electrons lost their kinetic energy through ionisation of the wall by their impact. The Lorentz force decreased by at least 1/100. The existence of a friction force between nanoparticles and a substrate carbon film could also be one of the causes. The effective cross-section of the nanoparticle might be considered, which decreases the impact of the electrons.

The Lorentz force for nanoparticle manipulation is also valid for W, Pt, and Cu, as shown in Figures 23, 24, and 26. Although the time to bond is different depending on the density of the weight, electron irradiation focusing to the localised region will be a candidate technology for fabricating circuits or functional dots by nanoparticle arrays.

7. Summary of the Nanostructure Evolution and Manipulations in the Electron Excited Field

Research conducted by my group on nanostructure evolution by electron beam irradiation from 1995 to 2005 was reviewed. I have utilised electron beams in TEM to synthesise nanomaterials and manipulate their nanostructures, in addition to observing and analysing nanostructures. An overview of the effects of electron irradiation is presented in Figures 28 and 29, where the abscissa is expressed as the electron irradiation intensity on a logarithmic scale. The electron beam was focused for synthesis and manipulation up to 10^{19}–10^{24} e/cm^{2}s, which is higher than 10^{16} e/cm^{2}s generally used for electron diffraction and 10^{18} e/cm^{2}s used for bright-field imaging.

Figure 28 shows that electron irradiation of metastable θ-Al$_2$O$_3$ provides oxide-free Al nanoparticles, rod-like α-Al$_2$O$_3$, and encapsulated nanoparticles, whereas flashing mode provides θ-, δ-Al$_2$O$_3$ nanoball/nanowire complexes. The formation of W nanoparticles from WO$_3$ requires a higher intensity of more than 10^{23} e/cm^{2}s. Electrons traveling in a spiral trajectory in the magnetic field of the pole piece transfer momentum to the Al nanoparticles enable various types of manipulation, such as migration, bonding, rotation, revolution, embedding, fullerene formation, and intercalation. The intensity is also more than 100 times higher than that of normal observation conditions, as shown in Figure 29. The combination of such syntheses and manipulation will provide more complicated nanostructures for future applications.

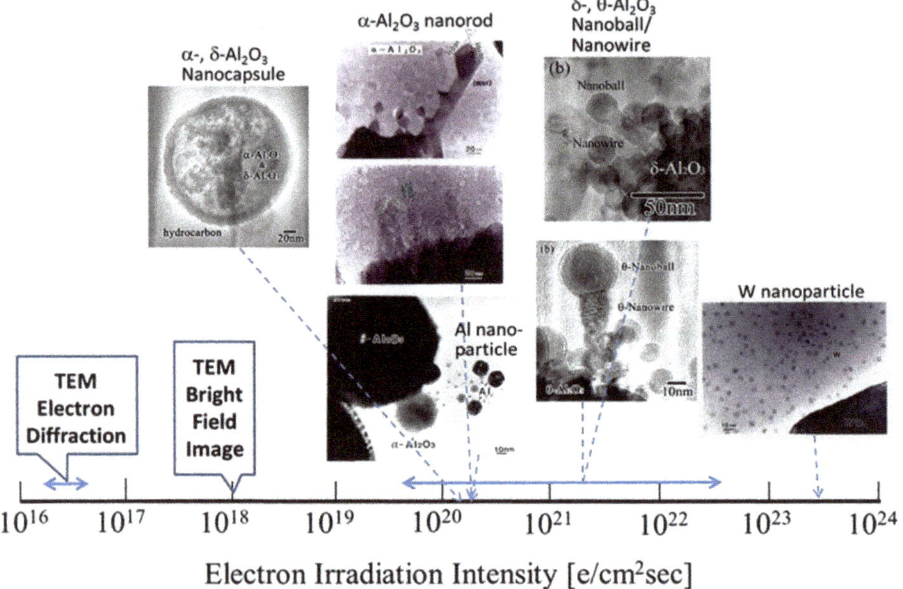

Figure 28. Nanostructured materials obtained by electron irradiation in TEM. Nanoparticles and nanosized oxides can be induced in an electron excited reaction field. The electron irradiation intensity ranged from 10^{20}–10^{23} e/cm^{2}s depending on the specific gravity of the materials and the metal-oxygen binding enthalpy of the starting oxide.

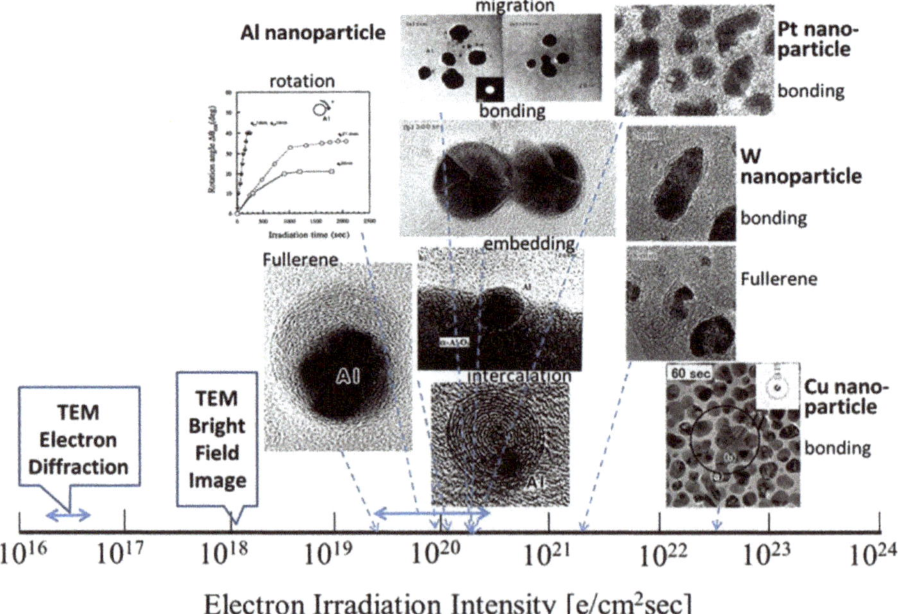

Figure 29. Manipulation of nanomaterials by electron irradiation in a TEM. Nanostructures can be controlled in an electron excited reaction field through migration, bonding, rotation, revolution, embedding, fullerene formation and intercalation. The electron irradiation intensity ranged between 10^{19}–10^{23} e/cm^{2}s depending on the size and weight of the material.

8. Recent Development in Control and Manipulation of Nanostructured Materials

In this review, pioneering works by the author's group published in 1995–2005 are summarised as a tool for nanomaterial control and manipulation at the TEM room temperature stage. In these works, mediate-accelerating keV was initially used, followed by accompanying magnetic field, and focusing electrons to the localised area. Although there are several works on the effects of electron irradiation, reviews on this topic are scarce. For example, lattice defects such as point defects or stacking faults are introduced as radiation damage in the region of MeV electron irradiation, for which an ultra-high voltage TEM has been used as an experimental simulation. Krasheninnikov et al. published an excellent review paper on the effects of ion and electron irradiation, collecting more than 680 papers [23], which contained derivation and simulation for nanostructured materials. Accompanying magnetic field and focusing electrons to the localised area in TEM are unique technologies for manipulation, which were partly covered in the papers by Zheng et al. [24] and Andres et al. [25]. Zheng et al. reported that the trapping force for one nanoparticle was on the order of 10^{-9} N in the electron density gradient of 10^{18-19} e/cm^{2}s [24] which is the same order of magnitude as discussed in Section 6.3. Simulation by first-principle theory using the density of state is important for predicting the formation, growth, and coalescence of nanoparticles [25].

9. Conclusions

The author's group succeeded in inducing the formation of Al nanoparticles by electron irradiation of metastable θ-Al$_2$O$_3$, followed by manipulation of the nanoparticles. A series of phenomena was observed without heating using high-resolution TEM (HRTEM), with an electron beam intensity as low as 10^{19-20} e/cm^{2}s. The typical morphology of the nanoparticles was that of a nanodecahedron surrounded by (111) surfaces with twins. Electron beam irradiation of a group of Al nanoparticles promoted their rotation, revolution, and migration to the irradiation centre, resulting in bonding and embedding into an α-Al$_2$O$_3$ matrix. The driving force is considered to be the momentum transfer from electrons spiralling across the pole piece of the HRTEM in a strong magnetic field to the Al nanoparticles. When nanoparticles were placed on an amorphous carbon film, onion-like fullerene nucleated and grew beneath them, and finally, a metallofullerene or Al-atom-intercalated structure was formed by electron irradiation.

To develop the manipulation technology for other types of nanoparticles, an electron beam was used to irradiate Cu nanoparticles of 10–50 nm in diameter at an irradiation intensity of 5.5×10^{22} e/cm^{2}s, Pt nanoparticles at 1.0–3.3×10^{20} e/cm^{2}s, and W nanoparticles derived from WO$_3$ at 9×10^{20} to 4×10^{23} e/cm^{2}s. The behaviour of Cu, Pt, and W nanoparticles under electron irradiation was similar to that of Al, and a nanofilm was finally formed. The CSL boundary structures at the bonded interface of Cu nanoparticles were found to be unstable Σ7 and Σ13b, which are different from the stable Σ3 obtained in Al and Pt with weaker electron beam irradiation.

The possible scientific contribution of electron irradiation is the synthesis of materials in a metastable state through a nonequilibrium reaction in vacuum, as well as the induction of hybridised nano-/mesostructures. It also enables the study of the nature of materials in a pristine and controlled environment, for example, without the formation of an oxide. From the viewpoint of application to devices, nanosized balls, dots, wires, and tube-forming three-dimensional structured circuits may be used as elements of nanodevices, and chemically active points embedded in the substrate for use as a catalyst can be achieved by the manipulation of electron irradiation. With respect to industrial applications, our technologies will contribute to the development of micro- and nanoelectromechanical systems, memories, photonics, battery electrodes, H$_2$ storage, and more.

Funding: This research partly received funding from ERATO, JST.

Acknowledgments: The author gratefully acknowledges the research scientists at Tanaka Solid Junction Project, JST, BingShe Xu (now: Shaanxi Univ. Sci. Tech.) and Yoshitaka Tamou (now Mitsubishi Materials), and graduate students at Nagoya Institute of Technology, Toru Kameyama (now: Sony GMO Corp.) and Yoshiki Miki (now Nippon Giant Tire Co., Ltd.).

Conflicts of Interest: The author declares no conflict of interest.

References

1. TANAKA Solid Junction. Available online: https://www.jst.go.jp/erato/en/research_area/completed/tky_P.html (accessed on 12 July 2021).
2. Abe, S.; Ogawa, M.; Kondo, K. Synthesis of Spherical Oxide Particles by Vaporized Metal Combustion Process. In Proceedings of the 4th Silicon for the Chemistry Industry; Norwegian University of Science and Technology: Trondheim, Norway, 1998; Volume 4, pp. 329–338.
3. Xu, B.S.; Tanaka, S.-I. Phase transformation and bonding of ceramic nanoparticles in a TEM. *Nanostruct. Mater.* **1995**, *6*, 727–730. [CrossRef]
4. Horiuchi, S. High resolution transmission electron microscope. *Kyoritsu* **1988**, *2*, 14.
5. Xu, B.S.; Tanaka, S.-I. Creation and Bonding of Al Nanodecahedron by an Electron Beam. In Proceedings of the Interface 6th Beijing Conference and Exhibition on Instrumental Analysis, Beijing, China, 24–27 October 1995; Volume 111, pp. A31–A32.
6. Xu, B.S.; Tanaka, S.-I. Control of Nanoscale Interphase Boundaries by an Electron Beam. *Mater. Sci. Forum* **1996**, *207*, 137–140. [CrossRef]
7. Tanaka, S.-I. Control of nanointerfaces by energy beam irradiation. *Mater. Sci. Forum* **2007**, *558–559*, 971–974.
8. Xu, B.S.; Tanaka, S.-I. Behavior and bonding mechanisms of aluminum nanoparticles by electron beam irradiation. *Nanostruct. Mater.* **1999**, *12*, 915–918. [CrossRef]
9. Xu, B.S.; Tanaka, S.-I. Embedding of Al nanoparticle in Al_2O_3 matrix by electron beam. *Supramol. Sci.* **1998**, *5*, 227–228. [CrossRef]
10. Xu, B.S.; Tanaka, S.-I. Formation of giant onion-like fullerenes under Al nanoparticles by electron irradiation. *Acta Mater.* **1998**, *46*, 5249–5257. [CrossRef]
11. Xu, B.S.; Tanaka, S.-I. Formation of a new electric material: Fullerene/metal polycrystalline film. *MRS Proc.* **1997**, *472*, 179–184. [CrossRef]
12. Tanaka, S.-I. Graphite Intercalation Formed by Electron Irradiation Having KeV Energy. In *Symposium of Spring Meeting*; The Japan Institute of Metals and Materials: Tokyo, Japan, 2019; p. 3.
13. Kameyama, T.; Tanaka, S.-I. Alumina nanostructures formed by electron beam irradiation. *J. Ceram. Soc. Jpn.* **2004**, *112-1*, S930–S932.
14. Kameyama, T.; Tanaka, S.-I. Nanostructure Derivation by Electron Beam Irradiation. In Proceedings of the International Conference on Novel Materials Processing by Advanced Electromagnetic Energy Sources (MAPEES'04), Osaka, Japan, 19–22 March 2004.
15. Kameyama, T. Control of Ceramic Nanostructures by Electron Beam Irradiation. Master's Thesis, Nagoya Institute of Technology, Nagoya, Japan, March 2004. (In Japanese).
16. Tamou, Y.; Tanaka, S.-I. Formation and coalescence of tungsten nanoparticles under electron beam irradiation. *Nanostruct. Mater.* **1999**, *12*, 123–126. [CrossRef]
17. Xu, B.S.; Tanaka, S.-I. Pt cluster bonding and fullerene formation in HRTEM. In Proceedings of the International 6th Beijing Conference and Exhibition on Instrumental Analysis, Beijing, China, 24–27 October 1995; pp. A33–A34.
18. Xu, B.S.; Iwamoto, C.; Tanaka, S.-I. Bonding Behavior of Platinum Nanoparticles under Electron Beam Irradiation in a High Resolution Transmission Electron Microscope. In Proceedings of the Interface Science and Materials Interconnection JIMIS-8, Toyama, Japan, 1–4 July 1996; pp. 395–398.
19. Xu, B.S.; Tanaka, S.-I. Formation and bonding of platinum nanoparticles controlled by high energy beam irradiation. *Scr. Mater.* **2001**, *44*, 2051–2054. [CrossRef]
20. Miki, Y.; Tanaka, S.-I. Migration and Bonding of Cu Nanoparticles Induced by Electron Beam Irradiation. In Proceedings of the International Conference on New Frontiers of Process Science and Engineering in Advanced Materials, Kyoto, Japan, 24–26 November 2004; pp. 52–57.
21. Fisher, S.B. On the temperature rise in electron irradiated foils. *Radiat. Effects* **1970**, *5*, 239–243. [CrossRef]
22. Yokota, T.; Murayama, M.; Howe, J.M. In situ transmission-electron-microscopy investigation of melting in submicron Al-Si alloy particles under electron-beam irradiation. *Phys. Rev. Lett.* **2003**, *91*, 265504. [CrossRef] [PubMed]
23. Krasheninnikov, A.V.; Nordlund, K. Ion and electron-induced effects in nanostructured materials. *J. Appl. Phys.* **2010**, *107*, 071301. [CrossRef]
24. Zheng, H.; Mirsaidov, U.M.; Wang, L.-W.; Matsudaira, P. Electron Beam Manipulation of Nanoparticles. *Nano Lett.* **2012**, *12*, 5644–5648. [CrossRef] [PubMed]
25. Andres, J.; Longo, E.; Gouveia, A.F.; Costa, J.P.C.; Gracia, L.; Oliveira, M.C. *In-situ* formation of metal nanoparticles through electron beam irradiation: Modeling real materials from First-Principles calculations. *J. Mater. Sci. Eng.* **2018**, *7*, 1000461. [CrossRef]

quantum beam science

Review

Self-Organized Nanostructures Generated on Metal Surfaces under Electron Irradiation

Keisuke Niwase

Department of Physics, Hyogo University of Teacher Education, Kato 673-1494, Hyogo, Japan; niwase@hyogo-u.ac.jp; Tel.: +81-795-44-2210

Abstract: Irradiation of high-energy electrons can produce surface vacancies on the exit surface of thin foils by the sputtering of atoms. Although the sputtering randomly occurs in the area irradiated with an intense electron beam of several hundred nanometers in diameter, characteristic topographic features can appear under irradiation. This paper reviews a novel phenomenon on a self-organization of nanogrooves and nanoholes generated on the exit surface of thin metal foils irradiated with high doses of 360–1250 keV electrons. The phenomenon was discovered firstly for gold irradiated at temperatures about 100 K, which shows the formation of grooves and holes with widths between 1 and 2 nm. Irradiation along [001] produces grooves extending along [100] and [010], irradiation along [011] gives grooves along [100], whereas no clear grooves have been observed for [111] irradiations. By contrast, nanoholes, which may reach depths exceeding 20 nm, develop mainly along the beam direction. The formation of the nanostructures depends on the irradiation temperatures, exhibiting an existence of a critical temperature at about 240 K, above which the width significantly increases, and the density decreases. Nanostructures formed for silver, copper, nickel, and iron were also investigated. The self-organized process was discussed in terms of irradiation-induced effects.

Keywords: electron irradiation; metal surface; sputtering; groove; hole; self-organization; pattern

Citation: Niwase, K. Self-Organized Nanostructures Generated on Metal Surfaces under Electron Irradiation. *Quantum Beam Sci.* **2021**, *5*, 4. https://doi.org/10.3390/qubs5010004

Received: 3 December 2020
Accepted: 14 January 2021
Published: 19 January 2021

Publisher's Note: MDPI stays neutral with regard to jurisdictional claims in published maps and institutional affiliations.

1. Introduction

Generation of nanosized structures is technologically important, and the controlled formation of structures in solids on a nanometer scale is critical to modern technology [1–3]. Such structures may have properties different from those of the bulk materials. Nanohole-based nanomaterials with a size less than the wavelength of an excitation laser beam, for example, are promised for applications such as chemical and biological sensing, membrane biorecognition, unique optical responses under laser excitation, etc.

It should be worthwhile to know the smallest size of nanostructured material. The scanning tunneling microscope has been employed to control the deposition of atoms on or extraction of atoms from the surface on an atomic scale [4–6]. However, there seems to be a minimum width when one tries to produce deep nanoholes or nanogrooves.

High-energy particle irradiation is one of method to generate interesting nanomaterials [7,8]. Aggregation of surface vacancies produced homogeneously by ion sputtering may produce pits that can become rather deep. Electron irradiation, which can induce back sputtering on the surface of thin foil specimens, is also one of the techniques used for nanometer scale etching, lithography and hole formation, and intense convergent electron beams have been utilized, so far. Deep nanoholes have been formed in MgO using a convergent nanosized electron beam, the smallest widths so far achieved being about 1 nm [9,10]. Parallel electron beams, on the other hand, of 500–1000 nm diameter have been shown to produce pits on the exit surface of Au(111) foils by sputtering over the electron energy range of 0.4–1.1 MeV [11].

Self-organization is one method to produce characteristic structures and several self-organization phenomena of defect clusters under high-energy particle irradiations such as voids, bubbles, and stacking fault tetrahedra have been reported so far [12–16]. Spontaneous well-ordered periodicity can be developed on a broad surface by ion beam sputtering and a numerical model has been proposed as the formation process [17].

The present paper reviews our studies on the evolution of nanosized structures resulting from the sputtering of atoms from the exit surface of thin metal specimens during homogeneous electron irradiation, focusing on a novel self-organization phenomenon, which can occur on the electron irradiated surface [18–22] and give additional data on the temperature dependence of the formation of nanoholes for gold.

2. Materials and Methods

In the present study, we used the wedge-shaped specimens produced from 99.998% Au, 99.9999% Ag, 99.999% Cu, 99.998% Ni, and 99.997% Fe foils with a thickness of about 100 μm. After eliminating lattice defects by annealing, they were thinned by jet-polishing to make wedge-shaped specimens. The electron irradiations were performed by using two transmission electron microscopes (TEM) of JEOL-ARM 1250 and JEOL-4000FX, both equipped with GATAN liquid-nitrogen cooling stages.

The irradiations were done along crystallographic directions near the [100], [110], or [111] direction of grains with surface orientations near {100}, using a beam of about 200–800 nm diameter as illustrated in Figure 1 and a flux density of about 10^{24} electrons m^{-2} s^{-1} with total fluences ranging from 10^{27} to 10^{28} electrons m^{-2}. The irradiation direction was adjusted by tilting the specimen. Although the surface orientation and the observation directions were not perfectly aligned with the crystallographic orientations in question, such crystallographic indices are used hereafter to show the surface orientation and observation directions. The irradiation time ranged from several 10 s to several 1000 s. A small fraction of incident electrons (10^{-6} for 1.0 MeV electrons) can give sufficient energy with target atoms on the electron exit surface to cause sputtering [23]. The electron microscopy observations were made with strongly reduced beam currents to suppress additional sputtering. Detailed experimental and theoretical study on the sputtering yield and angular distribution of sputtered atoms, among others, for gold foils in a high-voltage electron microscope have been given by Cherns et al. [23].

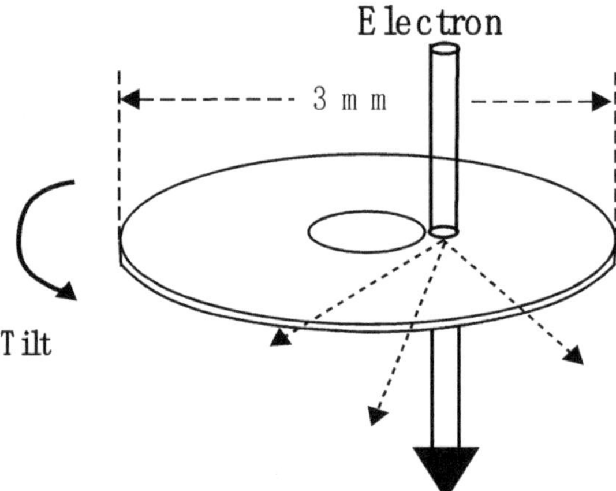

Figure 1. A schematic view of a disk specimen of 3 mm diameter. The irradiations were carried out at thin parts near a hole with an electron beam of about 200–800 nm diameter.

3. Results

3.1. Gold

3.1.1. Self-Organization of Nanostructure

Figure 2 shows a typical self-organized pattern observed for an Au(001) foil irradiated with 400 keV electrons along [001] direction at 95 K with an electron beam of about 300 nm in diameter. The micrograph was taken under a kinematic and slightly under-focus condition. Clear bright images extending along [100] and [010] with a width of 1–2 nm can be observed. Stereomicroscopy revealed that these bright lines are grooves formed on the electron exit surface.

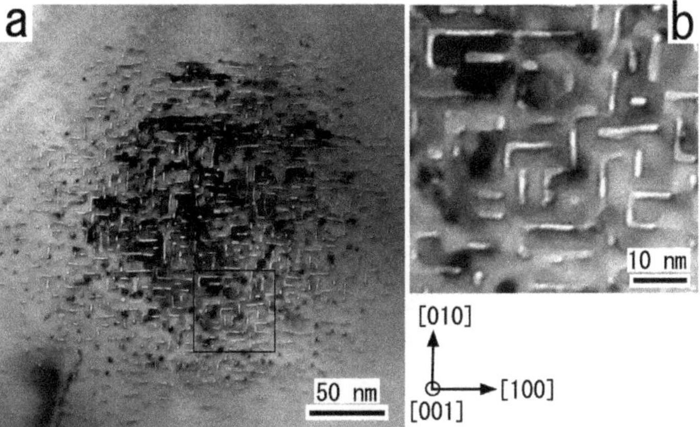

Figure 2. (**a**) TEM micrographs of an Au(001) foil irradiated along [001] with 400 keV electrons at 95 K; (**b**) a magnified view of black square area in photo (**a**).

Figure 3 shows an oblique observation of an Au(001) foil, which was irradiated along [112] with 400 keV electrons at 95 K and observed along [001] by tilting the specimen. Deep nanoholes and hillocks formed on the exit surface of an Au(001) foil can be seen. Aspect ratio on the nanohole is very high in spite of the small diameter of 1–2 nm.

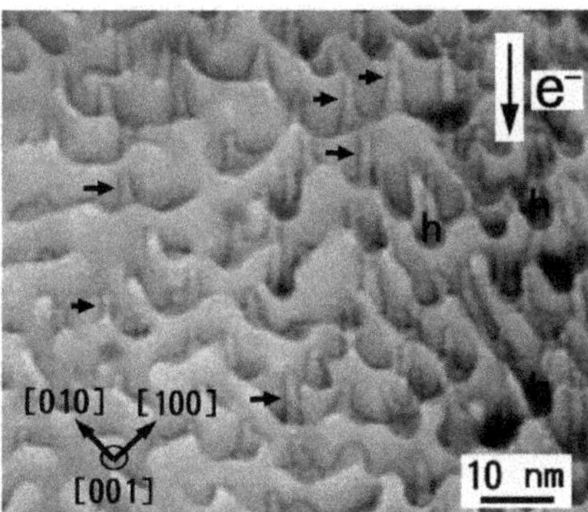

Figure 3. Nanoholes and hillocks generated on the exit surface of an Au(001) foil, which was irradiated along [112] with 400 keV electrons at 95 K and observed along [001]. Nanoholes and hillocks are denoted by arrows and by **h**, respectively.

3.1.2. Irradiation Directional Dependence on the Structure

Figure 4a–c exhibit TEM images obtained under kinematic and slightly under-focus conditions for a (001)-oriented foil after irradiation with 800 keV electrons at 110 K to a fluence of 8×10^{26} electrons m^{-2}, which were, respectively, irradiated along [001], [011] and [111] directions. Stereomicroscopy indicated that the elongated bright contrasts along [001] and [010] for the [001] irradiation were due to grooves at the electron exit surface of the foil. Different results were obtained after the irradiations along [011] or [111] directions, as seen in Figure 4b,c, respectively. We can find bright contrasts extending along [100] for [011] irradiation, which are due to grooves but cannot find groove formation after [111] irradiation. The white contrasts in Figure 4c are caused by pits. These results suggest strongly that the groove formation is affected by the angle between the beam direction and the surface normal.

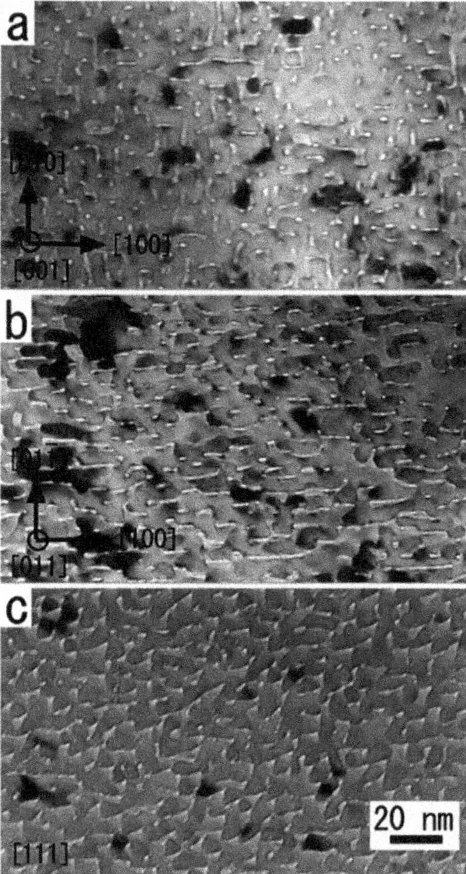

Figure 4. TEM micrographs of an Au(001) foil irradiated at 110 K with 800 keV electrons to a fluence of about 8×10^6 electrons m^{-2}. The directions of irradiations and observations are (**a**) [001], (**b**) [011], and (**c**) [111].

A detailed analysis has led us to schematic pictures on the nanostructure formed on the exit surface of gold films under electron irradiation as shown in Figure 5 [18]. The sputtered structure mainly consists of anisotropic nanogrooves, deep nanoholes, and hillocks. No clear grooves appear for the [111] irradiation. The nanoholes grow parallel to the irradiation direction both under [001] and [111] irradiations without much change in their widths but are zigzagged under [011] irradiation. Note the large aspect ratio of these nanoholes with extremely small dimension; the width and depth of the smallest nanoholes formed by the [111] irradiation, for example, are about 1.5 nm and more than 20 nm, respectively. In all three cases, the hillocks develop along the irradiation directions, although not indicated in Figure 5a for clarity.

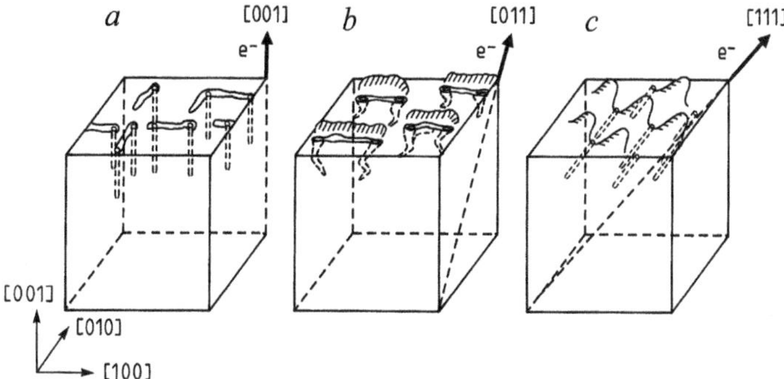

Figure 5. Schematic views of the sputtered rear surface of Au(001) foils irradiated with 800 keV electrons along (**a**) the [001] direction, (**b**) the [011] direction, and (**c**) the [111] direction.

Similar results were obtained for the groove and pit formation for 1250 keV electron irradiation of an Au(001) foil at 110 K. However, additionally, the formation of stacking-fault tetrahedra were observed, corresponding well to the estimated threshold electron energy of about 1150 keV for the Frenkel pair production in gold at 110 K [24].

3.1.3. Generation of Nanoslits and Nanoparticles

Successive sputtering due to electron irradiation induces thinning of the specimen, finally leading to the penetration of thin foils from the exit to the incident surfaces by nanogrooves and nanoholes. Figure 6a–d show the development of the nanostructure for an Au(011) foil irradiated with 400 keV electrons along [011] at 95 K. We can see that the width of nanogrooves in the pattern, which appeared in the initial stage of irradiation does not change significantly under irradiation. In Figure 6c, the penetration occurred in an area near the beam center, and then, nanoslits were formed as denoted by an arrow. In Figure 6d, the nanoslit denoted by the arrow in Figure 6c is closed, making the nanoslits beside it wider. This means the nanoslit became unstable under irradiation, probably due to its small size.

The self-organized pattern of nanoslits is seen to be determined in the initial stage of irradiation. Nanogrooves tend to develop along [100] direction, finally forming nanowires between the grooves. This method of the formation of nanowires using electron-beam irradiation in an electron microscope has been used to investigate characteristic nanostructures, which can appear in small size such as nanowires stabilized by the hcp lattices of the surfaces [25] and the ones consisting of helical atom rows coiled round the wire axis [26].

Figure 7a,b, respectively, show nanostructures of Au(001) foils irradiated with 1.25 MeV electrons at 110 K along [001] and [111] directions. The structures near the beam center denoted by X are seen to be pierced with a lot of holes produced by the irradiation and nanoparticles connected to each other. The shape of nanoparticles seen in Figure 7a,b reflects the self-organized nanostructures seen in Figure 4a,c, respectively. This method of particle formation may be useful for investigating the dependence of the characteristics of nanosized particle on size due to two advantages; one is the continuous preparation of a fresh surface under sputtering, and the other is the unique method of particle generation from thin foils.

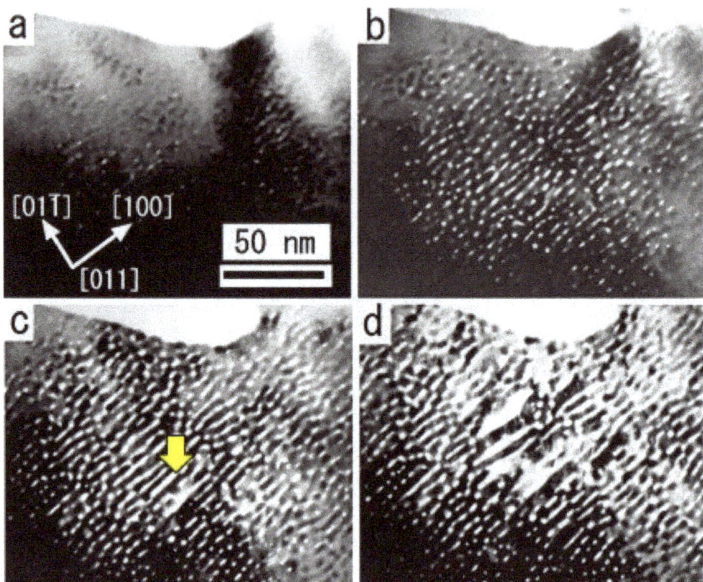

Figure 6. Development of nanostructure on the exit surface of an Au(011) foil irradiated with 400 keV electrons along [011] direction at 95 K: (**a**) 300 s; (**b**) 480 s; (**c**) 600 s; (**d**) 750 s.

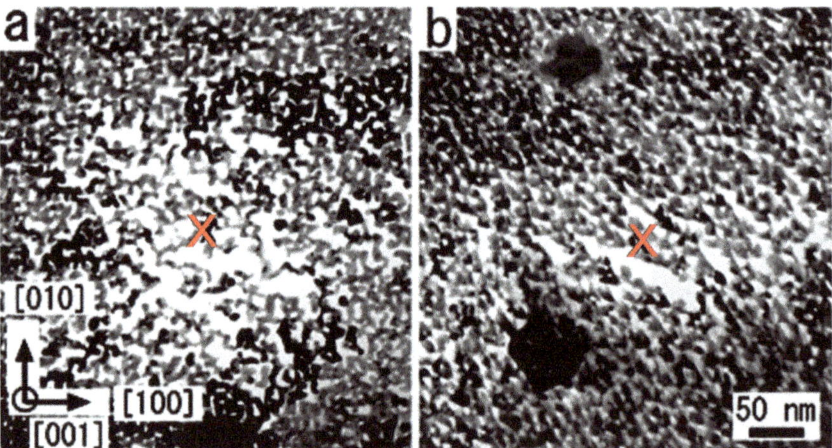

Figure 7. Nanostructures of Au(001) foils irradiated with 1.25 MeV electrons at 110 K along: (**a**) [001] direction, 1500 s; (**b**) [111] direction, 1700 s. The structures near the beam center are seen to be nanoparticles connected to each other and depend on the irradiation direction.

3.1.4. Annealing of the Nanostructure at Room Temperature

Figure 8a,b, respectively, show an area of Au foil irradiated along [111] at 800 keV with a fluence of about 8×10^{26} electrons m^{-2} and annealed at room temperature. Both the photos were observed obliquely from [001] direction, by tilting the specimen from the irradiation direction. The projection of the electron-beam direction during irradiation is indicated by an arrow in Figure 8a, and nanoholes are observed with parallel dark lines. They appear to develop rather randomly under [011] irradiation but grow along the irradiation direction under [111] irradiation [18]. The dark features are hillocks developing along the irradiation directions. On room-temperature annealing, the nanoholes become thermally unstable irrespective of the original irradiation direction. They transform to voids with large diameters, while the hillocks decrease in height, as seen in Figure 8b.

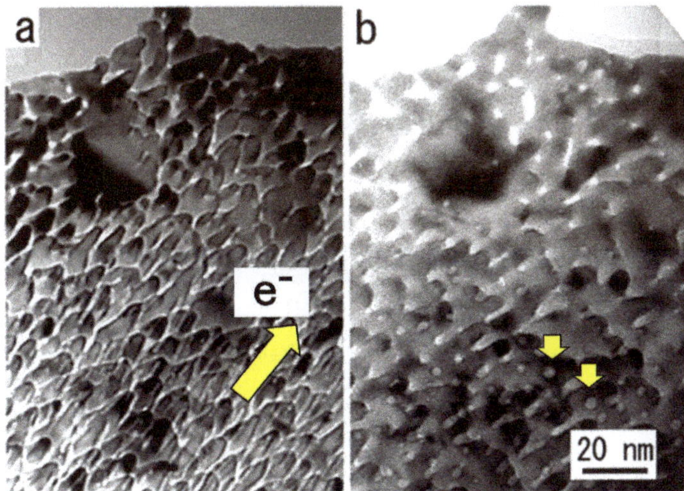

Figure 8. TEM micrographs of an area of Au foil (**a**) irradiated along [111] direction at 800 keV with a fluence of about 8×10^{26} electrons m^{-2} then (**b**) annealed at room temperature. Both the photos were taken along [001] direction. The projection of the electron-beam direction during irradiation is shown by a yellow arrow in photo (**a**). Voids transformed from nanoholes are indicated by arrows in photo (**b**).

3.1.5. Irradiation Temperature Dependence on the Structure

To know the irradiation temperature dependence on the formation of nanostructure, we investigated the structural change under electron irradiation at several different temperatures. Figure 9 shows bright-field images of Au(001) foils irradiated along [001] direction at four different temperatures. The electron irradiations were done with a beam of several hundred nm in diameter, and observation were done in areas where the structure looks homogeneous. The flux density is about 2×10^{24} electrons m^{-2} s^{-1}.

Nanoholes formed by the irradiations are observed as bright circular images in Figure 9a,b and dark circular images in Figure 9c,d. One should note that the density of nanohole is almost constant over the irradiated dose range at each irradiation temperature [11]. Then, the difference in the time indicated in Figure 9 does not significantly affect the investigation on the temperature dependence of the density of nanoholes.

Figure 10 shows the Arrhenius plot of the density of nanoholes. We can find two lines with apparent activation energy of 0.0046 eV below 240 K and 0.088 eV above 240 K. Similar studies were done several decades ago by Cherns [11]. He gave in situ observation of sputtering at 1.0 MeV with thin (111) gold films by using the Harwell EM7 high-voltage electron microscope. He measured the pit density, which corresponds to the nanohole density in the present study, against the irradiation temperature and found a temperature dependence on the pit density. However, he could not measure the pit densities greater than about 1–2 $\times 10^{16}$ m^{-2}, owing to a limit in the image resolution between adjacent pits. In the present study, we used JEOL-1250 transmission electron microscope, and the resolution is enough high to determine the density below 273 K, as seen in Figure 10.

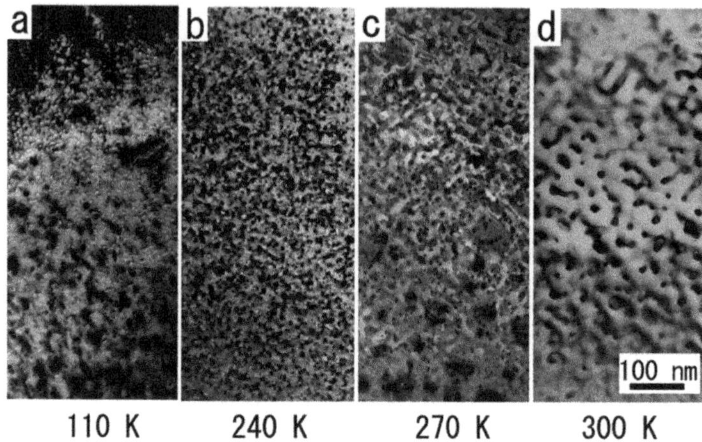

Figure 9. Au(001) foils irradiated along the [001] direction with 1250 keV electrons. (**a**) 110 K, 1070 s; (**b**) 240 K, 2130 s; (**c**) 270 K, 1180 s; (**d**) 300 K, 2680 s.

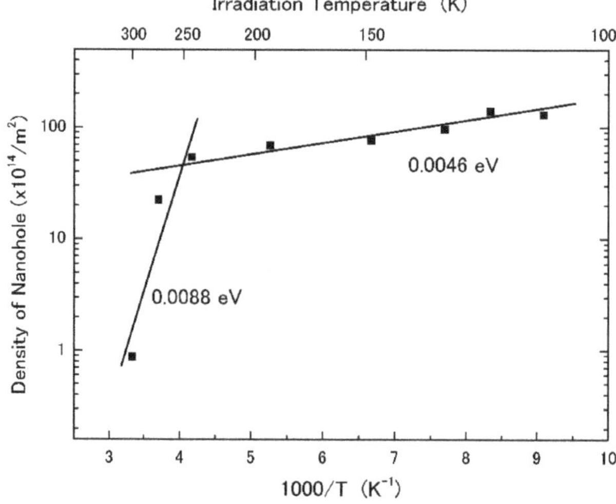

Figure 10. Arrhenius plot of the density of nanoholes. Two linear relations are seen, giving apparent activation energies of 0.0046 eV below 240 K and 0.088 eV above 240 K.

Cherns proposed a theoretical model in which pits form by the diffusion and agglomeration of surface vacancies produced by sputtering. The model explains the experimental results in some detail, and the apparent activation energy of pit density against $1/T$ is suggested to take a value of $E_m/3$, where E_m is the migration energy of surface vacancy. Utilizing Cherns' model in the present result on the density of pits on Au(001), the migration energy of a surface vacancy on Au(001) is derived to be 0.26 eV with the apparent activation energy of 0.088 eV shown in Figure 10. The value is lower than $E_m = 0.45$ eV given by Cherns, but almost in the range of error bar. If the apparent activation energy of 0.088 eV above 240 K were to relate to the migration of vacancy, a following question remains; "what is the activation process for the lower value of 0.0046 eV below 240 K?", which has not been found by Cherns. We can guess that the activation energy below 240 K may relate to the migration of adatom, etc., but further experimental and theoretical studies are awaited.

3.2. Comparison among Gold, Silver, and Copper

Figure 11 compares nanostructure, which appeared for Au(001), Ag(001), and Cu(001) foils irradiated with 400 keV electrons along [001] at 95 K [21]. Elongated bright images correspond to grooves formed on the electron exit surface of the foils. One can see that the patterns of grooves for both Ag and Cu are not as clear as the one for Au and extend weakly along [110] and [1$\bar{1}$0] directions. The width of grooves for Cu is about 2–4 nm, significantly larger than those for Au and Ag, which are about 1–2 nm.

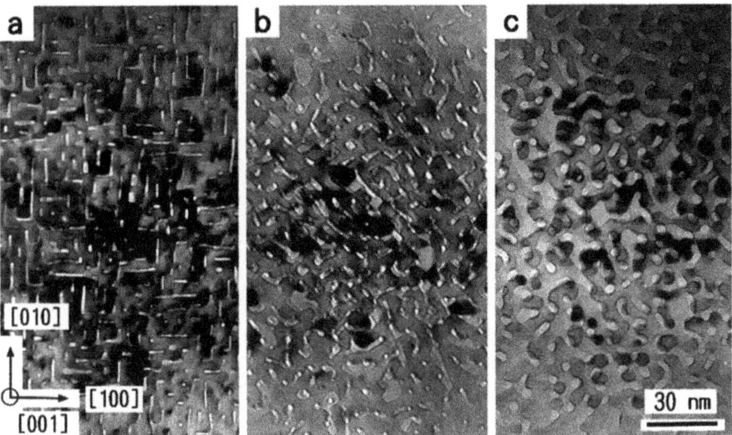

Figure 11. Nanostructures generated on (**a**) Au(001), (**b**) Ag(001), and (**c**) Cu(001) foils under 400 keV electron irradiation along [001] direction at 95 K. The patterns of grooves for both Ag and Cu are not as clear as the one for Au. The width of grooves for Cu is about 2–4 nm, significantly larger than those for Au and Ag, which are about 1–2 nm.

Contrary to the differences in the patterns for [001] irradiation, the groove formed by the oblique irradiations along [011] direction for Au, Ag, and Cu show a preferential elongation along [100], as denoted by arrows in Figure 12a–c. Note that the Ag specimen has a wavy surface. Then, the grooves formed on the surface are uneven and denuded in some areas, similar to the case of an Au foil with grain boundaries [11] but tend to grow along [100] in spite of the wavy surface. In addition to nanogrooves developed parallel to the exit surface of foils, nanoholes and hillocks are formed, as seen for Au, Ag, and Cu foils in Figure 13a–c, respectively [21]. The irradiations were done with 400 keV electrons at 95 K, and the photos were taken obliquely against the beam direction. Elongated images denoted by arrows correspond to nanoholes growing from the exit surface and dark contrasts denoted by **h** are hillocks. The directions of the growth of nanoholes and hillocks are along and opposite to the irradiation direction, respectively. One should note that the nanoholes grow deeper than 10 nm with an almost constant diameter of about 1–2 nm for Au and Ag and about 2–4 nm for Cu. Most of the nanoholes are formed in the areas of nanogrooves. The images of nanoholes observed along the irradiation direction can be found as extremely bright spots in Figure 12.

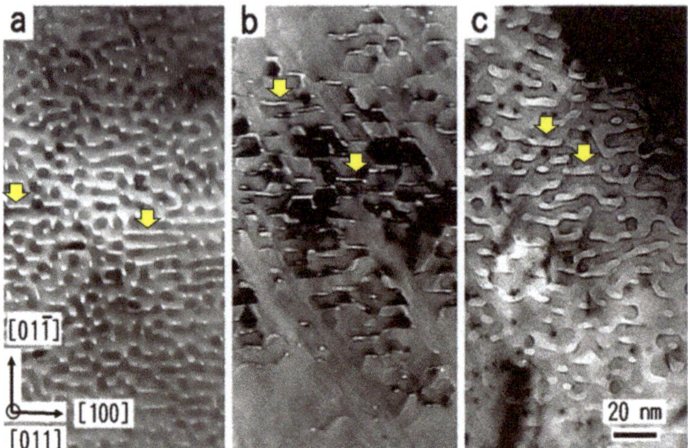

Figure 12. TEM images of nanostructure generated on the exit surface of Au, Ag, and Cu foils irradiated with 400 keV electrons along [011] direction at 95 K: (**a**) Au(001), 600 s; (**b**) Ag(001), 510 s; (**c**) Cu(001), 2160 s. Nanogrooves with bright contrast are seen to elongate along [100] direction for all the cases.

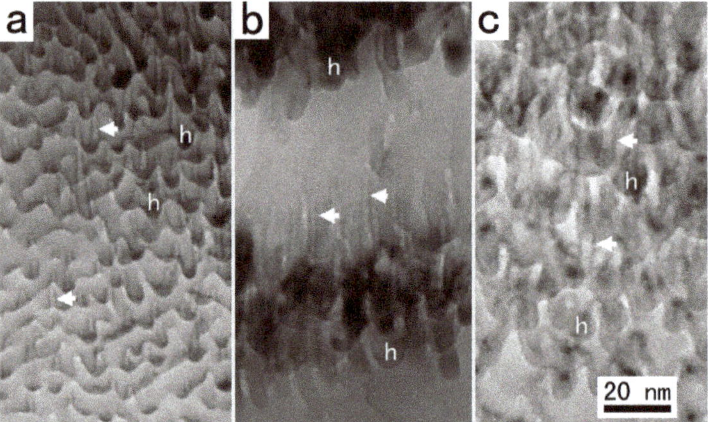

Figure 13. TEM images of deep nanoholes and hillocks generated on the exit surface of Au, Ag, and Cu foils irradiated with 400 keV electrons at 95 K: (**a**) Au(001), 600 s; (**b**) Ag(001) 600 s; (**c**) Cu(001), 3180 s. All images were taken oblique to the irradiation directions. Nanoholes and hillocks denoted by arrows and by **h** are seen to elongate opposite to and along the beam direction, respectively.

3.3. Ni

A nanostructure, which appeared for a Ni(001) foil irradiated with 400 keV electrons along [001] direction at 105 K to a dose of 7.0×10^{27} electrons m^{-2} [22], is shown in Figure 14. The diameter of electron beam for the irradiation was about 200 nm of which the center is near the center of Figure 14a. The micrograph was taken under a kinematic and slightly under-focus condition. One can see that the pattern near the center of the electron beam of photo (c) is rather different from that in the outer area of photo (c). The extension of bright images is not observed along [100] and [010] directions, but they extend slightly along [110] and [1$\bar{1}$0] instead. The pattern near the center of the electron beam is similar to those of Ag and Cu, which are, respectively, shown in Figure 11b,c. The results for Ni mentioned above are different from those of Au, where the pattern is almost the same in the whole area, although the extension of the grooves becomes longer near the beam center, as seen in Figure 2.

Figure 14. (**a**) TEM micrograph of nanostructures generated on the exit surface of Ni(001) foil irradiated with 400 keV electrons along [001] direction at 105 K to a dose of 7.0×10^{27} electrons m^{-2}. (**b**) A magnified view of the outer area of the electron beam. (**c**) A magnified view of the area near the beam center.

Figure 15 shows the Ni(001) specimen, which was tilted to the direction near [011] and irradiated obliquely with 400 keV electrons along [011] direction at 105 K to a dose of 8.4×10^{27} electrons m^{-2}. Bright images of nanogrooves tend to extend along [100], as shown in Figure 15a, similar to the cases of Au, Ag, and Cu (Figure 12). The irradiation was prolonged to a dose of 4.2×10^{28} electrons m^{-2}, then the penetration of nanogrooves from the electron exit surface to the incident surface occurred, as seen in Figure 15b. We can find large holes appear near the center of an electron beam, and several nanowires were generated. The nanowires tend to elongate along [100], reflecting the initial pattern of nanogrooves seen in Figure 15a.

3.4. Iron

A nanostructure formed for a Fe(111) foil irradiated with 400 keV electrons along [111] direction at 300 K to a dose of 4.4×10^{28} electrons m^{-2} [22] is shown in Figure 16. The TEM micrograph was taken under a kinematic and slightly under-focus condition. The beam center is near the edge of the specimen. The nanostructure generated by the electron irradiation can be roughly divided into two types; one is high density of nanoholes formed in the outer area of the electron beam, as seen in the magnified view, and the other is nanogrooves, which are observed near the beam center.

The appearance of the two types of structure should be due to the difference in the total dose, as the beam intensity is stronger near the beam center. The irradiation dose described above is an average value. In situ observation has revealed that nanoholes are generated at first and then developed to nanogrooves with increasing dose. The diameter of nanoholes is about 2–4 nm. Formation of high density of nanoholes under electron irradiation has not been observed for Au, Ag, and Cu [21] but has been reported for Si [27].

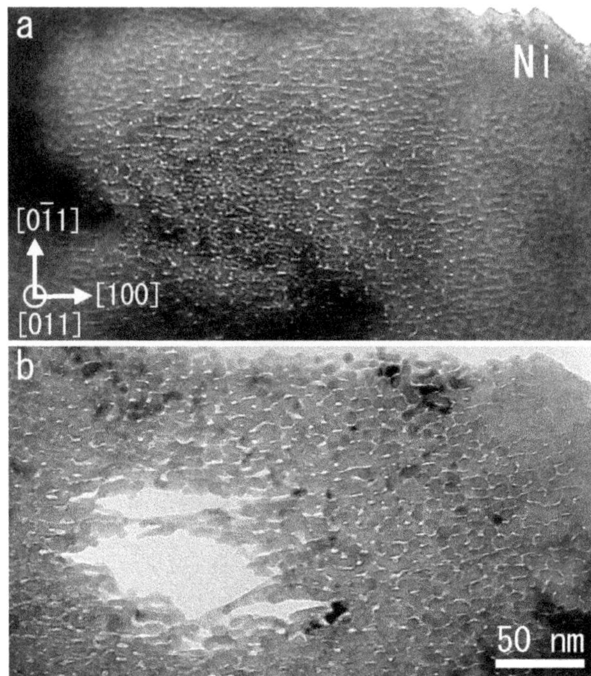

Figure 15. TEM micrographs of Ni(001) specimen tilted to the direction near [011] direction. Irradiation was done with 400 keV electrons along [011] direction at 105 K to doses of (**a**) 8.4×10^{27} e$^-$ m^{-2} and (**b**) 4.2×10^{28} e$^-$ m^{-2}. Penetration of the foil can be seen in photo (**b**).

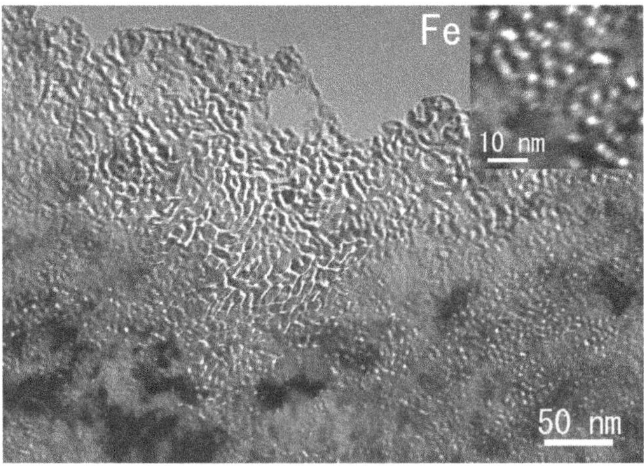

Figure 16. TEM micrographs of the Fe(111) foil irradiated with 400 keV electrons along [011] direction at 300 K to a dose of 4.4×10^{28} electrons m^{-2}. A magnified view of the nanoholes is given in the inset.

One should note that a nanostructure with several nanometer size appeared at 300 K in the case of Fe. In the case of Au, on the other hand, the nanostructure appeared at 110 K and became unstable on annealing at room temperature. Nanoholes are transformed into voids with larger diameters, while the hillocks decreased in height, as shown in Figure 8. Then, the stepwise annealing was carried out with 50 K/10 min steps from 373 to 823 K in the transmission electron microscope to know the thermal stability of the nanostructure generated for Fe.

The changes on annealing at 623 and 823 K are, respectively, shown in Figure 17a,b. The pattern formed on the Fe surface scarcely changed up to 573 K, but a clear change can be observed above 623 K. The steps denoted by arrows are seen to move downwards. In spite of such an apparent change in the specimen surface on annealing at 623 K (Figure 17a), we can still observe some nanogrooves and nanoholes whose size and pattern are similar to those before annealing (Figure 16). At 823 K, on the other hand, a remarkable change such as disappearance of the nanostructures can be observed (Figure 17b). However, there still remains some nanoholes whose diameters are almost the same as those before annealing. The result is clearly different from that of Au, where nanoholes were transformed into voids with larger diameters, and the surface became smoother. The changes in the Au specimen suggest the remarkable effect of surface diffusion. Therefore, we guess that thermal evaporation of Fe atoms from the surface becomes remarkable at a temperature above 623 K to form the steps and to change the structure. Moreover, material transport on Fe surface by surface diffusion is not so significant as to change the nanostructure below 823 K.

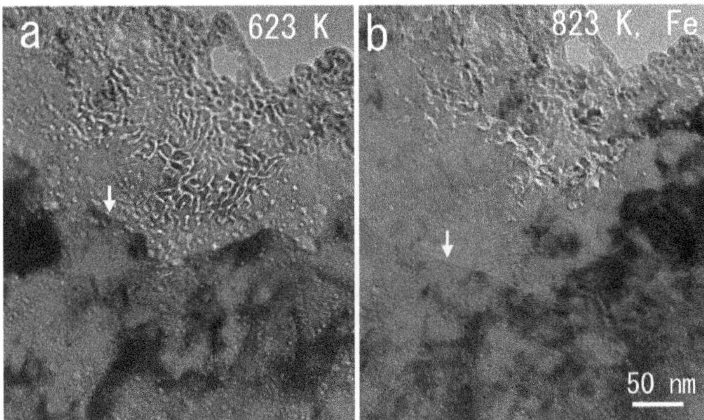

Figure 17. The change in nanostructure of the Fe(111) foil on annealing at (**a**) 623 K and (**b**) 823 K. The nanostructure was generated by the irradiation with 400 keV electrons along the [011] direction at 300 K to a fluence of 4.4×10^{28} electrons m^{-2}. Steps that appeared on annealing are denoted by arrows in photos (**a,b**).

However, residual O_2 gas in the conventional TEM may affect the formation or the behavior of nanostructure similarly to the case of silicon [27]. Nonetheless, the electron exit surface of Fe where the pattern developed should be rather clean due to sputtering. Further investigations on this point are desired.

4. Discussion

For the generation of aligned nanogrooves, we did not use a convergent nanosized electron beam as has been done for the generation of nanoholes [10] but a homogeneous electron beam of about 200–800 nm in diameter. In this case, the phenomena occurring on the electron exit surface may be as follows (Figure 18). Surface vacancies are formed at first in the first layer of the metal. Then, the surface vacancies agglomerate to produce monolayer islands of surface vacancies. Under prolonged irradiation, surface vacancies are produced in the area of islands, and multilayered pits are formed in several minutes. One cannot expect any anisotropic movement of surface vacancies on (001) and (111) surfaces in the process, but it may be expected on (011), judging from the nature of geometrical symmetry of the surfaces. Therefore, the present results on the pattern formation on the

(001) surface strongly suggest that we are dealing with a case of self-organized pattern formation controlled by some factor, which includes the anisotropy of material transfer.

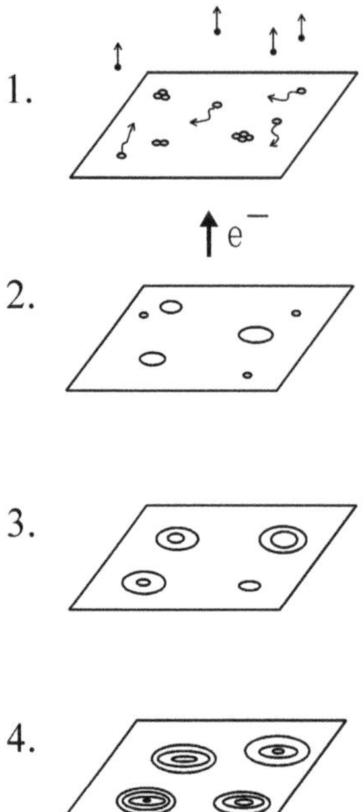

Figure 18. A sketch of the change in the electron exit surface at the initial stage of formation of pits. 1. At first, vacancies are formed on the surface layer of the foil by sputtering. 2. They then agglomerate to form monolayer islands of surface vacancies. 3. After prolonged irradiation, further surface vacancies are formed within islands. 4. Finally, multilayered pits will be formed.

We have found the formation of pits and the self-organization of grooves under electron irradiation. The growth process of the pits has been discussed theoretically by Cherns [11]. However, his model does not predict the generation of the deep nanoholes and the aligned nanogrooves found in the present work; probably, he used Au(111) foils on which the aligned nanogrooves do not appear. It is not clear when and why the aligned nanogrooves and the deep nanoholes are generated. In this section, we discuss the formation of deep nanoholes and the self-organization process of nanogrooves.

4.1. Formation Process of Deep Nanoholes

Figure 18 schematically shows the structural change in the electron exit surface at the initial stage of irradiation. At first, surface vacancies are formed in the first layer of the metal by sputtering. Then, the surface vacancies agglomerate to produce monolayer islands of surface vacancies. Under continuous irradiation, surface vacancies are formed in the area of islands, and finally, multilayered pits are generated. The growth process of the pits has been theoretically given by Cherns [11]. Nonetheless, his model does not predict the formation of the aligned nanogrooves and the deep nanoholes found in the present work.

In contrast with what has been said above for the formation of nanogrooves, the deepening of the nanoholes is not believed to involve self-organization. As may be seen from Figure 3, the diameters of nanoholes do not vary significantly along their depth, indicating one-dimensional growth along the irradiation direction. We can estimate that more than 2000 atoms must be removed to produce a nanohole of 1.5 nm width and 20 nm depth. We guess that irradiation-induced diffusion of surface vacancies plays a significant role, although there may be other processes leading to the deepening of nanohole. Note that the nanoholes are generated on the rear surface, growing opposite to the irradiation direction. Irradiation-induced diffusion of surface vacancies on the side walls of the holes can result in a vacancy migration opposite to the irradiation direction, which is towards the bottom of the nanohole. Thus, nanoholes can be deeper by this process.

How does the nanohole grow so deep? First, one can consider that a geometrical factor would be responsible for the deepening of pits. Surface vacancies created in a terrace with a width w, as shown in Figure 19, can contribute to the erosion of the descending step but not of the ascending step at temperatures below which the layer-by-layer removal starts [28]. We can easily see that the area A in the terrace becomes larger near the center if the length of the descending step and the width of the terrace w are the same. This means that the erosion rate of the descending step near the center should be faster in this case due to the large amount of incoming surface vacancies for the unit length of the descending step. Under prolonged irradiation, however, the pit profile should reach a steady state after a critical dose, because a faster erosion of the descending step would lead to a decrease in the terrace width, thereby leading to a decrease in the erosion rate of the descending step.

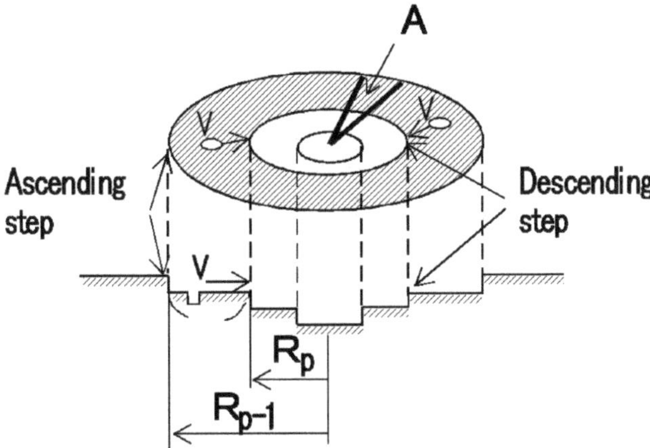

Figure 19. Oblique and cross-sectional views of a circular pit. Surface vacancies created in a terrace with width w can contribute to the erosion of the descending step but not of the ascending step at temperatures below that at which the layer-by-layer removal starts [28].

Here, we discuss the pit profile in the steady state. The growth rate of the circular step with an inner radius of R_p in Figure 19 is given [11] in the continuum approximation by

$$dR_p/dt = P(R_{p-1}^2 - R_p^2)/2N_0R_p \ (R_p > a_0),\tag{1}$$

where P and N_0 are the production rate of surface vacancies and the number of atoms per unit area of the surface, respectively, and a_0 is the atomic size.

As $R_{p-1} = R_p + w$, Equation (1) can be expressed as

$$\begin{aligned} dR_p/dt &= P[(R_p + w)^2 - R_p^2]/2N_0R_p \\ &= Pw/N_0 + Pw^2/2N_0R_p, \end{aligned}\tag{2}$$

When a pit grows steadily, Equation (2) should be constant (=k).

In the case where the terrace is far from the center ($R_p >> w$), the second term of Equation (2) is negligibly small. Then, the width of the terrace is approximately

$$w = kN_0/P,\tag{3}$$

In the case where the terrace is located beside the center (R_p = w), the width of the terrace is estimated, by applying R_p = w and dR_p/dt = k, as

$$w = (2/3)kN_0/P, \qquad (4)$$

The difference in the values of w in Equations (3) and (4) means that the width of the terrace near the center becomes narrower; i.e., the inclination of the face of the pit near the center becomes steeper. The shape of the steadily growing pit can be obtained by the numerical calculation for pit growth for finite intervals given by Cherns [11]. However, the result of these calculation does not reveal the significant deepening of pits observed in the present experimental study. Therefore, the deepening of the nanoholes should be due to some factors other than the geometrical factor.

Two factors can be mentioned as being responsible for the deepening, which are preferential sputtering at growth point and irradiation-induced diffusion. The sputtered surface quickly develops a high proportion of surface ledge sites, especially near the pit center, as seen in Figure 20. The sputtering rate would increase at the ledge sites due to the lower number of bonding. Therefore, it is probable that the preferential sputtering at the growth point leads to the deepening of pits. Irradiation-induced diffusion of surface vacancies and adatoms on the side walls of nanoholes, on the other hand, may play an important role. Surface vacancies and adatoms will move opposite to and along the irradiation direction, respectively, both movements enhancing the deepening of the holes.

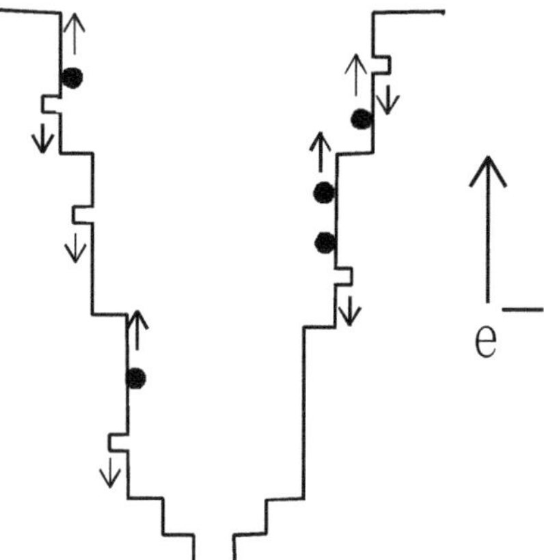

Figure 20. Growth process of nanohole. Surface vacancies on the wall of nanohole move to the tip of nanohole, but adatoms are evacuated from the nanohole. Thus, deep nanoholes can be formed.

4.2. Self-Organization of Nanogrooves

Interesting discovery on the development of nanostructure under electron irradiation in the present study is the pattern formation of nano-grooves, which can change depending on the irradiation directions, the irradiation temperatures, and the kinds of metals. The most outstanding feature appears for Au. The nanogrooves exhibit strong irradiation-direction dependencies in their growth. They grow along [100] and [010] directions for [001] irradiation and along [100] for [011] irradiation, whereas no clear grooves are formed for [111] irradiation. The widths of nanogrooves and holes are between about 1 and 2 nm, which are the smallest ones formed on metal surfaces so far. The formation of the groove pattern also has a surface orientation dependence.

To clarify the formation mechanism on the nanostructures, systematic investigations are needed. After finding the pattern formation for Au, we investigated several FCC metals of Au, Ag, Cu, Ni, and a BCC metal of Fe. For Ag and Cu, the pattern formation is not so clear. For Ni, the diffusion effect is not so high compared to the cases of Ag and Cu, but the pattern depended on the part of the specimen, as seen in Figure 11. For Fe, the pattern is not clear but rather random (Figure 14).

The generation of nanoholes and nanogrooves originally comes from the sputtering at the electron exit surface. Thus, the anisotropic growth of the nanogrooves and nanoholes should be attributed to the irradiation-induced anisotropic flow of point defects [29].

The fact that different groove patterns observed on the same Au(001) surface by irradiating along different directions (Figures 4 and 5) strongly suggests that the patterns are meta-stable structures, which has a very long lifetime at low temperatures.

The available experimental evidence demonstrates that we are dealing with a case of self-organized pattern formation, which satisfies the following conditions.

(1)   The groove patterns are generated far from thermal equilibrium in an open, highly dissipative system.

(2)   Only a small fraction of the total energy input is stored (e.g., as surface energy at the grooves).

Below 240 K, the temperature effect of surface vacancy, which will reduce the anisotropic growth of nanogrooves due to their random movements, should be low for Au, as seen in Figure 10. At the low temperature region, the migration of surface vacancies by thermal activation alone should be negligibly small. Thus, we may expect the phenomena observed to be controlled by some anisotropic mechanisms such as sputtering induced by the anisotropy of the sputtering yield [23], the surface reconstruction under irradiation [25], focused collision chains [30], or irradiation-induced diffusion [29].

We guess that the anisotropy of momentum transfer, either directly from the electrons by Rutherford scattering or indirectly through bulk collision sequences propagating along the densely packed directions, leads to the mobility of surface vacancies, finally leading to anisotropic groove patterns. One should note that there exists a difference in the anisotropy of collision sequence propagation between bulk Au and Ag, i.e., along the <100> direction for Au but <110> for Ag, which was experimentally supported by a difference in the anisotropy of the threshold energy for atom displacement in both materials [30].

Now, we draw a conceivable scenario, in the following section. Consider a surface atom of which momentum has been transferred either directly from the electrons by Rutherford scattering or indirectly through bulk collision sequences. In the case where the component of the transferred momentum normal to the surface is sufficiently large, the atom will be sputtered off the surface leaving a surface vacancy behind. Surface collision sequences propagating along close-packed directions in the surface shown in Figure 21 may be excited by momentum transfer to neighboring atoms—a process that is favored by the angular dependence of the Rutherford scattering cross-section. Under prolonged irradiation, the surface vacancies will move around by irradiation-induced diffusion and may agglomerate to form pits. Surface collision sequences ending at a descending surface step or at a hole or groove, on the other hand, may generate a surface adatom, leaving a surface vacancy at or near its point of departure. Adatoms created on the surfaces of grooves will in general reduce the groove surface area. Since the propagation of collision sequences is highly anisotropic, this may lead to a gradual alignment of the grooves and pits under prolonged irradiation. This process is similar to that of the formation of lattices of stacking-fault tetrahedra under high-dose electron irradiation [12,13]. Such a mechanism can work only if the mobility of the surface vacancies is not too large. This probably accounts for the fact that regular groove patterns have so far been observed for Au only after low-temperature irradiation but not under room-temperature irradiation conditions.

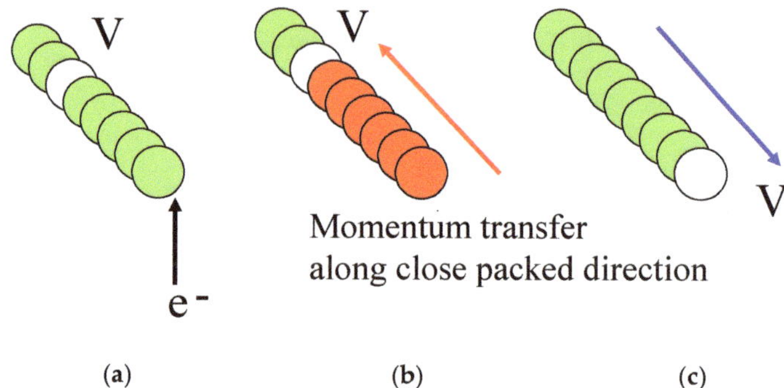

Figure 21. Movement of vacancy due to momentum transfer along close-packed direction. White circle indicates surface vacancy. (**a**) Collision of electron, (**b**) momentum transfer along close packed direction given by the collision of electron, (**c**) movement of surface vacancy opposite to the direction of momentum transfer along close packed direction.

Additionally, one should note that the pattern of aligned nanogroove for the 400 keV electron irradiation for Au (Figure 2) is clearer than the one for the 800 keV electron irradiation (Figure 4). The origin of the less clear pattern in the latter case should be attributed to some additional effects such as a complicated sputtering process; that is, at energies <600 keV, surface atoms are sputtered by direct recoils only, whereas at energies >600 keV, up to about 10% of the total yield comprises surface atoms sputtered by sub-surface recoils [23] and the formation of Frenkel pairs, which can occur at higher irradiation energy of electrons [24]. Angular distributions of sputtered atoms as a function of electron energy [23] may affect the formation of adatoms and the pattern formation. Moreover, the formation of aligned nanogrooves and nanoholes are significantly affected by the irradiation temperature (Figures 9 and 10), probably reflecting the mobility of surface vacancy.

5. Conclusions

We found a new type of self-organized nanostructure formation on the exit surface of thin gold foils irradiated with high doses of 360–1250 keV electrons at temperatures of about 100 K. The structure consists of aligned nanogrooves, which develop parallel to the surface, and nanoholes and hillocks, which grow parallel to the electron beam. The nanogrooves show strong irradiation-direction dependencies on their growth. They grow along [100] and [010] directions for [001] irradiation, along [100] for the [011] irradiation, whereas no clear grooves are formed for [111] irradiation. The widths of nanogrooves and holes are between about 1 and 2 nm, which are the smallest ones generated on metal surfaces so far. The final structures of the thin foils under electron irradiation are nanoparticles or nanowires. This method has been utilized to produce long gold nanowires for investigations of the interesting physics such as the electron transport properties and the multi-shell structure. Temperature dependence of the nanostructure for gold indicates that the effect of surface diffusion becomes significant above 240 K.

Furthermore, the self-organized structures for silver, copper, nickel, and iron are investigated. The formation of nanoholes and nanogrooves basically originates in the sputtering at the electron exit surface. The difference in the anisotropic growth of the nanogrooves and nanoholes among the kinds of metals should be attributed to the irradiation-induced anisotropic flow of point defects and some related factors, which are attributed to the nature of metals. Thus, we can use the present investigation for the studies of irradiation-induced effects and the surface diffusion effect, of which knowledge should be useful for the synthesis of new nanomaterials.

Funding: This study was supported by fellowships from the Ministry of Education of Japan and Max-Planck-Institut.

Acknowledgments: I express my sincere thanks to Wilfried Sigle, Fritz Phillipp, the late Alfred Seeger, and Hiroaki Abe for their collaboration. Experimental works were mostly done in Max-Planck-Institut für Metallforschung and partly Japan Atomic Energy Research Institute-Takasaki.

Conflicts of Interest: The authors declare no conflict of interest.

References

1. Wua, T.; Lin, Y.-W. Surface-enhanced Raman scattering active gold nanoparticle/nanohole arrays fabricated through electron beam lithography. *Appl. Surf. Sci.* **2018**, *435*, 1143. [CrossRef]
2. Tavakoli, M.; Jalili, Y.S.; Elahi, S.M. Rayleigh-Wood anomaly approximation with FDTD simulation of plasmonic gold nanohole array for determination of optimum extraordinary optical transmission characteristics. *Superlattices Microstruct.* **2019**, *130*, 454. [CrossRef]
3. Chen, X.; Wang, H.; Xu, N.-S.; Chen, H.; Deng, S. Resonance coupling in hybrid gold nanohole–monolayer WS2 nanostructures. *Appl. Mater. Today* **2019**, *15*, 145. [CrossRef]
4. Becker, R.S.; Golovchenko, J.A.; Swartzentruber, B.S. Atomic-scale surface modifications using a tunnelling microscope. *Nature* **1987**, *325*, 419. [CrossRef]
5. Eiger, D.M.; Schweizer, E.K. Positioning single atoms with a scanning tunnelling microscope. *Nature* **1990**, *344*, 524. [CrossRef]
6. Kobayashi, A.; Grey, F.; Williams, R.S.; Aono, M. Formation of nanometer-scale grooves in silicon with a scanning tunneling microscope. *Science* **1993**, *259*, 1724. [CrossRef] [PubMed]
7. Saif, M.J.; Naveed, M.; Asif, H.M.; Akhtar, R. Irradiation applications for polymer nano-composites: A state-of-the-art review. *J. Ind. Eng. Chem.* **2018**, *60*, 218. [CrossRef]
8. Córdoba, R.; Orús, P.; Strohauer, S.; Torres, T.E.; Teresa, J.M.D. Ultra-fast direct growth of metallic micro- and nano-structures by focused ion beam irradiation. *Sci. Rep.* **2019**, *9*, 14076. [CrossRef] [PubMed]
9. Turner, P.S.; Bullogh, T.J.; Devenish, R.W.; Maher, D.M.; Humphereys, C.J. Nanometre hole formation in MgO using electron beams. *Philos. Mag. Lett.* **1990**, *61*, 181. [CrossRef]
10. Kizuka, T.; Tanaka, N. Nanometre scale electron beam processing and in situ atomic observation of vacuum-deposited MgO films in high-resolution transmission electron microscopy. *Philos. Mag. A* **1995**, *71*, 631. [CrossRef]
11. Cherns, D. The surface structure of (111) gold films sputtered in the high voltage electron microscope a theoretical model. *Philos. Mag.* **1977**, *36*, 1429. [CrossRef]
12. Jin, N.Y.; Phillipp, F.; Seeger, A. Investigation of defect ordering in heavily irradiated metals by high-voltage electron microscopy. *Phys. Stat. Sol.* **1989**, *116*, 91. [CrossRef]
13. Seeger, A.; Jin, N.Y.; Phillipp, F.; Zaiser, M. The study of self-organization processes in crystals by high-voltage electron microscopy. *Ultramicroscopy* **1991**, *39*, 342. [CrossRef]
14. Ghoniem, N.M.; Walgraef, D.; Zinkle, S.J. Theory and experiment of nanostructure self-organization in irradiated materials. *J. Comput. Aided Mater. Des.* **2001**, *8*, 1. [CrossRef]
15. Kharchenko, D.O.; Kharchenko, V.O.; Bashtova, A.I. Modeling self-organization of nano-size vacancy clusters in stochastic systems subjected to irradiation. *Radiat. Eff. Defects Solids* **2014**, *169*, 418. [CrossRef]
16. Gao, Y.; Zhang, Y.; Schwen, D.; Jiang, C.; Sun, C.; Gan, J.; Bai, X.-M. Theoretical prediction and atomic kinetic Monte Carlo simulations of void superlattice self-organization under irradiation. *Sci. Rep.* **2018**, *8*, 6629. [CrossRef] [PubMed]
17. Vitral, E.; Walgraef, D.; Pontes, J.; Anjos, G.R.; Mangiavacchi, N. Nano-patterning of surfaces by ion sputtering: Numerical study of the anisotropic damped Kuramoto-Sivashinsky equation. *Comput. Mater. Sci.* **2018**, *146*, 193. [CrossRef]
18. Niwase, K.; Sigle, W.; Phillipp, F.; Seeger, A. Generation of nanosized grooves and holes on metal surfaces by low-temperature electron irradiation. *Philos. Mag. Lett.* **1996**, *74*, 167. [CrossRef]
19. Niwase, K.; Sigle, W.; Phillipp, F.; Seeger, A. Nanogrooves generated on Au surfaces by low-temperature electron irradiation. *J. Surf. Anal.* **1997**, *3*, 440.
20. Niwase, K.; Sigle, W.; Phillipp, F.; Seeger, A. Novel phenomena of irradiation damage in FCC metals irradiated with high-energy electrons at low temperatures. *J. Electron. Microsc.* **1999**, *48*, 495. [CrossRef]
21. Niwase, K.; Phillipp, F.; Seeger, A. Self-organized nanostructures generated on Au, Ag and Cu surfaces by low-temperature electron irradiation. *Jpn. J. Appl. Phys.* **2000**, *39*, 4624. [CrossRef]
22. Niwase, K.; Abe, H. Generation of nanosized structures on Ni and Fe surfaces by electron irradiation. *Mater. Trans.* **2002**, *43*, 646. [CrossRef]
23. Cherns, D. Sputtering of gold foils in a high voltage electron microscope a comparison of theory and experiment. *Philos. Mag.* **1977**, *35*, 693. [CrossRef]
24. Hohenstein, M.; Seeger, A.; Sigle, W. The anisotropy and temperature dependence of the threshold for radiation damage in gold—comparison with other FCC metals. *J. Nucl. Mater.* **1989**, *169*, 33. [CrossRef]
25. Kondo, Y.; Takayanagi, K. Gold nanobridge stabilized by surface structure. *Phys. Rev. Lett.* **1997**, *79*, 3455. [CrossRef]

26. Kondo, Y.; Takayanagi, K. Synthesis and characterization of helical multi-shell gold nanowires. *Science* **2000**, *289*, 606. [CrossRef] [PubMed]
27. Takeda, S.; Koto, K.; Iijima, S.; Ichihashi, T. Nanoholes on silicon surface created by electron irradiation under ultrahigh vacuum environment. *Phys. Rev. Lett.* **1997**, *20*, 2994. [CrossRef]
28. Michely, T.; Land, T.; Littmark, U.; Comsa, G. Morphological effects induced by the formation of a Pt-adatom lattice gas on Pt (111). *Surf. Sci.* **1992**, *272*, 204. [CrossRef]
29. Urban, K.; Seeger, A. Radiation-induced diffusion of point-defects during low-temperature electron irradiation. *Philos. Mag.* **1974**, *30*, 1395. [CrossRef]
30. Sigle, W. The anisotropy of the threshold energy for atom displacements in silver. *J. Phys Condens. Matter.* **1991**, *3*, 3921. [CrossRef]

quantum beam science

MDPI

Review of Swift Heavy Ion Irradiation Effects in CeO$_2$

William F. Cureton [1], Cameron L. Tracy [2] and Maik Lang [1,*]

[1] Department of Nuclear Engineering, University of Tennessee, Knoxville, TN 37966, USA; wcureton@vols.utk.edu

[2] Center for International Security and Cooperation, Stanford University, Stanford, CA 94305, USA; cltracy@umich.edu

* Correspondence: mlang2@utk.edu; Tel.: +1-865-974-8247

Abstract: Cerium dioxide (CeO$_2$) exhibits complex behavior when irradiated with swift heavy ions. Modifications to this material originate from the production of atomic-scale defects, which accumulate and induce changes to the microstructure, chemistry, and material properties. As such, characterizing its radiation response requires a wide range of complementary characterization techniques to elucidate the defect formation and stability over multiple length scales, such as X-ray and neutron scattering, optical spectroscopy, and electron microscopy. In this article, recent experimental efforts are reviewed in order to holistically assess the current understanding and knowledge gaps regarding the underlying physical mechanisms that dictate the response of CeO$_2$ and related materials to irradiation with swift heavy ions. The recent application of novel experimental techniques has provided additional insight into the structural and chemical behavior of irradiation-induced defects, from the local, atomic-scale arrangement to the long-range structure. However, future work must carefully account for the influence of experimental conditions, with respect to both sample properties (e.g., grain size and impurity content) and ion-beam parameters (e.g., ion mass and energy), to facilitate a more direct comparison of experimental results.

Keywords: cerium oxide; CeO$_2$; irradiation; swift heavy ions

Citation: Cureton, W.F.; Tracy, C.L.; Lang, M. Review of Swift Heavy Ion Irradiation Effects in CeO$_2$. *Quantum Beam Sci.* **2021**, *5*, 19. https://doi.org/10.3390/qubs5020019

Academic Editors: Akihiro Iwase and Hiro Amekura

Received: 30 April 2021
Accepted: 11 June 2021
Published: 16 June 2021

1. Introduction

Under ambient conditions, cerium dioxide (CeO$_2$) adopts the fluorite structure (space group *Fm-3m*), common to a variety of dioxides such as UO$_2$, ThO$_2$, PuO$_2$, and doped ZrO$_2$ [1,2]. In this phase, Ce cations arrange in a face-centered cubic array on 4a Wyckoff sites with a simple cubic substructure of oxygen anions occupying 8c sites (Figure 1). The flexibility of cerium's electronic structure allows for extensive transitions between the nominal $4f^0$–Ce^{4+} oxidation state and a $4f^1$–Ce^{3+} state without loss of the long-range fluorite structure. This transition is so prevalent over a wide range of conditions that materials containing cerium (including oxides) with monovalent Ce^{4+} cations are difficult to fabricate [3].

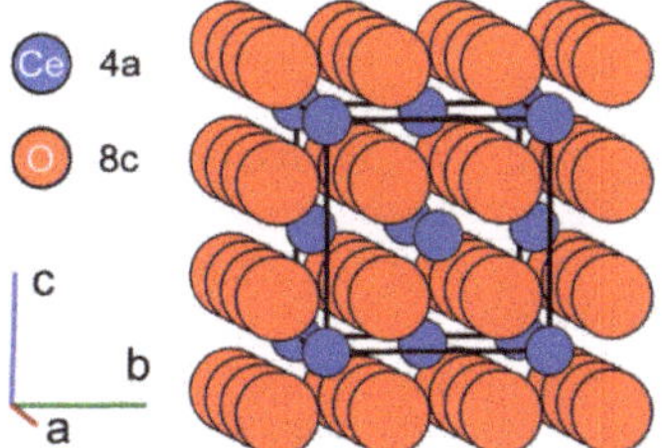

Figure 1. The cubic fluorite structure (space group *Fm-3m*) of CeO$_2$, where cations (blue spheres) occupy 4a Wykoff sites and anions (red spheres) occupy 8c sites.

The reduction in cation charge state is coupled to the production of oxygen vacancies (V_O) in the system; for every positively charged $V_{\ddot{O}}$, two Ce^{3+} cations are necessary for charge compensation. Under reducing conditions, oxygen vacancy formation is expressed in Kröger-Vink notation as:

$$2Ce_{Ce} + O_O \rightarrow \frac{1}{2}O_{2(gas)} + V_{\ddot{O}} + 2e^{/} + 2Ce_{Ce} \tag{1}$$

The two excess electrons ($e^{/}$) liberated by the formation of molecular oxygen can subsequently be captured by surrounding cerium atoms, causing a reduction in the oxidation state of tetravalent cations {Ce(IV) $\rightarrow$ Ce(III)} as expressed in:

$$2Ce_{Ce} + V_{\ddot{O}} + 2e^{/} \rightarrow V_{\ddot{O}} + 2Ce^{/}_{Ce} \tag{2}$$

The accommodation of these $V_{\ddot{O}}$ and Ce^{3+} defects in the fluorite structure yields an average expansion of the unit cell, due to the larger cationic radius of Ce^{3+} cations compared with Ce^{4+} cations (1.14 Å and 0.97 Å, respectively [4]) and local distortions around the V_O and aliovalent cations [5,6]. As these Ce^{3+}_{Ce}–V_O defect complexes accumulate to sufficiently large concentrations, the oxygen vacancies begin to order and arrange themselves according to Pauling's 1st and 2nd rules [7,8]. This leads to transformations into various trigonal fluorite-derivative phases such as $Ce_{11}O_{20}$ and Ce_7O_{12} as a function of decreasing oxygen content [9,10].

The flexibility of the fluorite structure and the complex redox behavior of cerium oxide compounds make them attractive for use in a number of engineering applications, such as oxygen sensors [11], catalysts in chemical processes [12], and electrolyte materials in solid oxide fuel cells [13]. Being isostructural with the nuclear fuel materials UO_2, ThO_2, and PuO_2, CeO_2 is also an important analogue for studies of radiation effects in actinide oxides, as it obviates the need for handling of radioactive or highly regulated compounds. Due to these applications involving operation under extreme conditions, CeO_2 has been studied in great detail in recent decades, particularly focusing on its behavior in harsh chemical environments, as well as during exposure to high temperature and intense irradiation. Perhaps the most extreme and least understood of these conditions is high energy heavy ion irradiation.

Swift heavy ions (SHIs), having energies >0.5 MeV/u, can be produced at large accelerator facilities and are used for a wide range of applications. The highly transient energy deposition associated with swift heavy ion irradiation leads to far-from-equilibrium material conditions, similar to those induced in nuclear fuels by the slowing down of energetic fission fragments. SHIs are therefore used to study radiation effects in nuclear materials under well-controlled experimental conditions [14]. SHIs are also used to mimic the effects of galactic cosmic rays in electronic materials [15], to study degradation mechanisms in accelerator components [16], and for medical treatment purposes [17]. Finally, SHIs can be harnessed to tailor materials by inducing modifications to their structures and properties that are not attainable by conventional processing techniques. Examples include the production of nanostructures [18–21] and the tuning of optical properties [22] for engineering applications.

Highly energetic heavy ions deposit energy to a material's electrons primarily via excitation and ionization processes [23]. This energy dissipates radially in the electronic subsystem and is then transferred to the atomic system through electron–phonon coupling, leading to high energy densities on the order of eV/atom. This results in rapid heating in the material over picosecond time scales [24,25] within a nanoscale region around the ion path (1–10 nm in the radial direction), which causes a thermal spike and possibly localized melting, followed by rapid quenching [26]. These processes result in complex structural and chemical modifications along the ion path, leaving behind cylindrical damage zones with radii of a few nm and lengths of 10 s of μm, known as ion tracks. The type of irradiation damage that is induced within ion tracks depends on the target material and can include the formation of defects [27], disorder [28], amorphization [29], and crystalline-to-crystalline phase transitions [30].

Fluorite-structured binary oxide materials are generally resistant to the structural modifications induced by swift heavy ion irradiation. Irradiation-induced defect formation is commonly limited to isolated point defects and extended defect clusters. This is often accompanied by the build-up of heterogeneous microstrain and, at sufficiently large defect concentrations, longer-range material modifications such as unit cell swelling. Still the long-range fluorite structure is commonly maintained.

As ion tracks accumulate with increasing ion fluence, the swelling and microstrain typically increase in a linear fashion at relatively low fluences, then saturate at higher fluences when tracks eventually overlap. This behavior is described by the so-called single impact mechanism [31]. The mathematical description of this mechanism is shown in Equation (3) for the increase in unit cell parameter as a function of increasing ion fluence:

$$\frac{\Delta a}{a_0} = \frac{a(\phi) - a_0}{a_0} = \frac{a_{sat} - a_0}{a_0} \left(1 - e^{-\sigma\phi}\right) \tag{3}$$

where a is the measured unit cell parameter at a given ion fluence ϕ, a_0 is the reference unit cell parameter of the unirradiated material, a_{sat} is the saturation value of the unit cell parameter at high ion fluence, and σ is the cross-sectional ion track area.

Understanding the intermediate mechanisms by which electronic excitation and the subsequent atomic displacement yield changes to the long-range structure and chemistry of CeO_2 is critical to its performance in various engineering applications. However, these processes are complex, multiscale, and difficult to fully characterize. To bridge the gaps between SHI irradiation-induced electronic excitation, atomic displacement, defect production, and bulk material modification, we review here in detail the unique response of CeO_2 to swift heavy ion irradiation, as revealed by a wide range of characterization techniques. After initially summarizing the most important characterization techniques, this article first reviews work investigating the fundamental structural and electronic modifications induced by swift heavy ions that lead to the formation of ion tracks in CeO_2. This is followed by a description of how these radiation effects are influenced by the irradiation conditions (ion energy, ion stopping power, temperature, and pressure) and sample characteristics (chemical composition and microstructure). The manuscript concludes with a brief comparison of swift heavy ion effects in CeO_2 with those in related fluorite-derivative oxides, followed by a summary and outlook.

2. Methods

2.1. Irradiation Conditions

The controlled production of SHIs requires large, dedicated user facilities, rather than the more prevalent and accessible laboratory-based tandem ion accelerators. SHI irradiation of CeO_2 materials has been performed at several large accelerator facilities worldwide [23]. However, these facilities cover different ranges of ion species and energies. The energy deposited in a material per unit length (known as energy loss, dE/dx) depends on ion mass and energy and is a key parameter in SHI irradiation experiments. For highly energetic ions, energy is deposited primarily to a material's electronic subsystem (electronic energy loss, dE/dx_e), with only minor contributions from nuclear collisions (nuclear energy loss, dE/dx_n).

The effects of SHI energy deposition parameters on the induced structural modifications in CeO_2 has been investigated over a broad dE/dx range between 15 and 45 keV/nm, as shown in Figure 2. Ions of specific energy ranging from 0.5–30 MeV/u (with u being the number of nucleons) and species ranging from ^{127}I–^{238}U have been utilized in these experiments [32–56]. The strong dependence of the energy loss on the ion energy leads to continuous change in dE/dx along the ion's penetration path in the material (Figure 2), which must be considered for characterization in order to accurately relate irradiation effects to the specific dE/dx within the volume probed by a particular characterization technique [57].

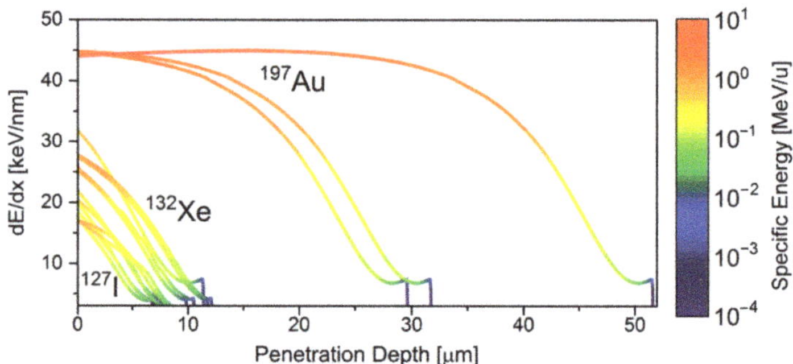

Figure 2. Energy loss profiles as a function of penetration depth determined with SRIM 2013 [57] for the ions used in the majority of studies [32–56] on SHI irradiated CeO_2, assuming 100% theoretical density. The color scale corresponds to the specific energy of ions and indicates the change in energy as ions penetrate into the material. The three ion species illustrate the range of masses and species used in these studies.

Other experimental parameters that are typically controlled in SHI irradiation studies include the ion flux, fluence, irradiation angle relative to the sample surface, and in situ environmental conditions (e.g., temperature and pressure). The many irradiation studies previously conducted on CeO_2 differ also with respect to sample properties. For example, various microstructures have been employed, ranging from single crystals [58] to polycrystalline pellets [32] to loose powder compacts [14]. As this review will show, all of these parameters greatly influence the response of CeO_2 to SHI irradiation. To gain further insight into radiation damage mechanisms, the impact of all ion-beam conditions and sample properties on the induced structural and chemical modifications must be understood at a holistic level.

2.2. Characterization

A wide range of characterization techniques have been employed to investigate the complex structural and chemical effects induced in CeO_2 by SHI irradiation. Such measurements are performed either ex situ (after ion irradiation) or in situ (directly at accelerator beamlines with various dedicated characterization infrastructures). Analytical methods used in prior work include various forms of scattering, spectroscopy, calorimetry, and electron microscopy. Some of these techniques provide direct insight into the damage structure within individual ion tracks (e.g., electron microscopy), while others are based on net damage accumulation and fluence-dependent measurements (e.g., X-ray diffraction). Characterization is commonly performed on materials after irradiation at ambient conditions to study damage formation, yet select works on irradiation at high temperatures and post-irradiation thermal annealing provide insight into damage recovery and defect dynamics. The following section gives a brief overview of the most prevalent characterization techniques used to investigate SHI irradiation effects in CeO_2 and other fluorite-structured oxide materials.

Scattering techniques provide information on the short-and long-range structural modifications to materials following SHI irradiation. Highly penetrating X-ray and neutron probes are commonly used for this purpose. X-ray scattering, which is sensitive to changes in the cation sublattice, is performed with either laboratory-based diffractometers or at large synchrotron facilities. Neutron probes are useful for investigating the structure of light (low-Z) atomic constituents in a material (e.g., O in CeO_2) because, unlike X-rays, neutrons scatter efficiently on atoms of low atomic mass (Z of the atomic constituent).

Two scattering techniques have been most extensively used in the study of irradiated CeO_2: diffraction and total scattering. Diffraction experiments allow for quantification of the long-range volumetric changes (unit cell swelling) caused by the production of defects [14,36,38,39,41,45–47,52,54,56,59,60], as well as heterogeneous microstrain and phase transformations, should they occur. Total scattering experiments are used to study the local defect structure using real-space analysis. Recently, intense spallation neutrons have become available for materials research at dedicated facilities, such as the Nanoscale Ordered Materials Diffractometer (NOMAD) at the Spallation Neutron Source, Oak Ridge National Laboratory (ORNL). High-resolution pair distribution function (PDF) analysis is utilized at NOMAD [61] to investigate short-range structural changes associated with irradiation-induced defects, which are inaccessible to conventional long-range diffraction methods [50,53].

Spectroscopic techniques provide insight into the local damage structure and changes to the chemistry of irradiated materials. X-ray absorption spectroscopy (XAS) has been used to probe cation oxidation state changes in irradiated CeO_2 and associated modifications to the local bonding environment [35,46,54,60]. X-ray photoelectron spectroscopy (XPS) provides insight into the electronic structure of cations by X-ray induced electron excitation and the consequent emission of characteristic photons during de-excitation [34,35,51,58]. Both XAS and XPS are valuable characterization tools for CeO_2 due to its tendency to chemically reduce under highly ionizing irradiation. Raman spectroscopy reveals the impact of defects on correlated atomic vibrations [38,49,55]. Structural modifications in fluorite-structured materials lead to a breakdown of selection rules and the appearance of so-called Raman forbidden modes in the spectra. Quantitative analysis of these forbidden modes provides information on defect concentrations. Raman spectroscopy further reveals the formation of $Ce-V_O$ defect complexes in irradiated CeO_2 [62]; thus, this technique is also useful for studying the material's redox response under SHI irradiation.

Electron microscopy (EM) is a powerful tool for the analysis of radiation effects in materials, as it provides direct imaging of the damage structure and its spatial distribution. High-resolution transmission electron microscopy (HRTEM), using state-of-the-art microscopes, has recently provided valuable insight into the atomic-scale nature of defects in irradiated CeO_2 [32,33,37,40,42,44,48]. This technique has provided new information on the size and damage morphology of individual SHI tracks. In modern microscopes, imaging capabilities are often coupled with other modes of operation such as electron diffraction and spectroscopy, which are utilized to determine complementary structural and electronic defect properties. Finally, heating stages in electron microscopes are useful to monitor changes in the damage structure and defect recovery as a function of increasing temperature.

Thermodynamic techniques like differential scanning calorimetry (DSC) provide information on the heat capacity of a material through comparison of its thermal behavior with that of an unirradiated reference sample under a precisely controlled high-temperature environment. With respect to irradiated samples, DSC is used to quantify the stored defect energy [53], providing insight into the nature of the initial defects and their kinetics during thermally induced recovery processes. When DSC is used in concert with complementary structural characterization techniques, the structure–energetics relationship of defects can be directly established. For example, in situ scattering measurements show how the defect structure recovers at high temperature, while DSC measurements yield the associated energetics of these processes [53].

3. Radiation Effects in CeO_2 Induced by Swift Heavy Ions

3.1. Impact on Structure and Chemistry

DSC measurements on SHI irradiated CeO_2 by Shelyug et al. [53] revealed that only ~1% of the energy deposited by ions is subsequently stored in irradiation-induced defects within the structure. XRD measurements of irradiated samples display distinct changes in Bragg peak position, intensity, and width in CeO_2 after SHI irradiation, indicative of unit cell expansion and an increase in structural distortions around defects (microstrain) (Figure 3a) [14,36,38,39,41,45–47,52,54,56,59,60]. The original fluorite structure peaks are retained, and no diffuse scattering is apparent, indicating a very high resistance to irradiation-induced amorphization. The observed changes in the XRD patterns are consistent with the formation of point defects and defect clusters. HRTEM measurements suggest that oxygen Frenkel pairs are the primary defect produced following SHI irradiation [42], a finding supported by molecular dynamics simulations [63]. Dislocation loops have been observed in SHI irradiated CeO_2 above a threshold stopping power of 12 keV/nm [40].

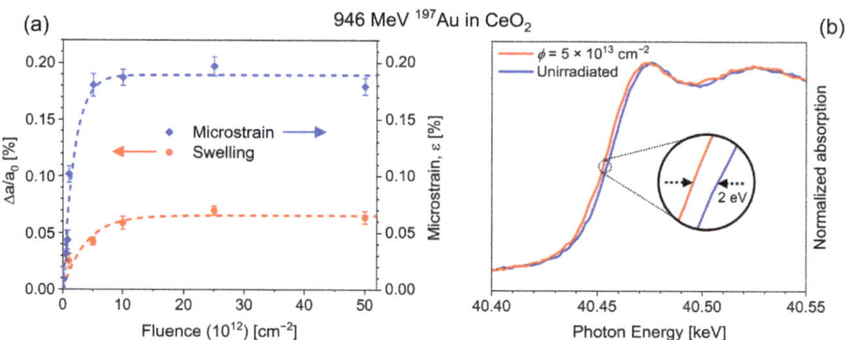

Figure 3. Structural and chemical changes in CeO_2 after irradiation with 946 MeV ^{197}Au ions, as adapted from Tracy et al. [46]. (**a**) Relative change in unit cell parameter (red) and heterogeneous microstrain (blue) based on X-ray diffraction measurements as a function of ion fluence and (**b**) corresponding X-ray absorption spectroscopy measurements of an unirradiated (blue) sample and after irradiation to a fluence of 5×10^{13} ions/cm^2 (red).

XAS and XPS measurements on SHI irradiated CeO_2 reveal that structural changes are accompanied by partial reduction of nominally Ce^{4+} cations to Ce^{3+} [34,39,46,51,54,58]. XAS spectra show a shift in the K-edge absorption energy of approximately -2 eV (Figure 3b), which suggests a partial reduction to the trivalent state rather than the transition of all cerium cations to the trivalent state, which corresponds to a shift of -7 eV [64]. This evidence of SHI irradiation-induced reduction is corroborated by magnetic measurements that display ferromagnetism in SHI irradiated CeO_2, indicative of the magnetic moments produced by the 4f electrons in Ce^{3+} cations [39,56]. Iwase et al. [34] observed a saturation of this redox behavior at a Ce^{3+}/Ce^{4+} ratio of ~12% at a maximum fluence of 6×10^{13} ions/cm^2 using 200 MeV ^{132}Xe ions. Coupled XRD and XAS measurements by Tracy et al. [46,59] revealed that the observed redox changes are directly linked with unit cell swelling and microstrain buildup, as the fluorite structure must distort to accommodate larger trivalent cations (1.14 Å versus 0.97 Å of initial tetravalent cations), as well as oxygen vacancies [4]. In certain cases, irradiation induced reduction occurs to a sufficient extent that the formation of a secondary phase is observed at high fluences [43,46,52]. This hypostoichiometric trigonal $Ce_{11}O_{20}$ phase consists of Ce^{4+} and Ce^{3+} cations along with ordered oxygen vacancies.

While changes in unit cell parameter and microstrain of CeO_2 (Figure 3a) follow a behavior that is consistent with a single impact model [31,46] (Equation (3)), it remains unclear whether or not the induced redox changes follow the same trend. XPS data from Iwase et al. [34] suggest a single impact mechanism for redox effects, based on the increase in Ce^{3+} cations as a function of ion fluence. Still, additional research is needed to accurately monitor the structural and chemical changes over a range of irradiation conditions, ideally using coupled XRD and XAS measurements to better understand the formation and accumulation of Frenkel-type defects (linked to structural changes) and redox-type defects (linked to chemical change).

3.2. Defect Formation and Ion Track Morphology

Irradiation induced unit cell expansion and the accumulation of heterogeneous microstrain are typically studied with XRD experiments designed to track net damage accumulation in a series of samples irradiated to increasing ion fluences. This is useful for the determination of ion track cross sections and effective track diameters, but provides no direct insight into the morphology of individual ion tracks. Electron microscopy, on the other hand, allows for direct imaging of tracks.

A recent HRTEM investigation by Takaki et al. [42] provided detailed insight into the size and damage morphology of 200 MeV ^{132}Xe ion tracks in CeO_2. This provides the basis for more fundamental understanding of the defect mechanisms leading to the formation of SHI tracks. A core-shell track morphology was observed, wherein the interior of the track is oxygen deficient and the annular shell oxygen rich (Figure 4a). This suggests that SHI traversal and associated energy deposition causes the radial expulsion of oxygen from the ion path region. Since cerium's electronic structure is flexible, the oxygen vacancies within a core region are stabilized by charge compensation from partial cation reduction (Ce^{4+} to Ce^{3+}). These measurements also showed that oxygen anions are displaced up to 17 nm from the center of the ion track [42]. Neutron total scattering measurements of CeO_2 irradiated with 2000 MeV ^{197}Au ions showed that a fraction of the displaced oxygen atoms form peroxide-like defect clusters (Figure 4b) [50]. These defect clusters may act as a structural and chemical compensation mechanism to balance the oxygen interstitials with their counterpart oxygen vacancies stabilized through cation reduction.

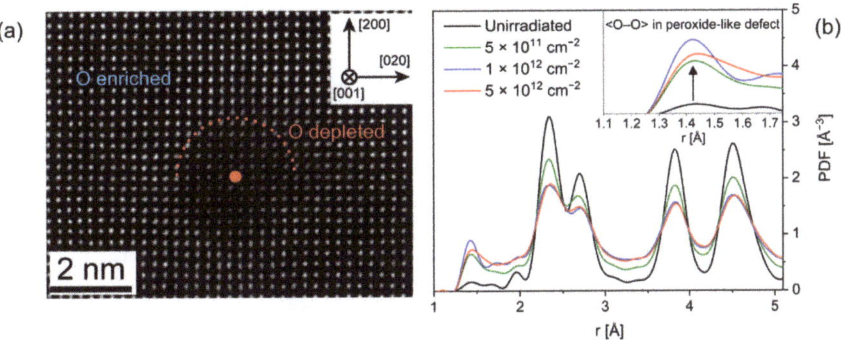

Figure 4. (a) HRTEM image of a single 200 MeV ^{132}Xe ion track in CeO_2 with a core-shell damage morphology, consisting of an oxygen depleted core (red) and oxygen interstitial rich shell (blue). (b) Neutron pair distribution functions (PDFs) of CeO_2 before and after irradiation with 2000 MeV ^{197}Au ions up to 5×10^{12} ions/cm^2. A loss of structural order is indicated by the symmetric peak broadening and the decrease in the intensity of correlation peaks. A structural feature at ~1.45 Å is observed after irradiation (inset) indicative of the formation of peroxide-like defects. Adapted from (a) Takaki et al. [42] and (b) Palomares et al. [50].

A combination of the results from XRD and EM measurements provides more comprehensive information on SHI tracks in CeO_2, including size, morphology, and internal damage structure. The areal extent of changes in unit cell parameters and microstrain within a single ion track can be deduced from the fitting of fluence-dependent XRD data (single-impact model, Equation (3)). The comparison of effective track diameters reveals a systematic discrepancy between those determined from unit cell expansion (3.9–5.8 nm) and microstrain (4.6–10.5 nm) over a range of irradiation conditions (^{132}Xe and ^{197}Au ions of 167 and 946 MeV energies) [46,52]. Relating these XRD-based results with diameters obtained by HRTEM investigation of SHI tracks (core diameter: ~4 nm and core + shell: ~17 nm) suggests that most of the swelling induced in CeO_2 occurs in the core region, since this matches well with the track diameter determined from analysis of unit cell expansion data. Thus, the unit cell expansion can be attributed to defects within the track core, which are predominantly oxygen vacancies and reduced Ce^{3+} cations. Microstrain can instead be attributed to distortions arising from all defects within the core and shell, such that the effective diameter associated with the region of increased microstrain represents the total ion track size, including both the core and shell periphery.

Besides complex ion tracks, SHIs have been shown to also induce interesting surface damage morphologies in CeO_2, as described by Ishikawa et al. [44]. For ions incident in oblique directions relative to sample surfaces, the formation of hillocks was observed. These hillocks are spherical in shape and crystalline, with an ideal fluorite structure of similar atomic spacing to that of the unirradiated matrix. The spherical hillocks are located above ion tracks and have a mean diameter of 10.6 ± 1.3 nm for irradiation with 200 MeV ^{197}Au ions. Currently, CeO_2 and $Gd_2Zr_2O_7$ are the only materials in which SHI irradiation-induced hillocks have been shown to exhibit a fully crystalline structure with no amorphous component [44,65]. This is consistent with the exceptionally high radiation tolerance of CeO_2.

The spherical, droplet-like shapes of these hillocks imply the influence of surface tension in a liquid phase, such that the observation of hillocks on the surface of SHI irradiated CeO_2 supports the conclusion that a thermal spike and localized melting occur within ion tracks. In this scenario, all atoms are displaced from their original sites over picosecond time frames, with oxygen anions moving further away from the location of the original ion path than cerium cations. Rapid quenching restores the initial crystal structure, but some defects and defect clusters remain. The core-shell damage morphology is therefore a remnant of these highly transient processes, and the separation of oxygen anions from the track core and incomplete recovery result in the observed oxygen defect clusters and cation oxidation state reduction. This is supported by the molecular dynamics modeling of Devanathan et al. [63], who showed that the rapid increase in temperature within ion tracks in CeO_2 and the subsequent quenching process do not result in complete restoration of the initial atomic arrangement.

4. Effects of Varying Irradiation Conditions

The types of modifications that are induced in CeO_2 by SHI irradiation depend on a number of parameters that are individually adjusted in each irradiation experiment. These can be grouped into (i) ion beam settings: ion species, energy, energy loss, fluence, and flux; (ii) environmental conditions: irradiation temperature and pressure; and (iii) sample properties: grain size and impurity content. The following sections briefly summarize the effects that each parameter can have on the radiation response of CeO_2.

4.1. Ion Beam Conditions

The irradiation response of CeO_2 has been studied using a wide range of ion species (^{127}I–^{238}U), energies (0.5–30 MeV/u), dE/dx (~15–45 keV/nm), ion fluxes (10–10^{10} ions/cm^{2}s), and ion fluences (10^{11}–10^{16} ions/cm^2) [32–56]. The ion energy and energy loss strongly influence the induced structural damage as seen in unit cell parameter changes with increasing fluence for various irradiation conditions (Figure 5a). For SHI irradiations within the electronic dE/dx regime, there typically exists a material dependent energy loss threshold above which ion tracks will form [23]. Track formation has been documented by TEM in SHI irradiated CeO_2 for an energy loss of ~16 keV/nm [37], which suggests a lower threshold when compared with other fluorite-structured materials. For example, UO_2 exhibits a threshold between ~22–29 keV/nm [66]. However, the dE/dx threshold for track formation has not been accurately determined for CeO_2 over a wide range of ion species and energies; this should be the subject of future research.

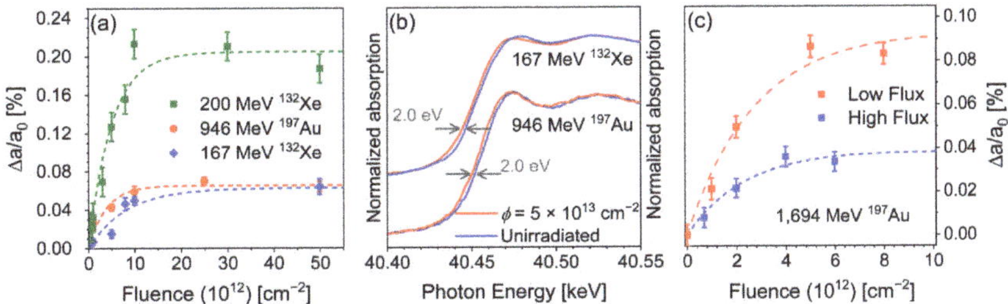

Figure 5. (**a**) Relative change in unit cell parameter based on XRD pattern analysis as a function of fluence for CeO_2 irradiated with 200 MeV ^{132}Xe (green, dE/dx = 27 keV/nm), 946 MeV ^{197}Au (red, dE/dx = 44 keV/nm), and 167 MeV ^{132}Xe (blue, dE/dx = 27 keV/nm). (**b**) X-ray absorption spectra of the cerium K-edge in CeO_2 before irradiation (blue) and after irradiation to a fluence of 5×10^{13} ions/cm^2 (red) with 167 MeV ^{132}Xe (top) and 946 MeV ^{197}Au (bottom). (**c**) Relative change in the unit cell parameter, determined by XRD pattern analysis, for CeO_2 irradiated with 1694 MeV ^{197}Au ions at ion fluxes of ~10^9 ions/cm^2/s (red) and ~10^{10} ions/cm^2/s (blue). Dashed lines in (**a**,**b**) are fits based on a single-impact model. These data are based on work published by Tracy et al. [46] (**a**,**b**) and unpublished work (**a**,**c**).

The energy loss governs the nature and extent of the radiation damage induced in a material. In most cases, a higher energy loss induces a higher energy density within the track region, which produces more defects and results in the formation of defect clusters [53]. Sonoda et al. [37] demonstrated that the track size increases with increasing dE/dx in CeO_2, which agrees well with thermal spike calculations [67] based on the Szenes model [68], indicative of a quadratic relation between track diameter and dE/dx value. While the amount and type of defects, as well as track size, show a clear dependence on energy loss, redox changes of the cerium cations appear to be only weakly dependent on the dE/dx (Figure 5b). This behavior is not well understood, and it remains unclear if cation reduction also has a critical dE/dx threshold akin to track formation and whether or not they are the same.

The energy loss alone cannot fully explain the observed radiation behavior, as demonstrated by the large change in unit cell parameter values in CeO_2 (Figure 5a) induced by two different ion irradiation experiments using nearly the same dE/dx (~27 keV/nm for both 167 MeV ^{132}Xe and 200 MeV ^{132}Xe ions). This is explained by the additional impact of ion energy, also known as the velocity effect [23,69]. SHIs lose their energy to the electron subsystem by collisions with electrons, exciting them to high-energy states that are sometimes sufficient for ionization from the target atoms. The maximum energy imparted to these so-called delta electrons is determined by the ion velocity: higher velocity ions yield electrons with higher kinetic energies, allowing them to travel further away from their initial positions. For irradiations with comparable energy loss values, ions with higher velocities (kinetic energies) will deposit their energy over a larger volume. This results in larger ion tracks, but lower energy densities within those tracks, yielding less pronounced in-track material modifications (Figure 5b) [46,54].

Differential scanning calorimetry measurements by Shelyug et al. [53] further illustrated this velocity effect. A comparison of the enthalpy of radiation damage (the energetic difference between pristine and irradiated samples) in CeO_2 irradiated with 1100 and 2200 MeV ^{197}Au ions revealed that the higher velocity ions produced tracks with lower defect concentrations, although the track diameters were larger than those produced by the lower velocity ions. These results were corroborated by neutron total scattering measurements and fitting of a single impact model to the measured damage accumulation. To date, no dedicated studies of the velocity effect have been performed on CeO_2. Future research should further investigate the influence of SHI velocity on the induced damage structure.

In addition to ion mass and energy (i.e., dE/dx), as well as ion velocity, the ion flux on the sample has a substantial effect on the observed radiation response. Irradiation-induced material modifications are often studied in CeO_2 by irradiation to high fluences, at which ion tracks overlap (e.g., ~10^{13} ions/cm^2). To reach these fluence values within reasonable irradiation times, high ion fluxes (fluence per unit time, given in ions/cm^2/s) are often used. The flux used in a given experiment also depends on the accelerator and the beam mode utilized (e.g., pulsed versus continuous); these can vary greatly among facilities.

During irradiation with a high ion flux, relatively large amounts of energy are deposited in the sample over a short time interval. For sufficiently high fluxes, the resulting increase in thermal energy can outrace its dissipation, yielding high sample temperatures. As shown in Section 4.3, the irradiation temperature can influence the radiation behavior in insulators like CeO_2. In contrast, a lower ion flux allows more time for the dissipation of thermal energy, and therefore produces less bulk sample heating and reduced mobilities of irradiation induced defects. This impedes the recovery processes, yielding higher defect concentrations and more extensive material modifications. This is demonstrated by the swelling behavior of CeO_2 irradiated with 1694 MeV Au ions at two different fluxes but otherwise identical irradiation conditions (Figure 5c). The higher flux irradiation leads to a faster unit cell parameter increase as a function of ion fluence, consistent with a larger ion track (ion track size is proportional to the slope of the initial linear region). However, the saturation value of swelling, which is caused by the concentration of defects within tracks, is greatly reduced compared to the low-flux irradiation (Figure 5c). This shows that the ion-beam flux is an important parameter that must be considered when comparing results from different irradiation experiments. It remains unclear how ion-matter interactions are impacted over a large range of fluxes, whether there is an effect beyond the increase in temperature, and whether or not there is a critical flux value below which thermal effects can be neglected. Systematic studies are needed to quantify the effect of ion-beam flux on defect formation and recovery processes.

4.2. Sample Microstructure

Swift heavy ion irradiation of CeO_2 has been performed on samples produced by various synthesis and processing procedures, resulting in a range of microstructures with grain sizes from the nm-scale to bulk. The microstructure of a material influences its properties, including mechanical behavior, transport properties, and radiation response. For example, grain boundaries serve as defect sinks during irradiation with low-energy ions, leading to an enhanced radiation tolerance of nanocrystalline materials due to their high grain boundary densities [70]. This is in contrast to the reduced stability of nanocrystalline CeO_2 reported for SHI irradiation experiments [43]. Tracy et al. [46] and Cureton et al. [52] have shown that upon irradiation with 946 MeV ^{197}Au ions, nanocrystalline CeO_2 (grain size: 20 nm) exhibits greater unit cell swelling [52] (Figure 6a) and redox changes [46] compared with microcrystalline CeO_2 (grain size: 2 μm).

The grain size also has an influence on the phase stability of ceria. The irradiation-induced formation of hypostoichiometric $Ce_{11}O_{20}$ has been observed for both micro- and nanocrystalline materials, but to a much larger extent for the latter (Figure 6b,d) [43,46,52]. This phase transition is linked to the reduction of cerium cations and associated oxygen expulsion processes described in Section 3.2 [52]. The unit cell parameter in $Ce_{11}O_{20}$ is larger than that of the initial fluorite phase and further increases with ion fluence, indicating that defects continue to accumulate in this hypostoichiometric phase (Figure 6a). The phase transition is apparent in nanocrystalline CeO_2 at much lower fluences, and $Ce_{11}O_{20}$ grows at a faster rate compared with microcrystalline CeO_2 (Figure 6d).

The greater unit cell swelling in nanocrystalline CeO_2 (Figure 6a) can be explained by the ion-track formation mechanism discussed in Section 3.2. SHI tracks form in CeO_2 via expulsion of oxygen anions in the radial direction, away from the ion path. This causes reduction of cerium cations in the track core where oxygen is depleted, while the track shell is enriched with oxygen point defects and small defect clusters. As shown by Takaki et al. [42], oxygen can be expelled by tens of nm, which corresponds to the grain size of samples used in the study shown in Figure 6. During quenching of the excited track region, some of the oxygen will diffuse back to the core and recombine with oxygen vacancies, eliminating defects and decreasing the number of reduced cerium cations. In nano-ceria, however, the fraction of oxygen that diffuses back can be assumed to be reduced due to losses to grain boundaries. For a given fluence, this leads to a larger unit cell increase as compared with microcrystalline CeO_2. This scenario is supported by previous X-ray absorption measurements (Figure 6c) [46], revealing increased redox changes in nanocrystalline ceria (edge shift of 4.3 eV compared with 2.0 eV in the microcrystalline material). It further explains the enhanced formation of hypostoichiometric $Ce_{11}O_{20}$ in nano-ceria. This discussion shows that the microstructure of CeO_2 must be considered when comparing results from different SHI irradiation experiments. Systematic research is needed to quantify the swelling and redox changes of CeO_2 as a function of grain size, particularly in the 1 nm–1 μm range, for which oxygen diffusion may control the radiation response.

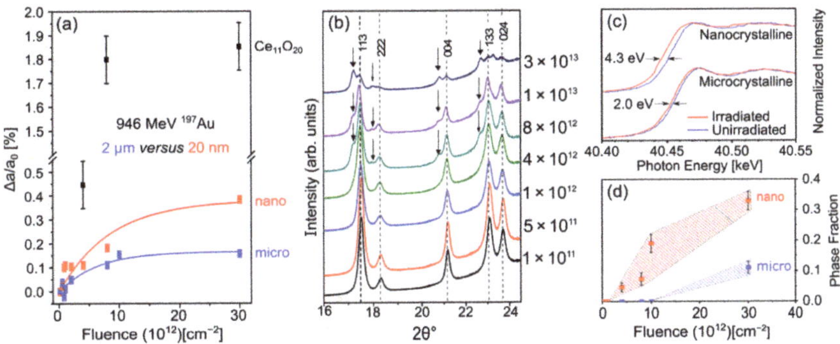

Figure 6. (a) Fluence-dependent change in unit cell parameter based on XRD pattern analysis for microcrystalline CeO_2 (blue, grain size = 2 μm), nanocrystalline CeO_2 (red, grain size = 20 nm), and the $Ce_{11}O_{20}$ phase produced in nanocrystalline samples (**black**), all relative to the unit cell parameter of unirradiated CeO_2. (b) X-ray diffraction patterns of irradiated nanocrystalline CeO_2 as a function of fluence, displaying the emergence of new peaks corresponding to the $Ce_{11}O_{20}$ phase, as indicated with arrows. (c) X-ray absorption spectra of the cerium K-edge in CeO_2 before irradiation (blue) and after irradiation to a fluence of 5×10^{13} ions/cm^2 (red) in nanocrystalline (**top**) and microcrystalline (**bottom**) samples. (d) Phase fraction of $Ce_{11}O_{20}$ relative to the fluorite phase as a function of ion fluence for microcrystalline (blue) and nanocrystalline (red) CeO_2. Irradiation was performed with 946 MeV Au ions. These data were adapted from Cureton et al. [52] (**a,b,d**) and Tracy et al. [46] (**c**).

4.3. High-Temperature Conditions

Irradiation temperature is a key parameter to consider for cerium dioxide as an analogue for nuclear fuels. Nuclear light water reactor (LWR) fuel operating conditions range from room temperature at reactor startup to a typical maximum of ~1200 °C under normal operation. In general, increased temperature enhances defect recovery due to higher defect mobility, but it might also enhance defect production and promote more complex defects due to higher initial temperatures within an ion-induced thermal spike. The profound effect of irradiation temperature on defect production in CeO_2 is demonstrated by the flux effect discussed in Section 4.1 (Figure 5c). During ion-beam experiments, high temperatures are typically achieved by mounting samples on stages with resistive coils that heat the material.

Prior SHI irradiation studies of CeO_2 have demonstrated a systematic decrease in the saturation level of swelling [54] and the ion track diameter [32,33] with increasing irradiation temperature (Figure 7a,b). Both changes are consistent with thermally-driven defect recovery due to enhanced defect mobility; however, a more comprehensive understanding is provided by consideration of the thermal-spike model of track formation. This mathematical model describes the interaction of swift heavy ions with a material in terms of a rapid increase in temperature over picosecond timescales induced by the high energy deposition within a nm-sized track region, followed by rapid quenching that freezes in structural damage [67,71]. The initial sample temperature is an input parameter in the thermal-spike model [49] that is added to the transient temperature spike and affects the final track diameter; higher temperatures typically yield larger ion tracks.

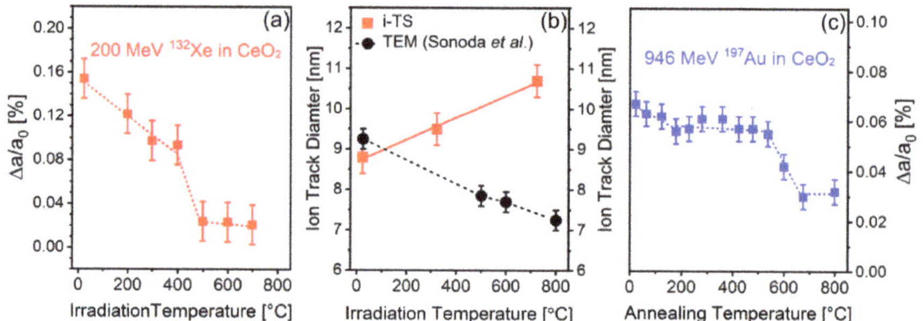

Figure 7. (**a**) Change in unit cell parameter from XRD pattern analysis as a function of irradiation temperature for CeO$_2$ irradiated with 200 MeV ^{132}Xe ions, relative to unirradiated reference samples heated under identical conditions [54]. (**b**) Ion track diameters determined by inelastic thermal-spike calculations [54] (red) and measured based on TEM images [32,33] (black) as a function of irradiation temperature. (**c**) Effect of thermal annealing on the relative change in unit cell parameter in CeO$_2$ previously irradiated with 946 MeV ^{197}Au ions [45]. Dashed and solid lines are to guide the eye. This figure was adapted from (**a**,**b**) Cureton et al. [54], (**b**) Sonoda et al. [32,33], and (**c**) Palomares et al. [45].

In CeO$_2$, the thermal spike model predicts an increase in track diameter as a function of increasing irradiation temperature up to 700 °C [54] (Figure 7b). This prediction is not supported by TEM characterization, which shows an opposite trend [32,33]. This discrepancy either indicates that the thermal-spike model does not fully capture all aspects of ion-matter interactions at elevated temperatures, or that the model and experiment describe two different track regions. The thermal spike approach accounts for the entire track region (i.e., the full range of delta electron pathways and associated electron-phonon coupling), and the deduced track diameter represents the size of the track core plus the shell (see Section 3.2). TEM characterization, on the other hand, is sensitive to changes in electron density, and the measured track diameter mostly represents the smaller track core, which is enriched in (high-Z) cerium cation defects [32,33]. Under these assumptions, the discrepancy between TEM data and thermal spike calculations could indicate that the track core shrinks with increasing irradiation temperature, while the shell thickness increases. This is consistent with XRD results and the reported reduced unit cell expansion at higher irradiation temperature, since the track core is primarily responsible for ion-induced swelling (oxygen vacancies and associated Ce^{3+} cations). Systematic ion-beam studies over a range of temperatures (including cryogenic) are required to gain further insight into the manner in which track formation is modified by increases in the irradiation temperature. For example, conventional scanning transmission EM imaging (which is more sensitive to changes at cation sublattice) coupled with annular bright-field (ABF) imaging (which more sensitive to changes at anion sublattice) could reveal changes in ion track core and shell sizes at various irradiation temperatures.

In addition to the in situ heating experiments described above, where temperature is applied during ion-beam exposure, samples can be thermally annealed after irradiation to elucidate defect kinetics and damage recovery mechanisms [45]. Synchrotron XRD measurements of CeO$_2$ irradiated with 946 MeV ^{197}Au ions and subsequently annealed within a hydrothermal diamond anvil cell [14] revealed a two-step defect recovery mechanism with corresponding activation energies of 1.0 and 2.1 eV (Figure 7c). These activation energies were attributed to O-interstitial migration and Ce-vacancy migration, respectively, but the reoxidation of Ce^{3+} to Ce^{4+} after oxygen vacancy annihilation was not considered [72]. Unlike isostructural ThO_2, the damage recovery in ceria remained incomplete up to 800 °C, with a unit cell parameter increase of ~0.03% relative to an unirradiated reference sample remaining at this highest achieved temperature [45]. This suggests that defect clusters with relatively high binding energies, such as $2Ce^{3+}–V_O^{2-}$ complexes, are stable up to 800 °C and require higher annealing temperatures for recovery.

High-temperature calorimetry measurements of previously irradiated CeO_2 (using 1100 MeV and 2200 MeV ^{197}Au ions) revealed that defect recovery is enhanced within an oxygen atmosphere compared with heating in an inert environment [53]. This suggests that recovery processes related to the reoxidation of ion induced Ce^{3+} cations to Ce^{4+} play an important role in the annealing of SHI damage in CeO_2 and must be considered [53]. Thus, ex situ annealing experiments are useful for identifying and characterizing the different defects that form in CeO_2 during SHI irradiation.

4.4. Combined Pressure and Ion Irradiation

Pressure is another parameter which can be adjusted during SHI irradiation. While irradiation is typically carried out in vacuum conditions, a limited number of investigations have focused on the combined effects of ion irradiation and high pressure. The use of SHIs, as opposed to ions of lower energies, is essential in such efforts, as energies on the order of 200 MeV/u are required to penetrate the mm-thick anvils of a conventional high-pressure cell. The synergistic effects of pressure and ion irradiatio often yield material modifications that cannot be obtained otherwise [73].

The high-pressure response of CeO_2, in the absence of irradiation, is characterized by a sluggish phase transformation above ~30 GPa to the $PbCl_2$-type cotunnite phase [74,75]. This transformation is typical of fluorite-structured materials, and in ceria it reaches completion at ~50 GPa. When irradiation is conducted at high pressure, this transformation is modified. Figure 8 illustrates the effects of separate and coupled pressure and irradiation, revealing that pressure significantly modifies the response of CeO_2 to SHI irradiation and vice versa.

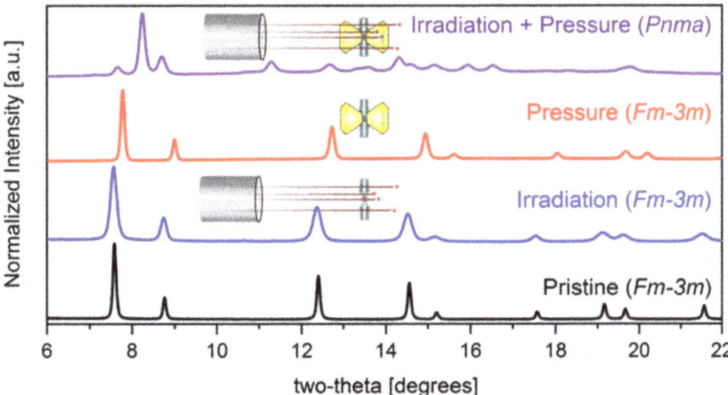

Figure 8. X-ray diffraction patterns of CeO_2 exposed to 4×10^{12} 946 MeV ^{197}Au ions/cm^2 at ambient conditions (blue), pressurized to 21.7 GPa in a diamond anvil cell (DAC) in the absence of irradiation (red), and exposed to a combination of pressure and irradiation, utilizing compression in a DAC to 21.8 GPa and irradiation with 7100 MeV ^{238}U ions to a fluence of 4×10^{12} ions/cm^2 (violet).

Irradiation with 7100 MeV ^{238}U ions to a fluence of 4×10^{12} ions/cm^2 at a pressure of only 21.8 GPa (roughly 10 GPa below the typical transformation onset pressure) triggers a complete fluorite-to-cotunnite transformation. The highly localized and dense electronic excitations produced by SHIs provide a means of overcoming the energy barrier for cotunnite high-pressure phase formation. According to ab initio calculations, hyper-

and hypo-stoichiometry increase and decrease, respectively, the critical phase transition pressure in fluorite-structured oxides [76,77]. The hypostoichiometry observed in CeO_2 ion track cores might therefore explain the transformation to the cotunnite phase at a lower-than-expected pressure. Moreover, the formation of the cotunnite phase is proposed to proceed through buckling of the (111) cation planes into adjacent anion planes in the fluorite structure, effectively increasing cation coordination [76]. Expelling anions from the track core during SHI irradiation at high pressure should reduce the resistance of cation-plane buckling and enhance the efficiency of cotunnite formation. An interplay of thermal and dynamic pressure effects during the highly transient ion irradiation process could also play a role in the observed transformation behavior. This was supported by compression experiments on pre-irradiated ceria samples that did not find any evidence of an accelerated fluorite-to-cotunnite transformation. More systematic research is needed to fully understand the effects of coupled extremes of irradiation, pressure, and temperature.

5. Effects of Chemical Composition

This review has so far focused on SHI irradiation effects in pure CeO_2 having ideal or near-ideal stoichiometry. However, the study of related materials that deviate from this ideal composition can also provide valuable insight into its radiation response. First, swift heavy ion irradiation has been shown to induce local nonstoichiometry in CeO_2, such that later ion impacts will interact not with ideal CeO_2, but rather with a nonstoichiometric phase. Second, since CeO_2 is used as a surrogate for nuclear fuel materials, doping with different atomic species (mimicking the accumulation of fission products) is an important aspect to consider in SHI irradiations. Finally, because the redox chemistry of Ce appears to play a key role in the radiation response of CeO_2, a comparative study of structurally-related materials featuring cations with distinct redox behavior can help to isolate the effects of cation chemistry on the response of this material to SHI irradiation.

5.1. Effects of Doping

In addition to trivalent cerium cations, CeO_2 can incorporate a number of dopant species while maintaining the fluorite structure [78]. To date, only a few studies have focused on the effects of doping on the response of CeO_2 to ion irradiation. These studies investigated CeO_2 doped with trivalent lanthanide cations, given their similarity to the lanthanide Ce. Doping with the lanthanide sesquioxides Gd_2O_3 and Er_2O_3 has been shown to enhance irradiation-induced swelling and structural disorder in CeO_2 irradiated with 200 MeV ^{132}Xe ions, compared with an undoped reference sample [60,79–81]. This was explained in terms of defect stabilization by localized strain fields around the dopant cations, but changes to the redox behavior of the cerium cations by chemical doping were not considered.

Ab initio modeling by Lucid et al. showed that incorporation of dopants alters the energetics of cerium cation redox processes in CeO_2, with certain dopants inhibiting reduction (e.g., Sm) while others (e.g., Eu) facilitate it [82]. Since redox behavior plays a crucial role in SHI induced material modifications (particularly in nanocrystalline ceria), doping with certain lanthanide sesquioxides may provide a means of tuning ion-matter interactions and radiation stability. This should be addressed in future studies by comparing, for example, SHI induced unit cell changes in CeO_2 containing a range of dopant species (e.g., varying dopant sizes, masses, charge states, and redox energetics).

5.2. Dependence of Radiation Response on the A-Site Cation Species

As CeO_2 is utilized as an analogue to study radiation effects in nuclear fuel materials (e.g., UO_2 and ThO_2), it is particularly important to compare the SHI irradiation responses of these materials. Despite all three having the same fluorite structure, each A-site cation has a distinct electronic structure, resulting in varying behavior under highly ionizing irradiation. As mentioned previously, the track formation is very different between CeO_2 and UO_2, with the latter exhibiting no observable ion tracks after irradiation with fission fragments (dE/dx ~18–22 keV/nm) [66]. This suggests that UO_2 is able to dissipate the energy deposited by a SHI much more efficiently than CeO_2 under similar irradiation conditions. This behavior may be related to differences in the types of defects formed, as suggested by a molecular dynamics (MD) investigation [64]. In MD simulations 99% of SHI irradiation-induced defects were produced on the oxygen sublattice in UO_2, in contrast to CeO_2, where cerium and oxygen defects are produced in stoichiometric quantities after ion impact (redox changes not considered). The production of appreciable quantities of both cation and anion defects in CeO_2 led to a larger quantity of surviving defects after track quenching [64].

While the radiation responses of ThO_2 and CeO_2 are more similar, with both exhibiting a core-shell type track morphology, the extent of structural changes within individual ion tracks differs between these materials due to the accessible cation valence states (monovalent Th^{4+} versus Ce^{4+}/Ce^{3+}) [83]. The formation of Ce^{3+} cations with larger ionic radius and the correspondingly complex oxygen defect structure in CeO_2 yields more pronounced swelling and microstrain build-up in this material, relative to ThO_2, for which this redox-driven defect mechanism is inactive (Figure 9) [46,52,83]. Instead, the SHI radiation response of ThO_2 appears to be based solely on the accumulation of point defects and small, simple defect clusters [83]. The situation is more complex in UO_2, given its multiple accessible U cation oxidation states (U^{3+}, U^{4+}, U^{5+}, and U^{6+}), enabling cation reduction or oxidation. Raman spectroscopy and XRD measurements have shown that UO_2 undergoes some oxidation during SHI irradiation, which causes minor unit cell contraction (Figure 9) [52].

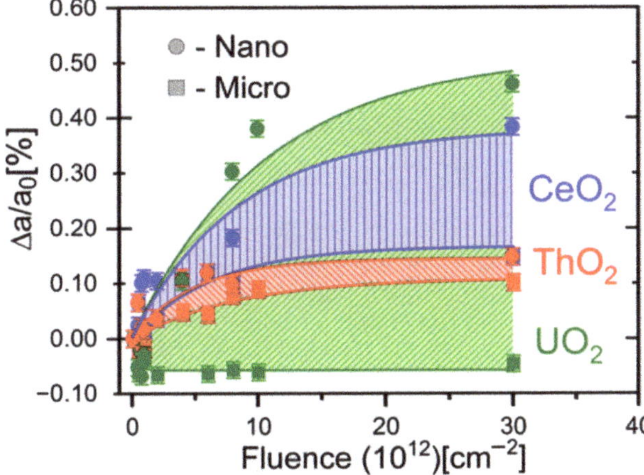

Figure 9. Relative change in unit cell parameter from XRD measurements as a function of fluence for microcrystalline (squares) and nanocrystalline (circles) UO_2 (green), ThO_2 (red), and CeO_2 (blue) irradiated with 946 MeV ^{197}Au ions. The shading displays the variation in irradiation-induced unit cell parameter changes between microcrystalline (2 μm) and nanocrystalline (20 nm) samples for each material. Adapted from Cureton et al. [52].

Changes in radiation behavior among CeO_2, ThO_2, and UO_2 are particularly evident when structural modifications are compared between microcrystalline and nanocrystalline materials. As shown in Figure 6c, ion-induced redox processes are more efficient in nanocrystalline CeO_2 (as discussed in Section 4.2), leading to a larger saturation level of swelling at high ion fluences (Figure 9). Since such redox effects are absent in ThO_2, its overall swelling behavior is very similar for both nano- and microcrystalline materials. The largest discrepancy in unit cell parameter changes after SHI irradiation is found in microcrystalline and nanocrystalline UO_2 (Figure 9). The former oxidizes under SHI irradiation, yielding unit cell contraction, while the latter exhibits a significant degree of swelling.

Characterization by X-ray diffraction and Raman spectroscopy suggest that the increase in structural disorder in nanocrystalline UO_2 results from oxygen loss to grain boundaries, implying that pronounced redox changes are induced during SHI irradiation [53].

These results highlight the manner in which the electronic structure of the cation dramatically changes the response of fluorite-structured oxides to swift heavy ion irradiation. This raises concerns about the use of CeO_2 as an analogue material for UO_2 in radiation-damage studies, at least for swift heavy ion (or fission-fragment) irradiation. The data shown in Figure 10 again emphasize that the grain size of a material plays a crucial role in ion-matter interactions, with its specific effects showing a strong dependence on composition.

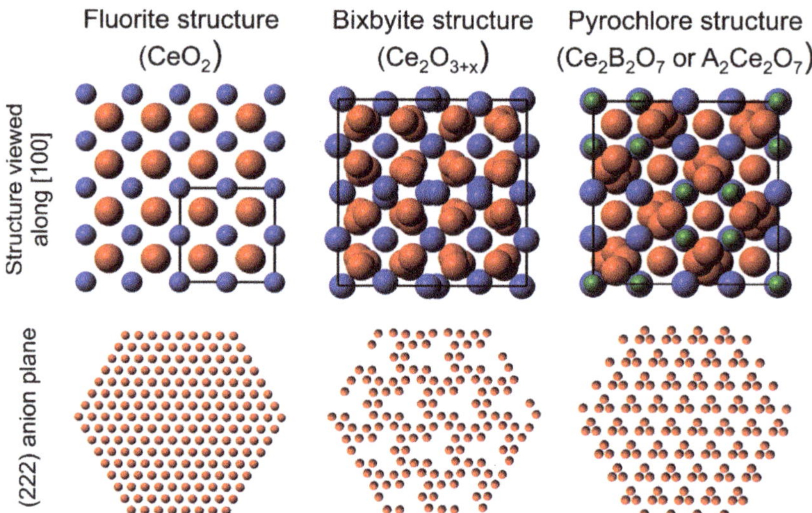

Figure 10. Ce-bearing oxides with fluorite, bixbyite, and pyrochlore structures viewed along the (100) direction (**top**) and the corresponding (222) plane anion layers (**bottom**). Blue and green spheres represent cations, while red spheres represent oxygen anions. Unit cells are delineated with black lines. The bixbyite structure can be characterized as a fluorite-derivative with 1/8 of the anions replaced by constitutional vacancies, while the pyrochlore structure is a fluorite-derivative with ordering of two cation species and $\frac{1}{4}$ of the anions replaced by constitutional vacancies.

5.3. Radiation Effects in Structurally-Related Lanthanide Oxides

If aliovalent dopants are added to CeO_2 in sufficiently high concentrations, ordering of these new cations can occur alongside ordering of the defects produced on the anion sublattice to maintain charge neutrality. Even in undoped CeO_2, defects can order and change the symmetry of the material if they accumulate in large quantities. Thus, dopant and defect ordering can yield new fluorite-derivative structures [84], with bixbyite-structured lanthanide sesquioxides and pyrochlore-structured lanthanide/transition metal oxides being two commonly-studied examples. These materials can be considered as defect-rich (sesquioxides) and heavily doped (pyrochlores) variants of CeO_2, and both exhibit anion-deficient fluorite-derivative structures (Figure 10) [85]. The following section summarizes the main SHI irradiation effects observed in these materials, which provide further insight into the radiation response of CeO_2.

The bixbyite structure is characteristic of most lanthanide oxides for which the lanthanide element is in the trivalent oxidation state (Ln^{3+}). This structure is a derivative of the fluorite structure, but with ordered constitutional vacancies replacing $\frac{1}{4}$ of the anion sites and the remaining atoms relaxed towards these vacant sites (Figure 10) [86]. The responses of bixbyite materials to SHI irradiation have been characterized using a range of ion energies and masses [30,87–90]. This prior work revealed a strong dependence of the radiation response on the cation ionic radius. The magnitude of the induced structural changes generally increases with cation size (and therefore decreases with cation mass, due to the contraction that occurs across the lanthanide series). Sesquioxides with small cations tend to retain their bixbyite structures under swift heavy ion irradiation, those with medium cations tend to undergo transformations to high temperature polymorphs, and those with large cations tend to amorphize. These modifications, which are generally proportional in magnitude to dE/dx, are attributed to the displacement of anions into constitutional vacancies which, when sufficiently extensive, can yield collective atomic relaxation to form accessible polymorphic or amorphous structures [30].

These processes provide insight into the likely behavior of anion-deficient CeO_{2-x} materials under irradiation. While Ce_2O_3 preferentially adopts a hexagonal structure, unlike most of the lanthanide sesquioxides, it is stable in cubic bixbyite-like phases starting at slightly higher oxygen contents [91,92]. Radiation damage mechanisms similar to those observed in the lanthanide sesquioxides have been reported for CeO_2, with the displacement of oxygen being a dominant mode of defect production (see Section 3.2) [50]. To date, the response of Ce_2O_3 to swift heavy ion irradiation has not been characterized. However, based on the ionic radius of Ce, which is large among the lanthanides, amorphization is expected to be the dominant response of this material. This suggests that cerium reduction and the concomitant introduction of oxygen vacancies will reduce the radiation tolerance of CeO_2 by making possible the oxygen displacement-driven transformation mechanisms previously observed in several lanthanide sesquioxides. To clarify this behavior, a detailed study of the swift heavy ion irradiation response of CeO_{2-x} materials as a function of x would be useful.

Like the bixbyite-structured oxides, the responses of pyrochlore-structured materials to SHI irradiation have been extensively studied, due in large part to their potential applications in the immobilization of nuclear wastes [93]. These materials exhibit the general formula $A_2B_2O_7$, where A is a large trivalent cation and B is a smaller tetravalent cation [94]. While Ce most often occupies the A-site position due to its relatively large ionic radius, it can occupy the B-site if paired with a larger A-site cation such as La [95]. Pyrochlore materials adopt a fluorite-derivative superstructure, differing from the fluorite structure in that two cations are ordered on the face-centered cubic cation sublattice, while 1/8 of the anions are replaced with constitutional vacancies, and the remaining anions are relaxed towards these vacant sites (Figure 10). The potential utility of these materials for nuclear waste immobilization arises, in large part, from their chemical flexibility, since a wide range of aliovalent cations of various sizes can be incorporated onto the two cation sites. Due to this compositional variability, research on the radiation responses of pyrochlore materials proves instructive with respect to the possible effects of extensive chemical doping on the radiation responses of cerium oxides.

Pyrochlore materials show clear compositional trends in radiation tolerance. Like the lanthanide sesquioxides, cation ionic radii are the primary determinant of these trends [28,96–99]. For pyrochlore materials the ratio of the A- and B-site cation radii, r_A/r_B, governs radiation tolerance. When this ratio is large, due to the inclusion of relatively large A-site cations or small B-site cations, the energy of cation antisite defect formation is large and the irradiation-induced disorder on the cation sublattice cannot easily be incorporated into the material's crystalline structure [100]. This typically yields amorphization in response to SHI irradiation. In contrast, for materials with small cation radius ratios, cation antisite defects are relatively easily accommodated by the structure and disordering to a highly radiation tolerant defect-fluorite structure is typical [100]. This order-disorder transformation entails the mixing of A- and B-site cations onto a single face-centered cubic cation site and the mixing of oxygen and constitutional vacancies onto a fluorite-like anion sublattice.

Since Ce is relatively large among the lanthanides, pyrochlore materials that include this element on the A-site usually feature large cation ionic radius ratios, making amorphization a likely response to swift heavy ion irradiation. This indicates that doping of CeO_2 with additional cations, particularly smaller transition metal elements, such as the fission fragments found in nuclear fuels and nuclear wastes, might reduce the radiation tolerance of this material. This doping, if sufficiently extensive, could make accessible irradiation-induced phase transformation pathways to amorphous phases, as well as short-range structural modifications resulting from local defect ordering [98,101,102]. Thus, deviation from the ideal CeO_2 chemical composition due to the introduction of other atomic species appears likely to have a deleterious effect on radiation tolerance in this system.

5.4. Implications for CeO_2

Comparison of the SHI irradiation response of CeO_2 to those of closely-related fluorite and fluorite-derivative materials suggests that deviation from the ideal fluorite chemistry and crystallography tends to reduce the radiation tolerance of cerium oxides. Both Ce-bearing, bixbyite-structured materials (which can be considered anion-deficient fluorite derivatives) and Ce-bearing, pyrochlore-structured materials (which can be considered heavily doped, anion-deficient fluorite derivatives) are susceptible to irradiation-induced amorphization, a radiation response that has not previously been observed in pristine CeO_2. Likewise, doping with relatively low levels of lanthanide cations has been shown to reduce the radiation tolerance of CeO_2.

Substantial alterations in chemistry and crystallography are likely to occur in many of the harsh nuclear environments in which the use of CeO_2 has been proposed. For example, use of this material as a nuclear fuel matrix will necessarily entail the introduction of new cation species, including a diverse array of fission fragment elements. The accompanying exposure to highly ionizing radiation of fission fragments will induce redox changes and modification from ideal stoichiometry. As discussed in Section 5.2, substitution of Ce with Th or U greatly alters the redox response under SHI irradiation, and therefore impacts the induced structural changes. Similarly, doping with cation species that modify the energetics of redox processes [82] may yield a radiation response that is strongly dependent on the electronic structure of the dopant element.

6. Summary and Outlook

Fluorite-structured CeO_2 is generally resistant to structural modification by SHI irradiation. Defect production accompanied by unit cell expansion and microstrain constitute the dominant observable radiation responses. Ion tracks with core-shell morphologies are observed in this material, a remnant of the induced atomic-scale structural and chemical changes. The highly transient conditions along the SHI path lead to oxygen movement in radial directions. This results in an oxygen-depleted core region, where Ce^{4+} cations are partially reduced to Ce^{3+}. This core is surrounded by an oxygen rich shell that contains small peroxide-like oxygen defect clusters.

These redox driven processes under SHI irradiation are an important characteristic of CeO_2 and are very sensitive to alterations in sample microstructure and chemistry (grain size, stoichiometry, and dopants), as well as ion-beam parameters (ion mass, energy, fluence, flux, irradiation temperature, and pressure). This makes CeO_2 a suitable model material to study fundamental aspects of ion-matter interactions over a wide range of conditions. On the other hand, this complex behavior poses a challenge in terms of disentangling the individual contributions of each parameter and comparing results from different research groups and experiments.

Future ion irradiation studies should systematically examine the radiation response of CeO_2 under a wide range of experimental conditions, including coupled effects that are of relevance for nuclear applications. Such efforts should also cover the mostly unexplored irradiation regime between low energy ions (predominately nuclear dE/dx) and swift heavy ions (predominately electronic dE/dx) to gain insight into the material response under simultaneous contributions from both types of energy deposition.

Author Contributions: Conceptualization, W.F.C., C.L.T. and M.L.; Data curation, W.F.C., C.L.T. and M.L.; writing—original draft preparation, W.F.C., C.L.T. and M.L.; writing—review and editing, W.F.C., C.L.T. and M.L.; supervision, M.L.; project administration, M.L.; funding acquisition, M.L. All authors have read and agreed to the published version of the manuscript.

Funding: This work was funded by the Department of Energy (DOE) Office of Nuclear Energy's Nuclear Energy University Program under US-DOE, contract DE-NE0008895. Synchrotron XRD measurements were performed at HPCAT (Sector 16), Advanced Photon Source (APS), Argonne National Laboratory. HPCAT operations are supported by DOE-NNSA's Office of Experimental Sciences. The Advanced Photon Source is a U.S. Department of Energy (DOE) Office of Science User Facility operated for the DOE Office of Science by Argonne National Laboratory under Contract No. DE-AC02-06CH11357. HPCAT beam time was provided by the Chicago/DOE Alliance Center. The research at ORNL's Spallation Neutron Source was sponsored by the Scientific User Facilities Division, Office of Basic Energy Sciences, U.S. Department of Energy. W.F.C. was funded by an Integrated University Program Graduate Fellowship.

Data Availability Statement: No new data were created or analyzed in this study. Data sharing is not applicable to this article.

Acknowledgments: The authors gratefully acknowledge technical support by scientists at the different ion-beam, neutron, and X-ray user facilities: Christina Trautmann and Daniel Severin (GSI Helmholtz Center, Darmstadt, Germany), Maxim Zdorovets (Institute of Nuclear Physics, Astana, Kazakhstan), Vladimir Skuratov (Joint Institute for Nuclear Research, Dubna, Russia), Michelle Everett and Jörg Neuefeind (Spallation Neutron Source, Oak Ridge National Laboratory, Oak Ridge, USA) and Changyong Park (Advanced Photon Source, Argonne National Laboratory, Argonne, USA).

Conflicts of Interest: The authors declare no conflict of interest.

References

1. Eyring, L. Chapter 27 The Binary Rare Earth Oxides. In *Handbook on the Physics and Chemistry of Rare Earths*; Elsevier: Amsterdam, The Netherlands, 1979; Volume 3, pp. 337–399.
2. Haire, R.G.; Eyring, L. Chapter 125 Comparisons of the Binary Oxides. In *Handbook on the Physics and Chemistry of Rare Earths*; Elsevier: Amsterdam, The Netherlands, 1994; Volume 18, pp. 413–505.
3. Schelter, E.J. Cerium under the lens. *Nat. Chem.* **2013**, *5*, 348–348. [CrossRef] [PubMed]
4. Shannon, R.D. Revised effective ionic radii and systematic studies of interatomic distances in halides and chalcogenides. *Acta Crystallogr. Sect. A* **1976**, *32*, 751–767. [CrossRef]
5. Tsunekawa, S.; Ishikawa, K.; Li, Z.Q.; Kawazoe, Y.; Kasuya, A. Origin of Anomalous Lattice Expansion in Oxide Nanoparticles. *Phys. Rev. Lett.* **2000**, *85*, 3440–3443. [CrossRef] [PubMed]
6. Bishop, S.; Marrocchelli, D.; Chatzichristodoulou, C.; Perry, N.; Mogensen, M.B.; Tuller, H.; Wachsman, E. Chemical expansion: Implications for electrochemical energy storage and conversion devices. *Annu. Rev. Mater. Res.* **2014**, *44*, 205–239. [CrossRef]
7. Shoko, E.; Smith, M.F.; McKenzie, R.H. Mixed valency in cerium oxide crystallographic phases: Valence of different cerium sites by the bond valence method. *Phys. Rev. B* **2009**, *79*, 134108. [CrossRef]
8. Shoko, E.; Smith, M.F.; McKenzie, R.H. Charge distribution near bulk oxygen vacancies in cerium oxides. *J. Phys. Condens. Matter* **2010**, *22*, 223201. [CrossRef]
9. Kümmerle, E.A.; Heger, G. The Structures of C–Ce$_2$O$_{3+\delta}$, Ce$_7$O$_{12}$, and Ce$_{11}$O$_{20}$. *J. Solid State Chem.* **1999**, *147*, 485–500. [CrossRef]
10. Hull, S.; Norberg, S.T.; Ahmed, I.; Eriksson, S.G.; Marrocchelli, D.; Madden, P.A. Oxygen vacancy ordering within anion-deficient Ceria. *J. Solid State Chem.* **2009**, *182*, 2815–2821. [CrossRef]
11. Beie, H.J.; Gnörich, A. Oxygen gas sensors based on CeO$_2$ thick and thin films. *Sens. Actuators B Chem.* **1991**, *4*, 393–399. [CrossRef]
12. Trovarelli, A.; de Leitenburg, C.; Boaro, M.; Dolcetti, G. The utilization of ceria in industrial catalysis. *Catal. Today* **1999**, *50*, 353–367. [CrossRef]
13. Tuller, H.L. Mixed ionic-electronic conduction in a number of fluorite and pyrochlore compounds. *Solid State Ion.* **1992**, *52*, 135–146. [CrossRef]
14. Lang, M.; Tracy, C.L.; Palomares, R.I.; Zhang, F.; Severin, D.; Bender, M.; Trautmann, C.; Park, C.; Prakapenka, V.B.; Skuratov, V.A.; et al. Characterization of ion-induced radiation effects in nuclear materials using synchrotron x-ray techniques. *J. Mater. Res.* **2015**, *30*, 1366–1379. [CrossRef]
15. Liu, T.; Liu, J.; Xi, K.; Zhang, Z.; He, D.; Ye, B.; Yin, Y.; Ji, Q.; Wang, B.; Luo, J.; et al. Heavy Ion Radiation Effects on a 130-nm COTS NVSRAM Under Different Measurement Conditions. *IEEE Trans. Nucl. Sci.* **2018**, *65*, 1119–1126. [CrossRef]
16. Severin, D.; Ensinger, W.; Neumann, R.; Trautmann, C.; Walter, G.; Alig, I.; Dudkin, S. Degradation of polyimide under irradiation with swift heavy ions. *Nucl. Instrum. Methods Phys. Res. Sect. B Beam Interact. Mater. At.* **2005**, *236*, 456–460. [CrossRef]
17. Schardt, D.; Elsässer, T.; Schulz-Ertner, D. Heavy-ion tumor therapy: Physical and radiobiological benefits. *Rev. Mod. Phys.* **2010**, *82*, 383–425. [CrossRef]

18. Jain, I.P.; Agarwal, G. Ion beam induced surface and interface engineering. *Surf. Sci. Rep.* **2011**, *66*, 77–172. [CrossRef]
19. Ridgway, M.C.; Giulian, R.; Sprouster, D.J.; Kluth, P.; Araujo, L.L.; Llewellyn, D.J.; Byrne, A.P.; Kremer, F.; Fichtner, P.F.P.; Rizza, G.; et al. Role of Thermodynamics in the Shape Transformation of Embedded Metal Nanoparticles Induced by Swift Heavy-Ion Irradiation. *Phys. Rev. Lett.* **2011**, *106*, 095505. [CrossRef] [PubMed]
20. Akcöltekin, E.; Peters, T.; Meyer, R.; Duvenbeck, A.; Klusmann, M.; Monnet, I.; Lebius, H.; Schleberger, M. Creation of multiple nanodots by single ions. *Nat. Nanotechnol.* **2007**, *2*, 290–294. [CrossRef] [PubMed]
21. Ridgway, M.C.; Djurabekova, F.; Nordlund, K. Ion–solid interactions at the extremes of electronic energy loss: Examples for amorphous semiconductors and embedded nanostructures. *Curr. Opin. Solid State Mater. Sci.* **2015**, *19*, 29–38. [CrossRef]
22. Jubera, M.; Villarroel, J.; García-Cabañes, A.; Carrascosa, M.; Olivares, J.; Agullo-López, F.; Méndez, A.; Ramiro, J.B. Analysis and optimization of propagation losses in $LiNbO_3$ optical waveguides produced by swift heavy-ion irradiation. *Appl. Phys. B* **2012**, *107*, 157–162. [CrossRef]
23. Lang, M.; Djurabekova, F.; Medvedev, N.; Toulemonde, M.; Trautmann, C. 1.15—Fundamental Phenomena and Applications of Swift Heavy Ion Irradiations. In *Comprehensive Nuclear Materials*, 2nd ed.; Konings, R.J.M., Stoller, R.E., Eds.; Elsevier: Oxford, UK, 2020; pp. 485–516. [CrossRef]
24. Lankin, A.V.; Morozov, I.V.; Norman, G.E.; Pikuz, S.A.; Skobelev, I.Y. Solid-density plasma nanochannel generated by a fast single ion in condensed matter. *Phys. Rev. E* **2009**, *79*, 036407. [CrossRef] [PubMed]
25. Norman, G.E.; Starikov, S.V.; Stegailov, V.V.; Saitov, I.M.; Zhilyaev, P.A. Atomistic Modeling of Warm Dense Matter in the Two-Temperature State. *Contrib. Plasma Phys.* **2013**, *53*, 129–139. [CrossRef]
26. Duffy, D.M.; Daraszewicz, S.L.; Mulroue, J. Modelling the effects of electronic excitations in ionic-covalent materials. *Nucl. Instrum. Methods Phys. Res. Sect. B: Beam Interact. Mater. At.* **2012**, *277*, 21–27. [CrossRef]
27. Baldinozzi, G.; Simeone, D.; Gosset, D.; Monnet, I.; Le Caër, S.; Mazerolles, L. Evidence of extended defects in pure zirconia irradiated by swift heavy ions. *Phys. Rev. B* **2006**, *74*, 132107. [CrossRef]
28. Lang, M.; Zhang, F.; Zhang, J.; Wang, J.; Lian, J.; Weber, W.J.; Schuster, B.; Trautmann, C.; Neumann, R.; Ewing, R.C. Review of $A_2B_2O_7$ pyrochlore response to irradiation and pressure. *Nucl. Instrum. Methods Phys. Res. Sect. B Beam Interact. Mater. At.* **2010**, *268*, 2951–2959. [CrossRef]
29. Cusick, A.B.; Lang, M.; Zhang, F.; Sun, K.; Li, W.; Kluth, P.; Trautmann, C.; Ewing, R.C. Amorphization of Ta2O5 under swift heavy ion irradiation. *Nucl. Instrum. Methods Phys. Res. Sect. B Beam Interact. Mater. At.* **2017**, *407*, 25–33. [CrossRef]
30. Tracy, C.L.; Lang, M.; Zhang, F.; Trautmann, C.; Ewing, R.C. Phase transformations in Ln_2O_3 materials irradiated with swift heavy ions. *Phys. Rev. B* **2015**, *92*, 174101. [CrossRef]
31. Weber, W.J. Models and mechanisms of irradiation-induced amorphization in ceramics. *Nucl. Instrum. Methods Phys. Res. Sect. B Beam Interact. Mater. At.* **2000**, *166–167*, 98–106. [CrossRef]
32. Sonoda, T.; Kinoshita, M.; Chimi, Y.; Ishikawa, N.; Sataka, M.; Iwase, A. Electronic excitation effects in CeO_2 under irradiations with high-energy ions of typical fission products. *Nucl. Instrum. Methods Phys. Res. Sect. B Beam Interact. Mater. At.* **2006**, *250*, 254–258. [CrossRef]
33. Sonoda, T.; Kinoshita, M.; Ishikawa, N.; Sataka, M.; Chimi, Y.; Okubo, N.; Iwase, A.; Yasunaga, K. Clarification of the properties and accumulation effects of ion tracks in CeO_2. *Nucl. Instrum. Methods Phys. Res. Sect. B: Beam Interact. Mater. At.* **2008**, *266*, 2882–2886. [CrossRef]
34. Iwase, A.; Ohno, H.; Ishikawa, N.; Baba, Y.; Hirao, N.; Sonoda, T.; Kinoshita, M. Study on the behavior of oxygen atoms in swift heavy ion irradiated CeO_2 by means of synchrotron radiation X-ray photoelectron spectroscopy. *Nucl. Instrum. Methods Phys. Res. Sect. B Beam Interact. Mater. At.* **2009**, *267*, 969–972. [CrossRef]
35. Ohno, H.; Iwase, A.; Matsumura, D.; Nishihata, Y.; Mizuki, J.; Ishikawa, N.; Baba, Y.; Hirao, N.; Sonoda, T.; Kinoshita, M. Study on effects of swift heavy ion irradiation in cerium dioxide using synchrotron radiation X-ray absorption spectroscopy. *Nucl. Instrum. Methods Phys. Res. Sect. B Beam Interact. Mater. At.* **2008**, *266*, 3013–3017. [CrossRef]
36. Kinoshita, M.; Yasunaga, K.; Sonoda, T.; Iwase, A.; Ishikawa, N.; Sataka, M.; Yasuda, K.; Matsumura, S.; Geng, H.Y.; Ichinomiya, T.; et al. Recovery and restructuring induced by fission energy ions in high burnup nuclear fuel. *Nucl. Instrum. Methods Phys. Res. Sect. B Beam Interact. Mater. At.* **2009**, *267*, 960–963. [CrossRef]
37. Sonoda, T.; Kinoshita, M.; Ishikawa, N.; Sataka, M.; Iwase, A.; Yasunaga, K. Clarification of high density electronic excitation effects on the microstructural evolution in UO_2. *Nucl. Instrum. Methods Phys. Res. Sect. B Beam Interact. Mater. At.* **2010**, *268*, 3277–3281. [CrossRef]
38. Varshney, M.; Sharma, A.; Kumar, R.; Verma, K.D. Swift heavy ion irradiation induced nanoparticle formation in CeO_2 thin films. *Nucl. Instrum. Methods Phys. Res. Sect. B Beam Interact. Mater. At.* **2011**, *269*, 2786–2791. [CrossRef]
39. Shimizu, K.; Kosugi, S.; Tahara, Y.; Yasunaga, K.; Kaneta, Y.; Ishikawa, N.; Hori, F.; Matsui, T.; Iwase, A. Change in magnetic properties induced by swift heavy ion irradiation in CeO_2. *Nucl. Instrum. Methods Phys. Res. Sect. B Beam Interact. Mater. At.* **2012**, *286*, 291–294. [CrossRef]
40. Yasuda, K.; Etoh, M.; Sawada, K.; Yamamoto, T.; Yasunaga, K.; Matsumura, S.; Ishikawa, N. Defect formation and accumulation in CeO_2 irradiated with swift heavy ions. *Nucl. Instrum. Methods Phys. Res. Sect. B Beam Interact. Mater. At.* **2013**, *314*, 185–190. [CrossRef]
41. Kishino, T.; Shimizu, K.; Saitoh, Y.; Ishikawa, N.; Hori, F.; Iwase, A. Effects of high-energy heavy ion irradiation on the crystal structure in CeO_2 thin films. *Nucl. Instrum. Methods Phys. Res. Sect. B: Beam Interact. Mater. At.* **2013**, *314*, 191–194. [CrossRef]

42. Takaki, S.; Yasuda, K.; Yamamoto, T.; Matsumura, S.; Ishikawa, N. Atomic structure of ion tracks in Ceria. *Nucl. Instrum. Methods Phys. Res. Sect. B Beam Interact. Mater. At.* **2014**, *326*, 140–144. [CrossRef]
43. Grover, V.; Shukla, R.; Kumari, R.; Mandal, B.P.; Kulriya, P.K.; Srivastava, S.K.; Ghosh, S.; Tyagi, A.K.; Avasthi, D.K. Effect of grain size and microstructure on radiation stability of CeO$_2$: An extensive study. *Phys. Chem. Chem. Phys.* **2014**, *16*, 27065–27073. [CrossRef]
44. Ishikawa, N.; Okubo, N.; Taguchi, T. Experimental evidence of crystalline hillocks created by irradiation of CeO$_2$ with swift heavy ions: TEM study. *Nanotechnology* **2015**, *26*, 355701. [CrossRef] [PubMed]
45. Palomares, R.I.; Tracy, C.L.; Zhang, F.; Park, C.; Popov, D.; Trautmann, C.; Ewing, R.C.; Lang, M. In situ defect annealing of swift heavy ion irradiated CeO$_2$ and ThO$_2$ using synchrotron X-ray diffraction and a hydrothermal diamond anvil cell. *J. Appl. Crystallogr.* **2015**, *48*, 711–717. [CrossRef]
46. Tracy, C.L.; Lang, M.; Pray, J.M.; Zhang, F.X.; Popov, D.; Park, C.Y.; Trautmann, C.; Bender, M.; Severin, D.; Skuratov, V.A.; et al. Redox response of actinide materials to highly ionizing radiation. *Nat. Commun.* **2015**, *6*, 9. [CrossRef] [PubMed]
47. Pakarinen, J.; He, L.F.; Hassan, A.R.; Wang, Y.Q.; Gupta, M.; El-Azab, A.; Allen, T.R. Annealing-induced lattice recovery in room-temperature xenon irradiated CeO$_2$: X-ray diffraction and electron energy loss spectroscopy experiments. *J. Mater. Res.* **2015**, *30*, 1555–1562. [CrossRef]
48. Yablinsky, C.A.; Devanathan, R.; Pakarinen, J.; Gan, J.; Severin, D.; Trautmann, C.; Allen, T.R. Characterization of swift heavy ion irradiation damage in ceria. *J. Mater. Res.* **2015**, *30*, 1473–1484. [CrossRef]
49. Costantini, J.-M.; Miro, S.; Gutierrez, G.; Yasuda, K.; Takaki, S.; Ishikawa, N.; Toulemonde, M. Raman spectroscopy study of damage induced in cerium dioxide by swift heavy ion irradiations. *J. Appl. Phys.* **2017**, *122*, 205901. [CrossRef]
50. Palomares, R.I.; Shamblin, J.; Tracy, C.L.; Neuefeind, J.; Ewing, R.C.; Trautmann, C.; Lang, M. Defect accumulation in swift heavy ion-irradiated CeO$_2$ and ThO$_2$. *J. Mater. Chem. A* **2017**, 12193–12201. [CrossRef]
51. Maslakov, K.I.; Teterin, Y.A.; Popel, A.J.; Teterin, A.Y.; Ivanov, K.E.; Kalmykov, S.N.; Petrov, V.G.; Petrov, P.K.; Farnan, I. XPS study of ion irradiated and unirradiated CeO$_2$ bulk and thin film samples. *Appl. Surf. Sci.* **2018**, *448*, 154–162. [CrossRef]
52. Cureton, W.F.; Palomares, R.I.; Walters, J.; Tracy, C.L.; Chen, C.-H.; Ewing, R.C.; Baldinozzi, G.; Lian, J.; Trautmann, C.; Lang, M. Grain size effects on irradiated CeO$_2$, ThO$_2$, and UO$_2$. *Acta Mater.* **2018**, *160*, 47–56. [CrossRef]
53. Shelyug, A.; Palomares, R.I.; Lang, M.; Navrotsky, A. Energetics of defect production in fluorite-structured CeO$_2$ induced by highly ionizing radiation. *Phys. Rev. Mater.* **2018**, *2*, 093607. [CrossRef]
54. Cureton, W.F.; Palomares, R.I.; Tracy, C.L.; O'Quinn, E.C.; Walters, J.; Zdorovets, M.; Ewing, R.C.; Toulemonde, M.; Lang, M. Effects of irradiation temperature on the response of CeO$_2$, ThO$_2$, and UO$_2$ to highly ionizing radiation. *J. Nucl. Mater.* **2019**, *525*, 83–91. [CrossRef]
55. Costantini, J.-M.; Gutierrez, G.; Watanabe, H.; Yasuda, K.; Takaki, S.; Lelong, G.; Guillaumet, M.; Weber, W.J. Optical spectroscopy study of modifications induced in cerium dioxide by electron and ion irradiations. *Philos. Mag.* **2019**, *99*, 1695–1714. [CrossRef]
56. Yamamoto, Y.; Ishikawa, N.; Hori, F.; Iwase, A. Analysis of Ion-Irradiation Induced Lattice Expansion and Ferromagnetic State in CeO$_2$ by Using Poisson Distribution Function. *Quantum Beam Sci.* **2020**, *4*, 26. [CrossRef]
57. Ziegler, J.F.; Ziegler, M.D.; Biersack, J.P. SRIM—The stopping and range of ions in matter (2010). *Nucl. Instrum. Methods Phys. Res. Sect. B Beam Interact. Mater. At.* **2010**, *268*, 1818–1823. [CrossRef]
58. Kumar, A.; Devanathan, R.; Shutthanandan, V.; Kuchibhatla, S.V.N.T.; Karakoti, A.S.; Yong, Y.; Thevuthasan, S.; Seal, S. Radiation-Induced Reduction of Ceria in Single and Polycrystalline Thin Films. *J. Phys. Chem. C* **2012**, *116*, 361–366. [CrossRef]
59. Park, C.; Popov, D.; Ikuta, D.; Lin, C.; Kenney-Benson, C.; Rod, E.; Bommannavar, A.; Shen, G. New developments in micro-X-ray diffraction and X-ray absorption spectroscopy for high-pressure research at 16-BM-D at the Advanced Photon Source. *Rev. Sci. Instrum.* **2015**, *86*, 072205. [CrossRef]
60. Tahara, Y.; Zhu, B.; Kosugi, S.; Ishikawa, N.; Okamoto, Y.; Hori, F.; Matsui, T.; Iwase, A. Study on effects of swift heavy ion irradiation on the crystal structure in CeO$_2$ doped with Gd$_2$O$_3$. *Nucl. Instrum. Methods Phys. Res. Sect. B Beam Interact. Mater. At.* **2011**, *269*, 886–889. [CrossRef]
61. Neuefeind, J.; Feygenson, M.; Carruth, J.; Hoffmann, R.; Chipley, K.K. The Nanoscale Ordered MAterials Diffractometer NOMAD at the Spallation Neutron Source SNS. *Nucl. Instrum. Methods Phys. Res. Sect. B Beam Interact. Mater. At.* **2012**, *287*, 68–75. [CrossRef]
62. Schmitt, R.; Nenning, A.; Kraynis, O.; Korobko, R.; Frenkel, A.I.; Lubomirsky, I.; Haile, S.M.; Rupp, J.L.M. A review of defect structure and chemistry in ceria and its solid solutions. *Chem. Soc. Rev.* **2020**, *49*, 554–592. [CrossRef]
63. Devanathan, R. Molecular Dynamics Simulation of Fission Fragment Damage in Nuclear Fuel and Surrogate Material. *MRS Adv.* **2017**, *2*, 1225–1230. [CrossRef]
64. Skanthakumar, S.; Soderholm, L. Oxidation state of Ce in Pb2Sr2Ce{1-x}Ca{x}Cu3O8. *Phys. Rev. B* **1996**, *53*, 920–926. [CrossRef]
65. Zhang, J.; Lang, M.; Lian, J.; Liu, J.; Trautmann, C.; Della-Negra, S.; Toulemonde, M.; Ewing, R.C. Liquid-like phase formation in Gd2Zr2O7 by extremely ionizing irradiation. *J. Appl. Phys.* **2009**, *105*, 113510. [CrossRef]
66. Matzke, H.; Lucuta, P.G.; Wiss, T. Swift heavy ion and fission damage effects in UO$_2$. *Nucl. Instrum. Methods Phys. Res. Sect. B Beam Interact. Mater. At.* **2000**, *166–167*, 920–926. [CrossRef]
67. Toulemonde, M.; Paumier, E.; Dufour, C. Thermal spike model in the electronic stopping power regime. *Radiat. Eff. Defects Solids* **1993**, *126*, 201–206. [CrossRef]
68. Szenes, G. Ion-induced amorphization in ceramic materials. *J. Nucl. Mater.* **2005**, *336*, 81–89. [CrossRef]

69. Meftah, A.; Brisard, F.; Costantini, J.M.; Hage-Ali, M.; Stoquert, J.P.; Studer, F.; Toulemonde, M. Swift heavy ions in magnetic insulators: A damage-cross-section velocity effect. *Phys. Rev. B* **1993**, *48*, 920–925. [CrossRef]

70. Rose, M.; Gorzawski, G.; Miehe, G.; Balogh, A.G.; Hahn, H. Phase stability of nanostructured materials under heavy ion irradiation. *Nanostruct. Mater.* **1995**, *6*, 731–734. [CrossRef]

71. Toulemonde, M.; Dufour, C.; Meftah, A.; Paumier, E. Transient thermal processes in heavy ion irradiation of crystalline inorganic insulators. *Nucl. Instrum. Methods Phys. Res. Sect. B Beam Interact. Mater. At.* **2000**, *166*, 903–912. [CrossRef]

72. Weber, W.J. Alpha-irradiation damage in CeO_2, UO_2 and PuO_2. *Radiat. Eff.* **1984**, *83*, 145–156. [CrossRef]

73. Lang, M.; Zhang, F.; Zhang, J.; Wang, J.; Schuster, B.; Trautmann, C.; Neumann, R.; Becker, U.; Ewing, R.C. Nanoscale manipulation of the properties of solids at high pressure with relativistic heavy ions. *Nat. Mater.* **2009**, *8*, 793–797. [CrossRef]

74. Kourouklis, G.A.; Jayaraman, A.; Espinosa, G.P. High-pressure Raman study of CeO_2 to 35 GPa and pressure-induced phase transformation from the fluorite structure. *Phys. Rev. B* **1988**, *37*, 4250–4253. [CrossRef]

75. Duclos, S.J.; Vohra, Y.K.; Ruoff, A.L.; Jayaraman, A.; Espinosa, G.P. High-pressure x-ray diffraction study of CeO_2 to 70 GPa and pressure-induced phase transformation from the fluorite structure. *Phys. Rev. B* **1988**, *38*, 7755–7758. [CrossRef]

76. Wang, J.; Ewing, R.C.; Becker, U. Electronic structure and stability of hyperstoichiometric UO_{2+x} under pressure. *Phys. Rev. B* **2013**, *88*, 024109. [CrossRef]

77. Ge, M.Y.; Fang, Y.Z.; Wang, H.; Chen, W.; He, Y.; Liu, E.Z.; Su, N.H.; Stahl, K.; Feng, Y.P.; Tse, J.S.; et al. Anomalous compressive behavior in CeO_2 nanocubes under high pressure. *New J. Phys.* **2008**, *10*, 123016. [CrossRef]

78. Chen, W.; Navrotsky, A. Thermochemical study of trivalent-doped ceria systems: CeO2–MO1.5 (M = La, Gd, and Y). *J. Mater. Res.* **2006**, *21*, 3242–3251. [CrossRef]

79. Tahara, Y.; Shimizu, K.; Ishikawa, N.; Okamoto, Y.; Hori, F.; Matsui, T.; Iwase, A. Study on effects of energetic ion irradiation in Gd_2O_3-doped CeO_2 by means of synchrotron radiation X-ray spectroscopy. *Nucl. Instrum. Methods Phys. Res. Sect. B Beam Interact. Mater. At.* **2012**, *277*, 53–57. [CrossRef]

80. Sasajima, Y.; Ajima, N.; Osada, T.; Ishikawa, N.; Iwase, A. Molecular dynamics simulation of fast particle irradiation to the Gd_2O_3-doped CeO_2. *Nucl. Instrum. Methods Phys. Res. Sect. B Beam Interact. Mater. At.* **2013**, *316*, 176–182. [CrossRef]

81. Zhu, B.; Ohno, H.; Kosugi, S.; Hori, F.; Yasunaga, K.; Ishikawa, N.; Iwase, A. Effects of swift heavy ion irradiation on the structure of Er_2O_3-doped CeO_2. *Nucl. Instrum. Methods Phys. Res. Sect. B Beam Interact. Mater. At.* **2010**, *268*, 3199–3202. [CrossRef]

82. Lucid, A.K.; Keating, P.R.L.; Allen, J.P.; Watson, G.W. Structure and Reducibility of CeO_2 Doped with Trivalent Cations. *J. Phys. Chem. C* **2016**, *120*, 23430–23440. [CrossRef]

83. Tracy, C.L.; McLain Pray, J.; Lang, M.; Popov, D.; Park, C.; Trautmann, C.; Ewing, R.C. Defect accumulation in ThO_2 irradiated with swift heavy ions. *Nucl. Instrum. Methods Phys. Res. Sect. B Beam Interact. Mater. At.* **2014**, *326*, 169–173. [CrossRef]

84. Preuss, A.; Gruehn, R. Preparation and Structure of Cerium Titanates Ce_2TiO_5, Ce_2TiO_7, and $Ce_4Ti_9O_{24}$. *J. Solid State Chem.* **1994**, *110*, 363–369. [CrossRef]

85. Uberuaga, B.P.; Sickafus, K.E. Interpreting oxygen vacancy migration mechanisms in oxides using the layered structure motif. *Comput. Mater. Sci.* **2015**, *103*, 216–223. [CrossRef]

86. Adachi, G.-y.; Imanaka, N. The Binary Rare Earth Oxides. *Chem. Rev.* **1998**, *98*, 1479–1514. [CrossRef]

87. Gaboriaud, R.J.; Jublot, M.; Paumier, F.; Lacroix, B. Phase transformations in Y_2O_3 thin films under swift Xe ions irradiation. *Nucl. Instrum. Methods Phys. Res. Sect. B Beam Interact. Mater. At.* **2013**, *310*, 6–9. [CrossRef]

88. Sattonnay, G.; Bilgen, S.; Thomé, L.; Grygiel, C.; Monnet, I.; Plantevin, O.; Huet, C.; Miro, S.; Simon, P. Structural and microstructural tailoring of rare earth sesquioxides by swift heavy ion irradiation. *Phys. Status Solidi (b)* **2016**, *253*, 2110–2114. [CrossRef]

89. Lang, M.; Zhang, F.; Zhang, J.; Tracy, C.L.; Cusick, A.B.; VonEhr, J.; Chen, Z.; Trautmann, C.; Ewing, R.C. Swift heavy ion-induced phase transformation in Gd_2O_3. *Nucl. Instrum. Methods Phys. Res. Sect. B Beam Interact. Mater. At.* **2014**, *326*, 121–125. [CrossRef]

90. Bilgen, S.; Sattonnay, G.; Grygiel, C.; Monnet, I.; Simon, P.; Thomé, L. Phase transformations induced by heavy ion irradiation in Gd_2O_3: Comparison between ballistic and electronic excitation regimes. *Nucl. Instrum. Methods Phys. Res. Sect. B Beam Interact. Mater. At.* **2018**, *435*, 12–18. [CrossRef]

91. Bevan, D.J.M. Ordered intermediate phases in the system CeO_2-Ce_2O_3. *J. Inorg. Nucl. Chem.* **1955**, *1*, 49–59. [CrossRef]

92. Da Silva, J.L.F. Stability of the Ce_2O_3 phases: A DFT+U investigation. *Phys. Rev. B* **2007**, *76*, 193108. [CrossRef]

93. Ewing, R.C.; Weber, W.J.; Lian, J. Nuclear waste disposal—pyrochlore ($A_2B_2O_7$): Nuclear waste form for the immobilization of plutonium and "minor" actinides. *J. Appl. Phys.* **2004**, *95*, 5949–5971. [CrossRef]

94. Subramanian, M.A.; Aravamudan, G.; Subba Rao, G.V. Oxide pyrochlores—A review. *Prog. Solid State Chem.* **1983**, *15*, 55–143. [CrossRef]

95. Zhang, F.X.; Tracy, C.L.; Lang, M.; Ewing, R.C. Stability of fluorite-type $La_2Ce_2O_7$ under extreme conditions. *J. Alloys Compd.* **2016**, *674*, 168–173. [CrossRef]

96. Lang, M.; Lian, J.; Zhang, J.; Zhang, F.; Weber, W.J.; Trautmann, C.; Ewing, R.C. Single-ion tracks in $Gd_2Zr_{2-x}Ti_xO_7$ pyrochlores irradiated with swift heavy ions. *Phys. Rev. B* **2009**, *79*, 224105. [CrossRef]

97. Lang, M.; Zhang, F.X.; Ewing, R.C.; Lian, J.; Trautmann, C.; Wang, Z. Structural modifications of $Gd_2Zr_{2-x}Ti_xO_7$ pyrochlore induced by swift heavy ions: Disordering and amorphization. *J. Mater. Res.* **2009**, *24*, 1322–1334. [CrossRef]

98. Tracy, C.L.; Shamblin, J.; Park, S.; Zhang, F.; Trautmann, C.; Lang, M.; Ewing, R.C. Role of composition, bond covalency, and short-range order in the disordering of stannate pyrochlores by swift heavy ion irradiation. *Phys. Rev. B* **2016**, *94*, 064102. [CrossRef]
99. Park, S.; Lang, M.; Tracy, C.L.; Zhang, J.; Zhang, F.; Trautmann, C.; Rodriguez, M.D.; Kluth, P.; Ewing, R.C. Response of $Gd_2Ti_2O_7$ and $La_2Ti_2O_7$ to swift-heavy ion irradiation and annealing. *Acta Mater.* **2015**, *93*, 1–11. [CrossRef]
100. Sickafus, K.E.; Minervini, L.; Grimes, R.W.; Valdez, J.A.; Ishimaru, M.; Li, F.; McClellan, K.J.; Hartmann, T. Radiation Tolerance of Complex Oxides. *Science* **2000**, *289*, 748. [CrossRef]
101. Shamblin, J.; Feygenson, M.; Neuefeind, J.; Tracy, C.L.; Zhang, F.; Finkeldei, S.; Bosbach, D.; Zhou, H.; Ewing, R.C.; Lang, M. Probing disorder in isometric pyrochlore and related complex oxides. *Nat. Mater.* **2016**, *15*, 507–511. Available online: http://www.nature.com/nmat/journal/v15/n5/abs/nmat4581.html#supplementary-information (accessed on 30 April 2021). [CrossRef]
102. Shamblin, J.; Tracy, C.L.; Palomares, R.I.; O'Quinn, E.C.; Ewing, R.C.; Neuefeind, J.; Feygenson, M.; Behrens, J.; Trautmann, C.; Lang, M. Similar local order in disordered fluorite and aperiodic pyrochlore structures. *Acta Mater.* **2018**, *144*, 60–67. [CrossRef]

quantum beam science

Article

Simulation of Two-Dimensional Images for Ion-Irradiation Induced Change in Lattice Structures and Magnetic States in Oxides by Using Monte Carlo Method

Akihiro Iwase * and Shigeru Nishio

The Wakasa Wan Energy Research Center (WERC), 64-52-1 Nagatani, Tsuruga, Fukui 914-0192, Japan; snishio@werc.or.jp
* Correspondence: aiwase@werc.or.jp or iwase@mtr.osakafu-u.ac.jp

Abstract: A Monte Carlo method was used to simulate the two-dimensional images of ion-irradiation-induced change in lattice structures and magnetic states in oxides. Under the assumption that the lattice structures and the magnetic states are modified only inside the narrow one-dimensional region along the ion beam path (the ion track), and that such modifications are affected by ion track overlapping, the exposure of oxide targets to spatially random ion impacts was simulated by the Monte Carlo method. Through the Monte Carlo method, the evolutions of the two-dimensional images for the amorphization of TiO_2, the lattice structure transformation of ZrO_2, and the transition of magnetic states of CeO_2 were simulated as a function of ion fluence. The total fractions of the modified areas were calculated from the two-dimensional images. They agree well with the experimental results and those estimated by using the Poisson distribution functions.

Keywords: high energy irradiation; ion track overlapping; oxides; Monte Carlo simulation for two-dimensional images; lattice structures and magnetic states; binomial and Poisson distribution functions

Citation: Iwase, A.; Nishio, S. Simulation of Two-Dimensional Images for Ion-Irradiation Induced Change in Lattice Structures and Magnetic States in Oxides by Using Monte Carlo Method. *Quantum Beam Sci.* **2021**, *5*, 13. https://doi.org/10.3390/qubs5020013

Academic Editor: Lorenzo Giuffrida

Received: 9 March 2021
Accepted: 10 May 2021
Published: 13 May 2021

Publisher's Note: MDPI stays neutral with regard to jurisdictional claims in published maps and institutional affiliations.

1. Introduction

It is well known that, in a lot of polymers, the energetic ion irradiation and the subsequent chemical etching produce one-dimensional holes with small diameters [1]. Resultant perforated membranes have been used as filters for small particles [1,2]. In some ceramics irradiated with swift heavy ions, one-dimensional areas, in which the lattice structures and the physical properties are strongly modified, are produced along the ion beam path. Such one-dimensional structures are called "ion-tracks" [3]. The ion-tracks originate from the energetic ion induced high-density electronic excitation, and their production mechanisms have been explained by using the thermal spike models [4–6] and by the Coulomb explosion model [7–9]. A lot of studies have been performed in order to investigate the individual ion-track structures and their dependence on the electronic stopping power, Se, of the irradiating ions [10–15]. The effects of the ion-track overlapping on lattice structures of target materials, which appear for the high fluence irradiation, have also been investigated so far [10,16–22].

Ishikawa et al. have explained the ion track overlapping effect on the TiO_2 amorphization by using the binomial distribution function [23]. In our recent study, we analyzed the ion-track overlapping effect on the magnetization of CeO_2 by using the Poisson distribution function, and succeeded in the reproduction of the experimentally observed ion-fluence dependence of the magnetization [24]. As shown later in this report, the Poisson distribution function, which is the approximated formula of the binomial distribution function under some extreme condition, can also describe the ion-track overlapping effects on the evolution of the amorphization in TiO_2 and the lattice structure change in ZrO_2. Such analytical methods, however, only reproduce the total fraction of the areas modified by high energy ions as a function of ion fluence. In the present study, we used the Monte Carlo method in order to simulate the two-dimensional images of the lattice structures and the magnetic states as a result of the ion -track overlapping for TiO_2, ZrO_2 and CeO_2. The total fractions of the modified area were calculated from the two-dimensional images, and were compared with the experimental results and those estimated by using the Poisson distribution function.

2. Binomial and Poisson Distribution Functions for the Analysis of the Ion Track Overlapping Effect

The binomial distribution function,

$$f(n,k,p) = \frac{n!}{k!(n-k)!}p^k(1-p)^{n-k},\tag{1}$$

generally presents the discrete probability distribution of the number of successes, k, in a sequence of n independent trials, where p is the probability of the success and $1-p$, that of the failure for each trial. The paper of Ishikawa et al. [23] and our previous report [24] have shown that the probability of the ion-track overlapping can be described by the binomial distribution function. When the fluence of ions is Φ and the total area of the target is S_0, the number of irradiating ions, n (=ΦS_0) corresponds to the trial number. The ratio of the cross section of each ion track, S, to S_0 ($s = S/S_0$) and the number of track impacts, r, correspond to p and k in Equation (1), respectively. The probability of r times track impacts is given as

$$a(n,r,s) = \frac{n!}{r!(n-r)!}s^r(1-s)^{n-r}\tag{2}$$

If the ion track overlapping is discussed for the unit irradiation area (S_0 = 1 cm^2), n is replaced by the ion fluence, Φ, and s is replaced by the track cross section itself, S. Then, the fraction of the area for the r times track impacts in the unit area of the target, $A(\Phi,r)$ is,

$$A(\Phi,r) = \frac{\Phi!}{r!(\Phi-r)!}S^r(1-S)^{\Phi-r} \ for \ r = 1,2,3,\cdots\Phi\tag{3}$$

Although the dimensions of Φ and S are cm^{-2} and cm^2 as experimental parameters, these parameters can be treated in Equation (3) and other equations in this paper as dimensionless.

For the usual irradiation experiments of the ion tracks in materials, as a value of Φ is very large and a value of S is very small, the binomial distribution function can be approximated by the following Poisson distribution function [25],

$$A(\Phi,r) = \frac{(\Phi S)^r \exp(-\Phi S)}{r!}\tag{4}$$

We have confirmed that the result calculated by the Poisson distribution function (Equation (4)) is completely the same as that calculated by the binomial distribution function (Equation (3)) in the case of the irradiation studies referred in the present report [16,21,24]. Equation (4) will be used later for the comparison of the analytical result with that by the Monte Carlo simulation.

3. Monte Carlo Method for Two-Dimensional Imaging of Ion Track Overlapping Effect

We have developed the following Monte Carlo algorism for the present study. The target square for the Monte Carlo calculation consisted of 1000 × 1000 grid of cells. Its dimension was defined to be 100 nm × 100 nm throughout this work; a single impact on the target square was equivalent to the ion fluence of 10^{10} cm^{-2}. The target was randomly bombarded with energetic ions until the ion fluence reached the given value, Φ.

Numbers of ion impacts for all cells were examined after every single bombardment to visualize the two-dimensional impact map and to calculate the fraction of ion overlapping area, $A(\Phi, r)$ for all the number of the impacts by the ion-track, r, to compare the results by using the Poisson distribution function. It is worth noting here that the analysis by using the Poisson distribution function ignores temporal decay and spatial distribution of an effect of an impact. Hence, these conditions can also be included in the Monte Carlo calculations.

All the Monte Carlo calculations were carried out with the homemade code written in LabVIEW 2019 (32 bit) for Windows.

4. Results of the Monte Carlo Simulation and the Analysis Using the Poisson Distribution Function

4.1. Amorphization of TiO$_2$ by High Energy Heavy Ion Irradiation

Here, we refer the result of anatase TiO$_2$ by Ishikawa et al. as an example of the ion irradiation induced amorphization [21]. They irradiated TiO$_2$ films with 230 MeV Xe^{+15} ions and measured the x ray diffraction (XRD) spectra. They have shown that the XRD peak intensity decreases in an exponential manner as a function of ion fluence. This experimental result can be explained as a result of the overlapping of amorphous ion tracks. Along the ion beam path in TiO$_2$, the crystal structure becomes amorphous, and the amorphous areas still remains amorphous irrespective of the number of impacts by ion-tracks. As only the non-amorphized area contributes to the x ray diffraction, the intensity of the XRD peak is proportional to the fraction of zero impact ($r = 0$) area, $A(\Phi, 0)$, which is not amorphized by the irradiation. By using Equation (4), the relative XRD peak intensity, $I(\Phi)/I(0)$, is related to the ion-fluence, Φ, by the following function,

$$I(\Phi)/I(0) = A(\Phi, 0) = \exp(-S\Phi) \tag{5}$$

or

$$\ln(I(\Phi)/I(0)) = \ln A(\Phi, 0) = -S\Phi \tag{6}$$

where $I(0)$ is the XRD peak intensity for the unirradiated target and $I(\Phi)$, the XRD peak intensity after the irradiation with the fluence of Φ. From the slope of $\ln(I(\Phi)/I(0))-\Phi$ plot, the cross section of the ion track, S, has been determined as 72.3 nm^2, which corresponds to the track diameter of 9.6 nm. Because the total fraction of amorphized and non-amorphized areas is unity, the fraction of the amorphized area is expressed as,

$$1 - A(\Phi, 0) = 1 - \exp(-S\Phi), \tag{7}$$

which is well known as "the Poisson law".

The above model that TiO$_2$ sample is amorphized only inside the ion -track, and the amorphized region never contributes to the XRD diffraction seems to be too simple. The previous TEM (transmission electron microscope) observations, however, clearly show that just inside the ion-track in TiO$_2$, the structure becomes amorphous, and outside the ion-track, the crystal structure is maintained [26,27]. The XRD peaks for TiO$_2$ samples which are almost completely amorphized by high energy heavy ion irradiation are much smaller than for the unirradiated crystalline TiO$_2$ and are scarcely observed [22]. These experimental results, therefore, justify the above model for the analysis of ion track overlapping in TiO$_2$, and the fraction of amorphized area can be estimated by the decrease in XRD peak intensity.

The result of the Monte Carlo simulation for the Xe ion induced amorphization of TiO$_2$ is shown in Figure 1. The figure represents the two-dimensional images of amorphized (yellow) and non-amorphized (blue) areas for the ion-fluences of 5×10^{11}, 1×10^{12}, 2×10^{12} and 2.5×10^{12} cm^{-2}. The track diameter is 9.6 nm which has been determined by the experiment.

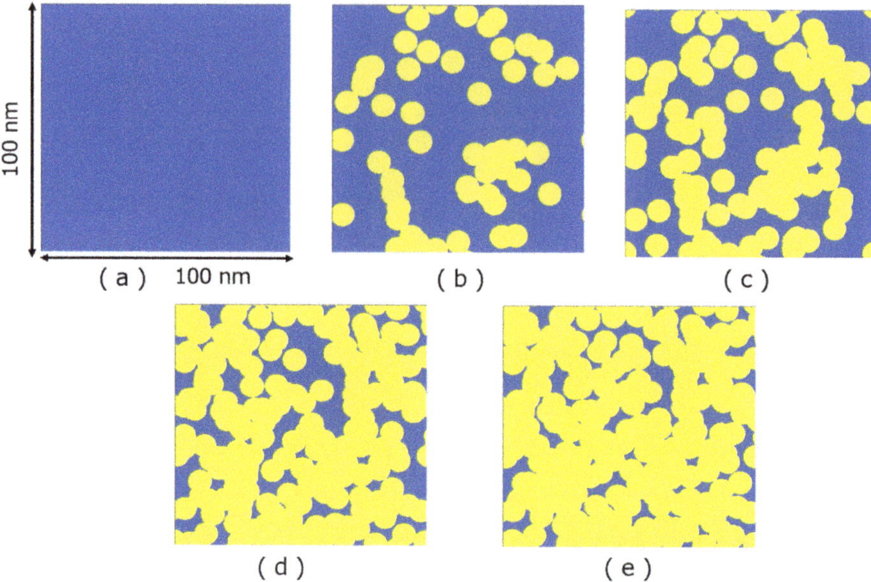

Figure 1. Two-dimensional images of amorphized (yellow) and non-amorphized (blue) areas for various ion fluences; (**a**) 0 cm^2 (unirrradiated), (**b**) 5×10^{11} cm^2, (**c**) 1×10^{12} cm^2, (**d**) 2×10^{12} cm^2 and (**e**) 2.5×10^{12} cm^2. The diameter of ion track is assumed to be 9.6 nm.

Figure 2 shows the fraction of the non-amorphized area as a function of ion-fluence, which has been calculated from the two-dimensional images. The figure also shows the experimental result [21] and the result calculated using Equation (6). The result of the Monte Carlo simulation well agrees with the experimental result and that estimated by the Poisson distribution function. Figure 3 confirms that the ion fluence dependence of the fraction of the amorphous area is expressed by the Poisson law (Equation (7)).

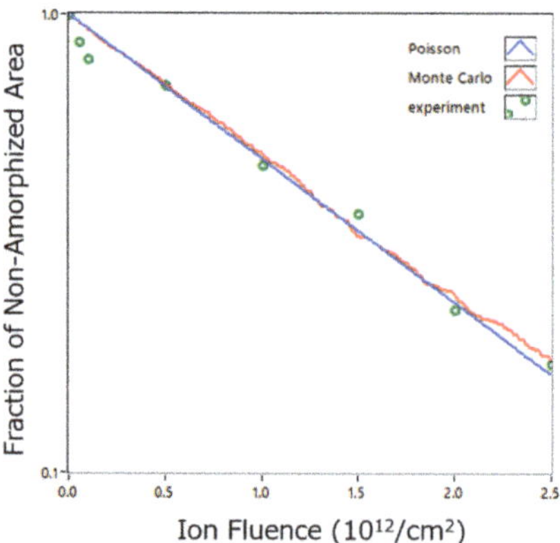

Figure 2. Fraction of non-amorphized area as a function of Xe ion fluence. Green circles, blue line, and red line represent the experimental result [21], result calculated by Equation (6) and the result from two-dimensional images, respectively. The logarithmic scale is used on the vertical axis.

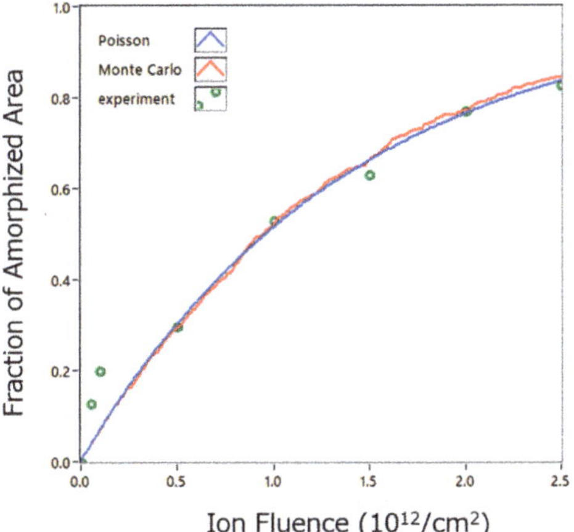

Figure 3. Fraction of amorphized area as a function of Xe ion fluence. Green circles, blue line, and red line represent the experimental result [21], result calculated by Equation (7) and the result from two-dimensional images, respectively.

4.2. Lattice Structure Change of ZrO_2 by High Energy Heavy Ion Irradiation

Next, one of the experimental results of the lattice structure transformation by high energy ion irradiation is referred. Benyagoub et al. irradiated monoclinic zirconia with 135 MeV Ni ions, and the lattice structures of the irradiated samples were characterized by XRD [16]. They have found that the lattice structure of monoclinic ZrO_2 gradually changes to the tetragonal structure by the ion irradiation. They have also shown that the evolution of the fraction of the tetragonal phase with the ion fluence shows a sigmoidal shape. This experimental result suggests that only one impact of the ion track does not cause the lattice structure transformation, but two or more impacts of the ion tracks are needed for the lattice structure transformation. This phenomenon can be explained by using the Poisson distribution function (Equation (4)) as follows. Since the total fraction of areas which are affected and not affected by ion tracks is unity:

$$\sum_{r=0}^{\Phi} A(\Phi,r) = A(\Phi,0) + A(\Phi,1) + \sum_{r=2}^{\Phi} A(\Phi,r) = 1. \tag{8}$$

According to Equation (4),

$$A(\Phi,0) = \exp(-S\Phi) \tag{9}$$

and

$$A(\Phi,1) = S\Phi\exp(-S\Phi) \tag{10}$$

Therefore, the fraction of two or more impacted areas, or the area of the tetragonal structure is,

$$\sum_{r=2}^{\Phi} A(\Phi,r) = 1 - \exp(-S\Phi) - S\Phi\exp(-S\Phi). \tag{11}$$

The result of the Monte Carlo simulation is shown in Figure 4 for the lattice structure transformation of ZrO_2. The figure represents the two-dimensional images of monoclinic (blue) and tetragonal (yellow) areas for the ion-fluences of 2.5×10^{12}, 5×10^{12}, 1.5×10^{13} and 3×10^{13} cm^{-2}. The track diameter is assumed to be 4.4 nm. This value was determined by the comparison of the experimental data and the result calculated by Equation (11).

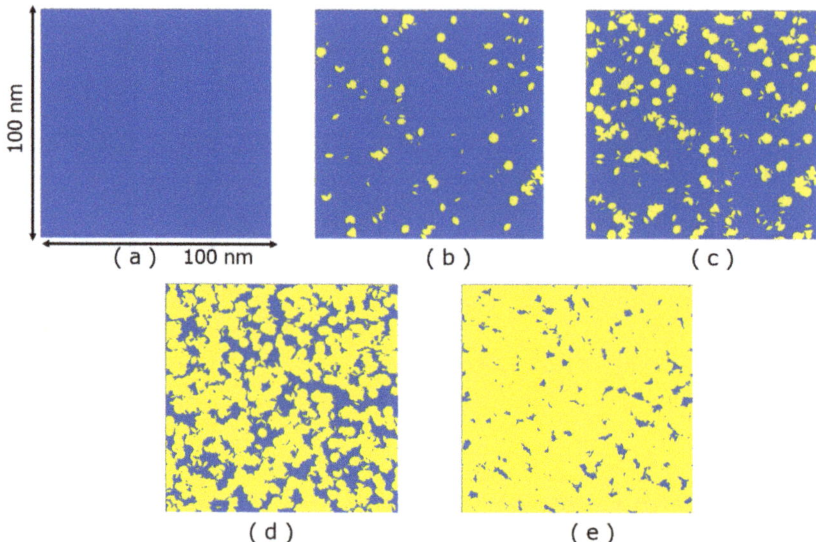

Figure 4. Two-dimensional images of tetragonal (yellow) and monoclinic (blue) areas for various ion fluences; (**a**) 0 cm^2 (unirrradiated), (**b**) 2.5×10^{12} cm^2, (**c**) 5×10^{12} cm^2, (**d**) 1.5×10^{13} cm^2 and (**e**) 3×10^{13} cm^2. The diameter of ion track is assumed to be 4.4 nm.

Figure 5 shows the fraction of the tetragonal structure area as a function of ion-fluence, which has been calculated from the two-dimensional images. The figure also shows the experimental result and the result estimated using Equation (11). The result of the Monte Carlo simulation well agrees with the experimental result and that calculated by using the Poisson distribution function with the track cross section, S, of 1.5×10^{-13} cm^2, or the track diameter of 4.4 nm. Benyagoub et al. have shown that the similar equation to Equation (11) can reproduce their experimental result for ZrO_2 [16]. The equation used in ref. [16] was deduced through a quite complicated manner [28]. Meanwhile, in our case, Equation (11) can be simply given as an approximated formula of the binomial distribution function.

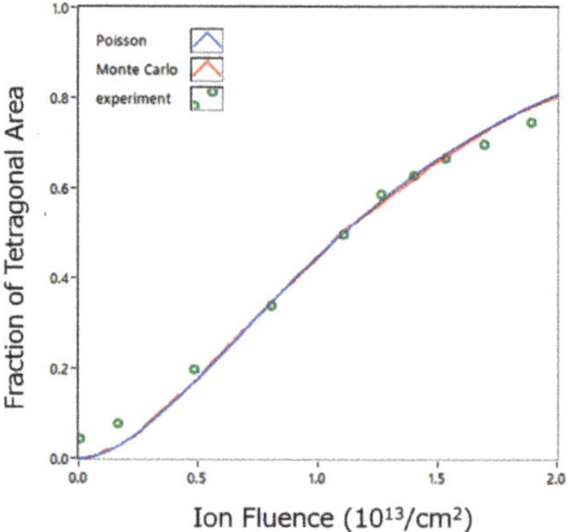

Figure 5. Fraction of tetragonal structure area of ZrO_2 as a function of ion fluence. Green circles, blue line, and red line represent the experimental result [16], result calculated by Equation (11) and the result from two-dimensional images, respectively. The value of S used in Equation (11) is 1.5×10^{-13} cm^2 which corresponds to the track diameter of 4.4 nm.

4.3. Change in Magnetic States of CeO$_2$ by High Energy Heavy Ion Irradiation

Concerning the appearance of magnetism in CeO$_2$ at room temperature, a lot of experimental and theoretical studies have been performed [29]. Although the mechanism of the appearance of the magnetism has never been fully clarified, previous studies have suggested that defects of O anions and Ce^{3+} state of cations somehow contribute to the magnetism of CeO$_2$ [29]. The measurements of EXAFS (extended x-ray absorption fine structure) and XPS (X-ray photoelectron spectroscopy) using synchrotron radiation facilities have revealed that 200 MeV Xe ion irradiation induces the oxygen deficiency around Ce cations in CeO$_2$, and the resultant change in valence state of cations from Ce^{4+} to Ce^{3+} [30,31]. The change in Ce valence state due to oxygen disorders has also been confirmed by the first principles calculation [32]. The SQUID (super quantum interference device) measurement shows that the irradiation with 200 MeV Xe ions induces the ferromagnetic state in CeO$_2$ [24,33]. These experimental and theoretical results imply that the appearance of the magnetism is attributed to the magnetic moment of localized 4f electrons on Ce^{3+} cations. Takaki and Yasuda have shown by the TEM observations that 200 MeV Xe ion irradiation produces one-dimensional defective regions (ion-tracks) in CeO$_2$ samples, and that only inside the ion-tracks, the arrangement of oxygen atoms is preferentially disordered [11]. Based on such previous results, we use the following model for the analysis of the ion-track overlapping effect on the magnetic state of Xe^{+14} ion irradiated CeO$_2$. Only inside the ion track, disorders of oxygen atoms and the resultant Ce^{3+} valence state are produced, and the ferro magnetic state appears. Outside the ion track, the sample is still nonmagnetic. With increasing Xe ion fluence, arrangements of not only oxygen atoms, but also cerium atoms become disordered by the ion track overlapping, resulting in the decreases in magnetization [24,33]. The effect of the ion track overlapping on the magnetic states for CeO$_2$ irradiated with 200 MeV Xe^{+14} ions is, therefore, more complicated than the cases of TiO$_2$ or ZrO$_2$. Our previous paper has shown that if the following effect of track overlapping on the saturation magnetization is assumed, the ion fluence dependence of the saturation magnetization calculated by using the Poisson distribution function well reproduces the experimental result [24]. The saturation magnetization is M$_0$ = 0 emu/g, M$_1$ = 0.1 emu/g, M$_2$ = 0.05 emu/g, M$_3$ = 0.025 emu/g and M$_4$ = 0.005 emu/g for the non-impacted area (r = 0), one-impacted area (r = 1), two-impacted area (r = 2), three-impacted area(r = 3), and the area for four or more impacts (r > =4), respectively. From Equation (4), $A(\Phi, r)$ for r = 0, 1, 2, 3 and for r of 4 or more impacts, are given by,

$$
\begin{aligned}
A(\Phi,0) &= \exp(-S\Phi) \\
A(\Phi,1) &= (S\Phi)\exp(-S\Phi) \\
A(\Phi,2) &= \frac{(S\Phi)^2}{2}\exp(-S\Phi) \\
A(\Phi,3) &= \frac{(S\Phi)^3}{6}\exp(-S\Phi) \\
A(\Phi,r>=4) &= 1 - \sum_{r=0}^{3} A(\Phi,r)
\end{aligned}
\tag{12}
$$

and the irradiation induced saturation magnetization as a function of ion fluence is given as,

$$
M(\Phi) = M_1 \cdot A(\Phi,1) + M_2 \cdot A(\Phi,2) + M_3 \cdot A(\Phi,3) + M_4 \cdot A(\Phi,r>=4)
\tag{13}
$$

Equation (13) will be used later for the comparison with the result of the Monte Carlo simulation.

The result of the Monte Carlo simulation is shown in Figures 6 and 7 for the transition of the magnetic states of CeO$_2$. The figure represents the two-dimensional images of areas having different saturation magnetization for the ion-fluence of 2.5×10^{12}, 5×10^{12}, 1.5×10^{13} and 3×10^{13} cm^{-2}. The track diameter is 4.7 nm which has been determined by the experiment [24].

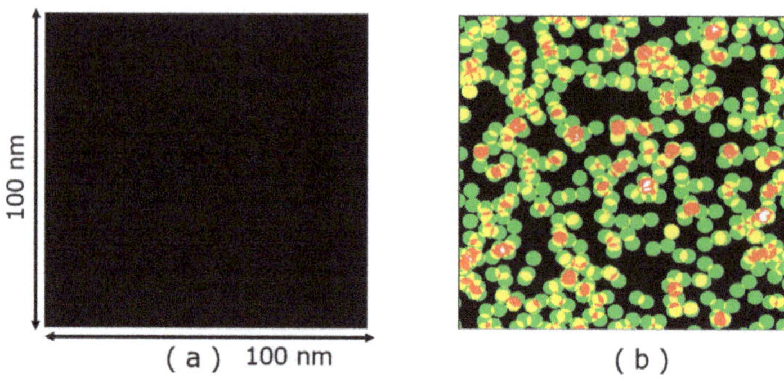

Figure 6. Two-dimensional images of irradiation-induced saturation magnetization of CeO_2 for ion fluences of (**a**) 0 cm^2 (unirradiated), and (**b**) 2.5×10^{12} cm^2. The diameter of ion track is assumed to be 4.7 nm. The correspondence relationship for colors and the values of magnetization is shown in Table 1.

Table 1. Correspondence relationship for colors in Figures 6 and 7, the number of ion track impacts and saturation magnetization.

Color		Number of Ion Track Impacts, r	Saturation Magnetization (emu/g) Mi
black		0	0
green		1	0.1
yellow		2	0.05
red		3	0.025
white		≥ 4	0.005

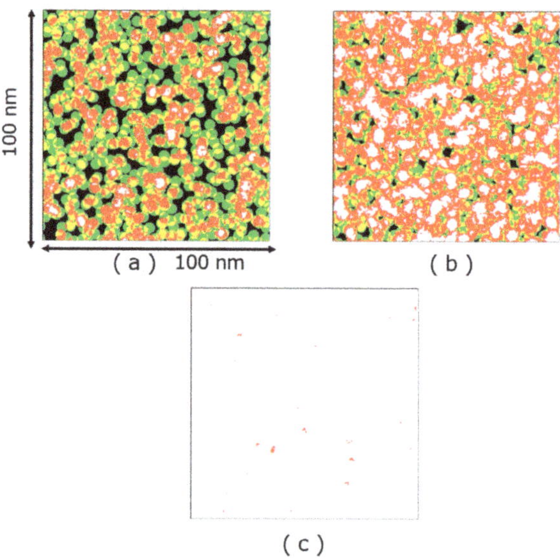

Figure 7. Two-dimensional images of irradiation-induced magnetization for ion fluences of (**a**) 5×10^{12} cm^2, (**b**) 1.5×10^{13} cm^2 and (**c**) 3×10^{13} cm^2. The diameter of ion track is assumed to be 4.7 nm. The correspondence relationship for colors and the values of magnetization is shown in Table 1.

The correspondence relationship for colors in Figures 6 and 7, the number of ion track impacts, and the saturation magnetization is shown in Table 1.

Figure 8 shows the saturation magnetization of CeO_2 as a function of ion-fluence, which has been calculated from the two-dimensional images. The figure also shows the experimental result [24] and the result calculated using Equation (13). The result of the Monte Carlo simulation well agrees with the experimental result and that calculated by using the Poisson distribution function.

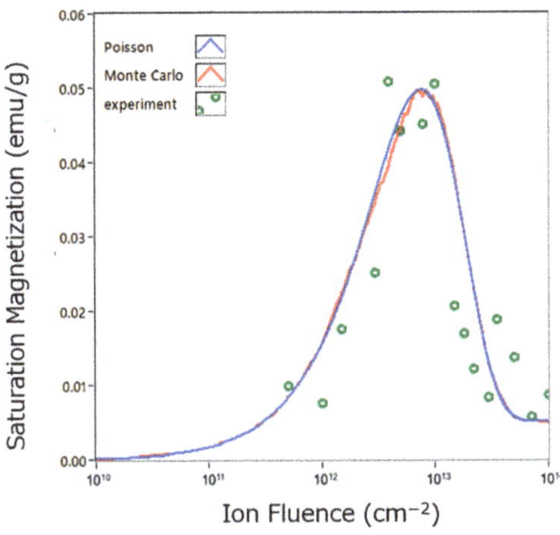

Figure 8. Saturation magnetization of CeO_2 as a function of ion fluence. Green circles, blue line, and red line represent the experimental result [24], result calculated by Equation (13) and the result from two-dimensional images, respectively. The value of S used in Equation (13) is 1.7×10^{-13} cm^2 corresponding to the track diameter of 4.7 nm.

5. Discussion

In the previous section, by using the Monte Carlo method, we have simulated the two-dimensional images for the three kinds of the ion track overlapping effects on the oxides irradiated with high energy heavy ions. In the case of the amorphization of TiO_2, as can be seen in Figure 1, the target is dotted with disk-shaped amorphized areas with the same diameter for small ion fluence, because only one impact of the ion-track can make a target amorphous. With increasing the ion fluence, the overlapping of amorphous tracks occurs more frequently, and for higher ion fluences, most part of the target becomes amorphous. For the crystal phase transformation of ZrO_2, Figure 4 shows that the areas of the tetragonal phase, which are surrounded by the monoclinic area, appear for the small ion fluence. The tetragonal phase areas have various shapes, which are far from disk-shape. This is due to the fact that only one ion-track impact does not cause the crystal phase transformation, but two or more ion-track impacts are needed for the crystal phase transformation. As can be seen in Figure 1, Figure 4, Figure 6, and Figure 7, the overlapping of ion -tracks leads to the modulated lattice and magnetic structures with a nanometer scale. Therefore, the Monte Carlo simulation provides a good first approach for understanding the nanometer-sized two-dimensional structures of oxides, which are produced by the overlapping of the ion-tracks. Moreover, the complementary usage of the Monte Carlo simulation with some experimental techniques of precise imaging, such as TEM, AFM (atomic force microscope), MFM (magnetic force microscope), and PEEM (photo-emission electron microscope), will also be useful in order to promote the study of the effect of high energy ion irradiation in materials.

In the present report, we have only mentioned the lattice structure and magnetic property changes by the ion-track overlapping. If the electrical conductivity appears inside the ion–tracks in insulators, and the overlapping of the ion–tracks affects their electrical conductivity, the formation of continuous electron paths or the conducting network will drastically change the macroscopic electrical property of the insulators. The two-dimensional images of ion track overlapping, which are simulated by the Monte Carlo method, may also be helpful for the understanding of such the percolation behavior in ion-irradiated insulators.

The ion-track overlapping effects have been analyzed so far by using complicated manners [20,28,34]. In the present study, the Monte Carlo method has confirmed that the effects of the ion-track overlapping can surely be described by a much simpler formula, the Poisson distribution function, which is the approximated formula of the binomial distribution function.

6. Summary

The two-dimensional images of the ion-track overlapping effects on the amorphization of TiO_2, the crystal phase transformation of ZrO_2, and the transition of the magnetic states of CeO_2 were simulated by the Monte Carlo method. From the two-dimensional images, the fractions of the areas which are modified by ion-tracks were calculated as a function of ion fluence. They agree well with the analytical result by using the Poisson distribution function.

Author Contributions: Monte Carlo simulation, S.N.; data analysis by Poisson distribution function, A.I.; data curation, S.N.; writing—original draft preparation, A.I.; writing—review and editing, A.I. and S.N. All authors have read and agreed to the published version of the manuscript.

Funding: This research received no external funding.

Data Availability Statement: The data presented in this study are available on request from the corresponding author.

Acknowledgments: One of the authors (A.I.) thanks N. Ishikawa and T. Yamaki for fruitful discussion.

Conflicts of Interest: The authors declare no conflict of interest.

References

1. Spohr, R. *Ion Tracks and Microtechnology*; Bethge, K., Ed.; Vieweg Verlag: Brauschweig, Germany, 1990.
2. Yamaki, T.; Nuryanthi, N.; Kitamura, A.; Koshikawa, H.; Sawada, S.; Voss, K.-O.; Severin, D.; Trautman, C. Fluoropoly-mer-based nanostructured membranes created by swift-heavy-ion irradiation and their energy and environmental applications. *Nucl. Instrum. Methods Phys. Res. Sect. B* **2018**, *435*, 162–168. [CrossRef]
3. Fleisher, R.L.; Price, P.B.; Walker, R.M. Nuclear Tracks in Solids (Principles & Applications). *Nucl. Technol.* **1981**, *54*, 126. [CrossRef]
4. Toulemonde, M. Nanometric phase transformation of oxide materials under GeV energy heavy ion irradiation. *Nucl. Instrum. Methods Phys. Res. Sect. B* **1999**, *156*, 1–11. [CrossRef]
5. Toulemonde, M.; Assmann, W.; Dufour, C.; Meftah, A.; Studer, F.; Trautmann, C. Experimental Phenomena and Thermal Spike Model Description of Ion Tracks in Amorphisable Inorganic Insulators. *Mat. Fys. Medd.* **2006**, *52*, 263–292.
6. Szenes, G. Information provided by a thermal spike analysis on the microscopic processes of track formation. *Nucl. Instrum. Methods Phys. Res. Sect. B* **2002**, *191*, 54–58. [CrossRef]
7. Fleischer, R.L.; Price, P.B.; Walker, R.M. Ion Explosion Spike Mechanism for Formation of Charged-Particle Tracks in Solids. *J. Appl. Phys.* **1965**, *36*, 3645–3652. [CrossRef]
8. Haff, P.K. Possible new sputtering mechanism in track registering materials. *Appl. Phys. Lett.* **1976**, *29*, 473–475. [CrossRef]
9. Bringa, E. Molecular dynamics simulations of Coulomb explosion. *Nucl. Instrum. Methods Phys. Res. Sect. B* **2003**, *209*, 1–8. [CrossRef]
10. Takaki, S.; Yasuda, K.; Yamamoto, T.; Matsumura, S.; Ishikawa, N. Structure of ion tracks in ceria irradiated with high energy xenon ions. *Prog. Nucl. Energy* **2016**, *92*, 306–312. [CrossRef]
11. Takaki, S.; Yasuda, K.; Yamamoto, T.; Matsumura, S.; Ishikawa, N. Atomic structure of ion tracks in Ceria. *Nucl. Instrum. Methods Phys. Res. Sect. B* **2014**, *326*, 140–144. [CrossRef]
12. Toulemonde, M.; Studer, F. Comparison of the radii of latent tracks induced by hjgh-energy heavy ions in Y3Fe5O12 by HREM, channeling Rutherford backscattering and Mossbauer spectrometry. *Philos. Mag. A* **1988**, *58*, 799–808. [CrossRef]
13. Jozwik-Biala, I.; Jagielski, J.; Thome, L.; Arey, B.; Kovarik, L.; Sattonnay, G.; Debelle, A.; Monnet, I. HRTEM study of track evolution in 120-MeV U irradiated $Gd_2Ti_2O_7$. *Nucl. Instrum. Methods Phys. Res. Sect. B* **2012**, *286*, 258–261. [CrossRef]
14. Kuroda, N.; Ishikawa, N.; Chimi, Y.; Iwase, A.; Ikeda, H.; Yoshizaki, R.; Kambara, T. Effects of defect morphology on the properties of the vortex system in $Bi_2Sr_2CaCu_2O_8+\delta$ irradiated with heavy ions. *Phys. Rev. B* **2001**, *63*, 224502. [CrossRef]

15. Ishikawa, N.; Iwase, A.; Chimi, Y.; Michikami, O.; Wakana, H.; Kambara, T. Defect Production Induced by Primary Ionization in Ion-Irradiated Oxide Superconductors. *J. Phys. Soc. Jpn.* **2000**, *69*, 3563–3575. [CrossRef]
16. Benyagoub, A.; Couvreur, F.; Bouffard, S.; Levesque, F.; Dufour, C.; Paumier, E. Phase transformation induced in pure zirconia by high energy heavy ion irradiation. *Nucl. Instrum. Methods Phys. Res. Sect. B* **2001**, *175-177*, 417–421. [CrossRef]
17. Bilgen, S.; Sattonnay, G.; Grygiel, C.; Monnet, I.; Simon, P.; Thomé, L. Phase transformations induced by heavy ion irradiation in Gd_2O_3: Comparison between ballistic and electronic excitation regimes. *Nucl. Instrum. Methods Phys. Res. Sect. B* **2018**, *435*, 12–18. [CrossRef]
18. Rymzhanov, R.; Medvedev, N.; Volkov, A.; O'Connell, J.; Skuratov, V. Overlap of swift heavy ion tracks in Al2O3. *Nucl. Instrum. Methods Phys. Res. Sect. B* **2018**, *435*, 121–125. [CrossRef]
19. Tracy, C.L.; Lang, M.; Zhang, F.; Trautmann, C.; Ewing, R.C. Phase transformations inLn2O3materials irradiated with swift heavy ions. *Phys. Rev. B* **2015**, *92*, 174101. [CrossRef]
20. Garcia, G.; Rivera, A.; Crespillo, M.; Gordillo, N.; Olivares, J.; Agulló-López, F.; Garcia, N.G. Amorphization kinetics under swift heavy ion irradiation: A cumulative overlapping-track approach. *Nucl. Instrum. Methods Phys. Res. Sect. B* **2011**, *269*, 492–497. [CrossRef]
21. Ishikawa, N.; Yamamoto, S.; Chimi, Y. Structural changes in anatase TiO2 thin films irradiated with high-energy heavy ions. *Nucl. Instrum. Methods Phys. Res. Sect. B* **2006**, *250*, 250–253. [CrossRef]
22. Rath, H.; Dash, B.N.; Benyagoub, A.; Mishra, N.C. Sensitivity of Anatase and Rutile Phases of TiO2 to ion irradiation: Ex-amination of the applicability of Coulomb Explosion and Thermal Spike Models. *Sci. Rep.* **2018**, *8*, 11774. [CrossRef]
23. Ishikawa, N.; Ohhara, K.; Ohta, Y.; Michikami, O. Binomial distribution function for intuitive understanding of fluence dependence of non-amorphized ion-track area. Nucl. Instrum. *Methods Phys. Res. Sect. B* **2010**, *268*, 3273–3276.
24. Yamamoto, Y.; Ishikawa, N.; Hori, F.; Iwase, A. Analysis of Ion-Irradiation Induced Lattice Expansion and Ferromagnetic State in CeO_2 by Using Poisson Distribution Function. *Quantum Beam Sci.* **2020**, *4*, 26. [CrossRef]
25. Simons, G.; Johnson, N.L. On the Convergence of Binomial to Poisson Distributions. *Ann. Math. Stat.* **1971**, *42*, 1735–1736. [CrossRef]
26. Awazu, K.; Wang, X.; Fujimaki, M.; Komatsubara, T.; Ikeda, T.; Ohki, Y. Structure of latent tracks in rutile single crystal of titanium dioxide induced by swift heavy ions. *J. Appl. Phys.* **2006**, *100*, 044308. [CrossRef]
27. O'Connell, J.; Skuratov, V.; Van Vuuren, A.J.; Saifulin, M.; Akilbekov, A. Near surface latent track morphology of SHI irradiated TiO_2. *Phys. Stat. Sol.* **2016**, *253*, 2144–2149. [CrossRef]
28. Benyagoub, A.; Thomé, L. Amorphization mechanisms in ion-bombarded metallic alloys. *Phys. Rev. B* **1988**, *38*, 10205–10216. [CrossRef] [PubMed]
29. Ackland, K.; Coey, J.M.D. Room temperature magnetism in CeO_2—A review. *Phys. Rep.* **2018**, *746*, 1–39. [CrossRef]
30. Ohno, H.; Iwase, A.; Matsumura, D.; Nishihata, Y.; Mizuki, J.; Ishikawa, N.; Baba, Y.; Hirao, N.; Sonoda, T.; Kinoshita, M. Study on effects of swift heavy ion irradiation in cerium dioxide using synchrotron radiation X-ray absorption spectroscopy. *Nucl. Instrum. Methods Phys. Res. Sect. B* **2008**, *266*, 3013–3017. [CrossRef]
31. Iwase, A.; Ohno, H.; Ishikawa, N.; Baba, Y.; Hirao, N.; Sonoda, T.; Kinoshita, M. Study on the behavior of oxygen atoms in swift heavy ion irradiated CeO_2 by means of synchrotron radiation X-ray photoelectron spectroscopy. *Nucl. Instrum. Methods Phys. Res. Sect. B* **2009**, *267*, 969–972. [CrossRef]
32. Iwasawa, M.; Ohnuma, T.; Chen, Y.; Kaneta, Y.; Geng, H.-Y.; Iwase, A.; Kinoshita, M. First-principles study on cerium ion behavior in irradiated cerium dioxide. *J. Nucl. Mater.* **2009**, *393*, 321–327. [CrossRef]
33. Shimizu, K.; Kosugi, S.; Tahara, Y.; Yasunaga, K.; Kaneta, Y.; Ishikawa, N.; Hori, F.; Matsui, T.; Iwase, A. Change in magnetic properties induced by swift heavy ion irradiation in CeO_2. *Nucl. Instrum. Methods Phys. Res. Sect. B* **2012**, *286*, 291–294. [CrossRef]
34. Khayat, O.; Afarideh, H. Heavy Charged Particle Nuclear Track Counting Statistics and Count Loss Estimation in High Density Track Images. *IEEE Trans. Nucl. Sci.* **2014**, *61*, 2727–2734. [CrossRef]

quantum beam science

MDPI

Article

Nanopore Formation in CeO$_2$ Single Crystal by Ion Irradiation: A Molecular Dynamics Study

Yasushi Sasajima [1,*], Ryuichi Kaminaga [1], Norito Ishikawa [2] and Akihiro Iwase [3]

[1] Department of Materials Science and Engineering, Graduate School of Science and Engineering, Ibaraki University, 4-12-1 Nakanarusawa, Hitachi 316-8511, Japan; kaminaga@advancesoft.jp
[2] Japan Atomic Energy Agency (JAEA), 2-4 Shirakata Shirane, Tokai 319-1195, Japan; ishikawa.norito@jaea.go.jp
[3] The Wakasa Wan Energy Research Center, 64-52-1 Nagatani, Tsuruga 914-0192, Japan; aiwase@werc.or.jp
* Correspondence: yasushi.sasajima.mat@vc.ibaraki.ac.jp

Abstract: The nanopore formation process that occurs by supplying a thermal spike to single crystal CeO$_2$ has been simulated using a molecular dynamics method. As the initial condition, high thermal energy was supplied to the atoms in a nano-cylinder placed at the center of a fluorite structure. A nanopore was generated abruptly at around 0.3 ps after the irradiation, grew to its maximum size at 0.5 ps, shrank during the time to 1.0 ps, and finally equilibrated. The nanopore size increased with increasing effective stopping power gSe (i.e., the thermal energy deposited per unit length in the specimen), but it became saturated when gSe was 0.8 keV/nm or more. This finding will provide useful information for precise control of the size of nanopores. Our simulation confirmed nanopore formation found in the actual experiment, irradiation of CeO$_2$ with swift heavy ions, but could not reproduce crystalline hillock formation just above the nanopores.

Keywords: nanopore structure; ceria; irradiation; molecular dynamics; simulation; structural analysis; defects

Citation: Sasajima, Y.; Kaminaga, R.; Ishikawa, N.; Iwase, A. Nanopore Formation in CeO$_2$ Single Crystal by Ion Irradiation: A Molecular Dynamics Study. *Quantum Beam Sci.* **2021**, 5, 32. https://doi.org/10.3390/qubs5040032

Academic Editor: Lorenzo Giuffrida

Received: 15 October 2021
Accepted: 9 November 2021
Published: 18 November 2021

Publisher's Note: MDPI stays neutral with regard to jurisdictional claims in published maps and institutional affiliations.

1. Introduction

When oxide metals are irradiated by high-energy heavy ions, nano-sized protrusions and pores are produced on the surface and inside the specimen, respectively. For example, in an irradiated NiO specimen, nano-sized protrusions and cylindrically shaped nanopores were generated simultaneously [1]. Using transmission electron microscopy (TEM), Jensen et al. [2] observed hillocks at both ends of tracks in Yttrium Iron Garnet (YIG) induced by high-energy -C$_{60}$ ions. They believed that the matter was emitted from the ion track, leaving the spherical hillocks on the sample surface at the entrance and exit. Recently, Ishikawa et al. [3] found that hemispherical protrusions and nanopores were also generated in CeO$_2$ by high-energy ion beam irradiation. Subsequently Ishikawa et al. [4] observed ion tracks and hillocks produced by swift heavy ions of different velocities in Y$_3$Fe$_5$O$_{12}$ by TEM. They found the dimensions of the hillocks increase as a function of stopping power, *Se*. The data can be interpreted by the lifetime of the melt region produced by irradiation. Ion-irradiated CeO$_2$ was studied extensively regarding its optical reflectivity [5] and spectroscopic characteristics [6], defects [7–10], grain size effects [11], X-ray Photoelectron Spectroscopy (XPS) [12], and effects of irradiation temperature on final structure [13].

Nanopore formation is an interesting phenomenon from the viewpoint of nano-order fabrication of materials and can lead to the realization of highly functional materials such as catalysts [14]. Therefore, it is crucial to clarify the formation mechanism of nano-sized protrusions and cylindrically shaped nanopores. Since high-energy beam irradiation is a non-equilibrium process and its relaxation time is very short, it is difficult to clarify such a mechanism only by an actual experiment. Molecular dynamics (MD) simulation provides a useful tool for the analysis of such a process because it can calculate the trajectories of individual atoms during a short time period.

In our previous papers, using computer-aided simulations, we studied the effects of high-energy, heavy-ion irradiation [15–19] as a theme in the development of a new generation of nuclear fuels with a high burn–up ratio. Employing MD simulations, we evaluated the disorder of a single crystal structure of specimens irradiated by fast particles. Our MD simulations elaborated the structural change followed by the high-energy dissipation of a nano-scale region in a single crystal of CeO_2 after irradiation [18,19]. Yablinsky et al. [20] analyzed the structure of ion tracks and investigated thermal spikes in CeO_2 with energy depositions using MD simulation. Medvedev et al. [21] developed the Monte-Carlo code TREKIS (Time-Resolved Electron Kinetics in swift heavy-ion Irradiated Solid) which models how a penetrating swift heavy ion (SHI) excites the electron subsystems of various solids, and the generated fast electrons spread spatially. Subsequently, the same group proposed a hybrid approach that consisted of the Monte-Carlo code TREKIS and the classical molecular dynamics code LAMMPS (Large-scale Atomic/Molecular Massively Parallel Simulator) for lattice atoms to simulate the formation process of a cylindrical track of about 2 nm diameter in Al_2O_3 irradiated by Xe 167 MeV ions [22]. Then, Rymzhanov et al. [23] simulated a structure in overlapping swift heavy-ion track regions in Al_2O_3 and found good coincidence with observation results of irradiated Al_2O_3 by high resolution TEM. Another paper by Rymzhanov et al. [24] examined swift heavy-ion irradiation of forsterite (Mg_2SiO_4) and the effects of the ion energy and its energy losses on the track radius were explained, and the track formation thresholds was determined [24]. Rymzhanov et al. [25] also clarified that different ion tracks were produced in MgO, Al_2O_3, and $Y_3Al_5O_{12}$ (YAG) by irradiation with Xe (176 MeV) ions, whereas no ion tracks in MgO, discontinuous distorted crystalline tracks in Al_2O_3 and continuous amorphous tracks in YAG were detected. Recently, Rymzhanov et al. [26] studied the irradiation process of MgO, CaF_2, and $Y_3Al_5O_{12}$ (YAG) with fast ions. They found MgO and CaF_2 showed recovery of transient damage in the surface region, forming a spherically shaped nano-hillock, whereas YAG showed almost no recovery of the transient disorder, forming an amorphous hillock. The movie they attached as supplemental material demonstrates nano-hillock formation of CaF_2 irradiated by 200 MeV Au. Some of the above same researchers teamed with Karlušić et al. [27] to research Al_2O_3 and MgO irradiated under grazing incidence with an I beam of 23 MeV. In this study, they found grooves surrounded with nano-hillocks on MgO surfaces and smoother, roll-like discontinuous structures on the surfaces of Al_2O_3.

In the present study, we used an MD method to simulate the nanopore structure formation process in a single crystal CeO_2 with two free surfaces by supplying a thermal spike. Structural analysis was done for the obtained specimens. We evaluated the number of Frenkel pairs as a function of the thermal energy deposited per unit length in the specimen, gSe, which represents the beam strength. We classified various types of oxygen Frenkel pairs by the distance between the vacancy and the corresponding oxygen atom.

2. Simulation Method

2.1. Molecular Dynamics

Concerning the MD simulation method, we adopted the same method as our previous work [19], except that the specimens have a free surface. Therefore, the calculation method is only outlined as follows. The Ce and O atoms were laid out to form the fluorite structure unit cell. Then, the unit cell was cloned 6 times each in <010> and <100> directions and 4 times in the <001> direction. The dimensions of the CeO_2 crystal were 3.25 nm × 3.25 nm × 2.33 nm, whereas those of the calculation region were 3.25 nm × 3.25 nm × 3.25 nm. There was a free space on the upper and lower sides of the CeO_2 crystal along the <001> direction to mimic the free surface. Figure 1a shows the unit structure of the CeO_2 fluorite structure and Figure 1b shows the whole system for calculation.

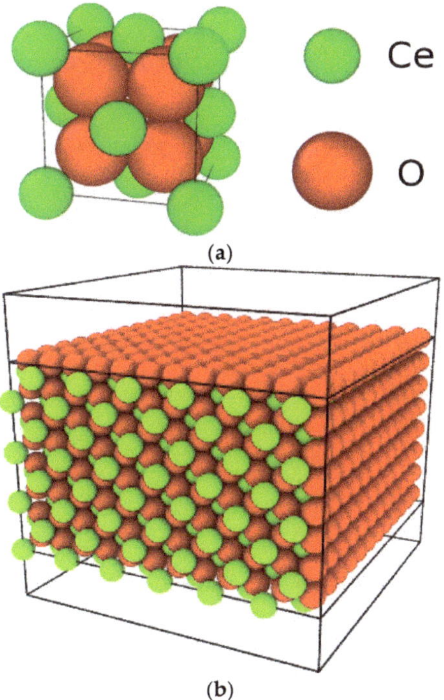

Figure 1. (**a**) The unit structure of CeO$_2$ fluorite. (**b**) The whole system for calculation.

In order to calculate the interaction between the atoms, we used the potential type proposed by Inaba et al. [28]. This is the Born–Mayer–Huggins potential type, which has a function form as

$$\phi_i(r_{ij}) = \frac{z_i z_j e^2}{r_{ij}} + f_0(b_i + b_j) exp\left(\frac{a_i + a_j - r_{ij}}{b_i + b_j}\right) - \frac{c_i c_j}{r_{ij}^6} \tag{1}$$

where r_{ij} is the distance between ions i and j, z_i is the effective valence of an ion i, e is the electron charge, f_0 is a constant to adjust the unit, c_i and c_j are the parameters of the molecular interaction term, and a_i and b_i are the parameters of the repulsion term. For the electrostatic interactions, the Ewald method was applied. The potential cut-off was 1.37 nm for short range interaction and for the real part of the Ewald summation. The potential parameters were determined to reproduce the lattice parameters at various temperatures and the bulk modulus of CeO$_2$. The fitted parameters for CeO$_2$ are presented in Table 1.

Table 1. Potential parameters for CeO$_2$ [28].

Parameters	Ce^{4+}	O^{2-}
z	2.700	−1.350
a (nm)	0.1330	0.1847
b (nm)	0.00454	0.0166
c ($J^{0.5}(nm)^3 mol^{-0.5}$)	0.00	1.294
f_0	4.07196	4.07196

A cylindrical region with a diameter of 1.0 nm located at the center of the calculation region in the <001> direction was considered as the irradiation beam trajectory. Using a MD method, we initially relaxed the structure at the temperature of 298 K before a high thermal energy was applied to the cylindrical region. This temperature was set as typical ambient temperature, or room temperature. Thermal energy is a part of the energy deposited by the heavy-ion irradiation. The present study names this thermal energy as an effective stopping power, and it is described as gSe. The parameter g in gSe signifies the ratio of thermal energy transferred from stopping power to the lattice as the vibration energy. In the MD simulation, the high thermal energy was applied to the cylindrical region by setting the velocity of the atoms in the region to values ranging from gSe = 0.0 to 1.6 keV/nm with 0.1 keV/nm energy bins according to the Maxwell distribution. According to the experiment by Ishikawa et al. [3], Se = 32.0 keV/nm for 200 MeV Au ion. The maximum value of gSe for our simulation is gSe = 1.6 keV/nm, thus the energy transfer ratio g = 0.05. According to TREKIS, a 167 MeV Xe ion gives Se = 21, 24.9 and 25 keV/nm for MgO, Al_2O_3 and YAG, respectively, and it gives a radial distribution of the excess lattice energy density around the trajectory of the irradiated ion (see Figure 2 in reference [25]). A rough estimation of gSe can be made from the excess lattice energy of MgO and Al_2O_3 as gSe = 2 keV/nm and the value of g is the same order as ours. The temperature of the region outside the cylindrical region with a diameter of 2.0 nm was kept at 298 K by the velocity scaling method [29]. In <010>, <001> and <100> directions, the periodic boundary condition was considered. The simulation duration was 10,000 molecular dynamic intervals, each taking 0.3 fs, summing up to 3 ps.

2.2. Structure Analysis

A vacancy is defined as the vacant site in the CeO_2 fluorite structure, such that the distance between the vacant site and the nearest atom from the site is larger than the atomic radius of O or Ce originally set at the vacant site, 0.097 nm (for O) and 0.138 nm (for Ce). This is what we call the "Lindemann criterion", and it may overestimate the number of vacancies. An oxygen Frenkel pair is defined as the pair of a vacant site and the oxygen atom which escaped from the site. Oxygen Frenkel pairs can be classified according to the distance between the vacant site and the oxygen atom that was originally located at the site, such as 1NN (1st nearest neighbor) Frenkel pair, 2NN Frenkel pair, and so on [30]. Figure 2 shows an example of a 1NN Frenkel pair, and Table 2 lists the distance between the vacant site and the oxygen atom for iNN Frenkel pairs (i = 1, 2, ... , 7). We evaluated the number of vacancies and Frenkel pairs in the irradiated specimen as a function of time and as a function of gSe.

1NN FP

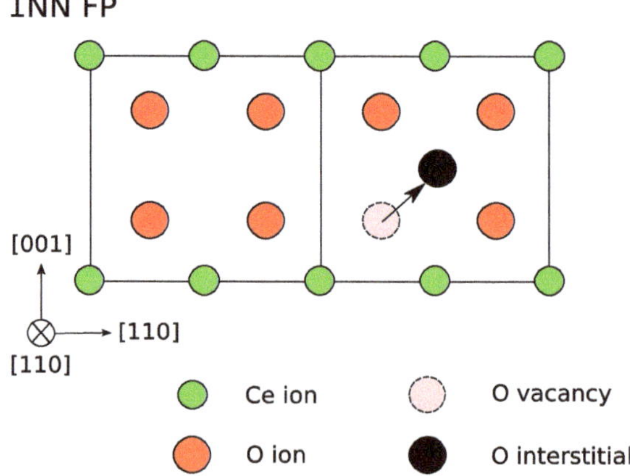

Figure 2. An example 1NN Frenkel pair.

Table 2. Distance between the vacant site and oxygen atom for iNN Frenkel pairs ($i = 1, 2, \ldots, 7$).

Type (NN: Nearest Neighbor)	Distance (nm)
1NN	0.234
2NN	0.449
3NN	0.590
4NN	0.703
5NN	0.800
6NN	0.887
7NN	0.966

3. Results and Discussion

Structural changes of the CeO_2 systems after the irradiation, as viewed from <001> and <100> directions, are shown in Figures 3 and 4 for the values of gSe = 0.0 and 0.8 keV/nm, respectively. The number of ejected atoms increased with increasing gSe value.

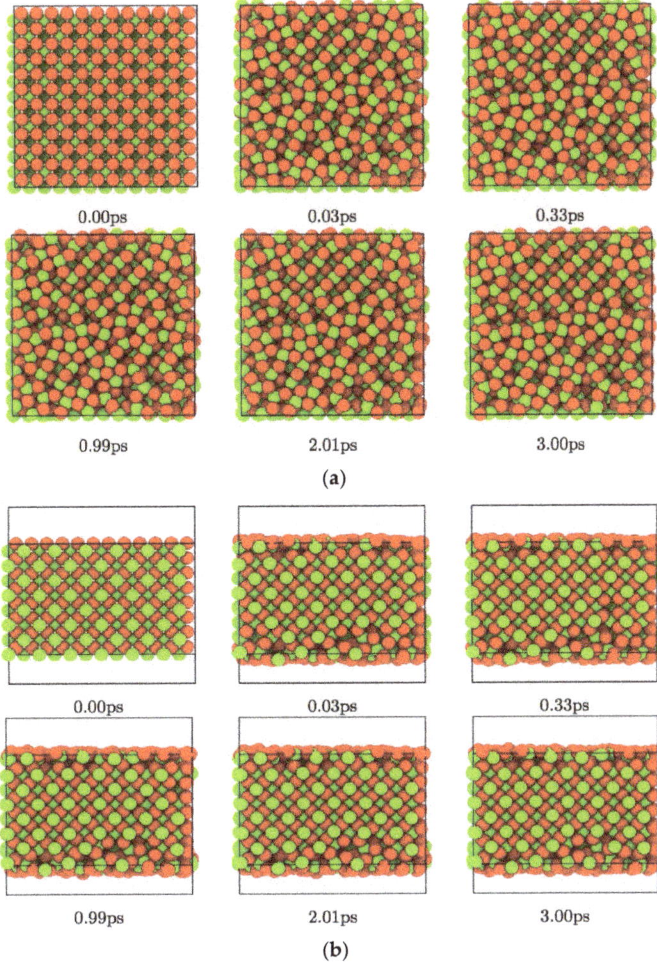

Figure 3. Structural change of the CeO_2 systems viewed from (**a**) <001> and (**b**) <100> directions after the irradiation and for gSe = 0.0 keV/nm.

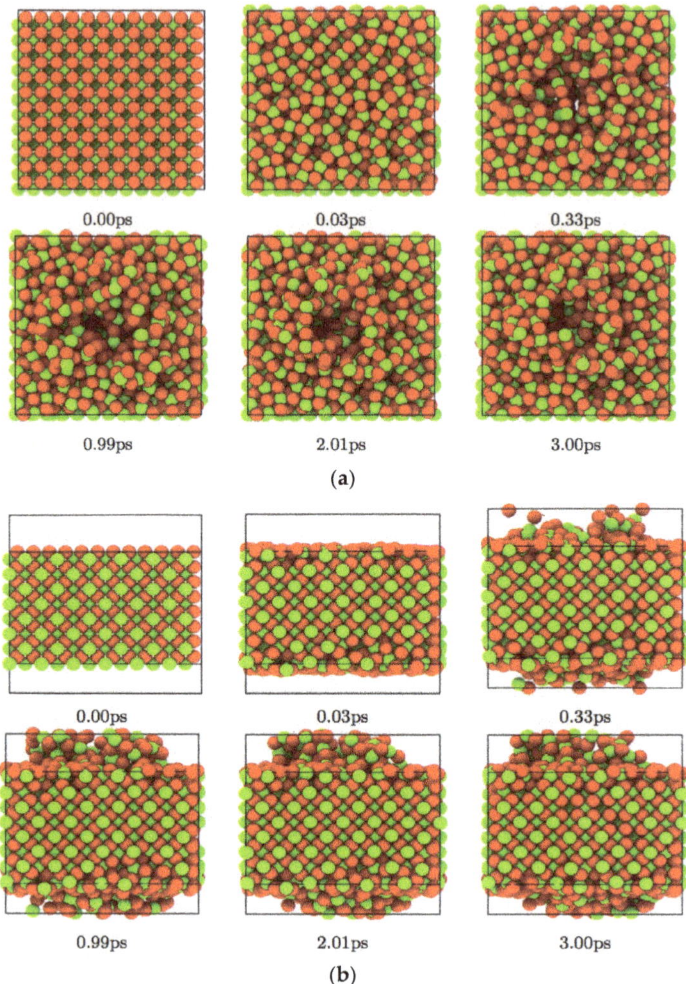

Figure 4. Structural change of the CeO_2 systems viewed from (**a**) <001> and (**b**) <100> directions after the irradiation and for gSe = 0.8 keV/nm.

Figure 3 indicates that the surface of the specimen showed disorder even when gSe = 0.0 keV/nm. This means that the disorder on the surface is not generated by irradiation. Therefore, we neglect the surface atoms for the structural analysis in order to clarify the formation process of the nanopore. In this case, no defects were found in the interior of the specimen; however, there is the possibility that defects are created at gSe = 0.0 keV/nm. The thermal energy provides the system necessary for the creation of defects. However, the ambient temperature is room temperature that is around 1/40 eV, which is low compared with the defect formation energy. Therefore, the defect creation event is rare, and no defects were found in the calculated system. We see the creation of defects among many specimens prepared, but this is out of our scope in the present study. As can be seen from Figure 4, a nanopore was formed in the specimen by ejection of the atoms located in the central region. In the case of gSe = 0.4 keV/nm, a nanopore was also produced but its diameter was smaller than that of gSe = 0.8 keV/nm.

It is interesting to compare this figure with our previous result for computer simulation of irradiation of CeO_2 single crystals with no free surface. The Ce sublattice was stable up to gSe = 1.5 keV/nm, and even at gSe = 2.0 keV/nm only several interstitial atoms were found as defects (see Figure 4 in reference [19]). It can be considered that the Ce sublattice damage is confined near the surface of the irradiated region compared with the O sublattice damage. To elaborate the nanopore formation process, the time change of the total number of vacancies for the various values of gSe was calculated.

Figure 5a,b show the results for gSe = 0.1–0.8 and 0.8–1.6 keV/nm, respectively. (No defect was found in the interior of the specimen at gSe = 0.0 keV/nm because of the low temperature 298 K.) For all these gSe cases, the number of vacancies increased abruptly up to 0.2 ps after the irradiation. It increased further and reached the maximum, then gradually decreased until 1.5 ps and equilibrated to a constant value after that. The equilibrated number of vacancies increased as gSe increased for the low values of gSe less than 0.8 keV/nm and converged to a constant value around 200 when gSe was larger than 0.8 keV/nm. It is noteworthy that the number of atoms in the region where vacancies were counted is 1296. In the range of gSe = 0.1–0.8 keV/nm, the more thermal energy given, the more disordered the structure. However, in the range of 0.8–1.6 keV/nm, the additional thermal energy cannot change the structure drastically because the structure is already disordered. If the same amount of energy is given to an ordered and a disordered structure, the increment of entropy of the ordered structure is much larger than that of the disordered structure.

The distributions of vacancies in the CeO_2 systems viewed from <001> and <100> directions are shown in Figures 6 and 7 for the values of gSe = 0.2 and 0.8 keV/nm, respectively. The nanopore was formed abruptly at around 0.3 ps after the irradiation and grew to its maximum size at 0.5 ps, then it shrank during the time to 1.0 ps and finally became equilibrated. The size of the nanopore increased as gSe increased, but it saturated when gSe was 0.8 keV/nm or more. This finding will provide useful information for precise control of the size of the nanopore.

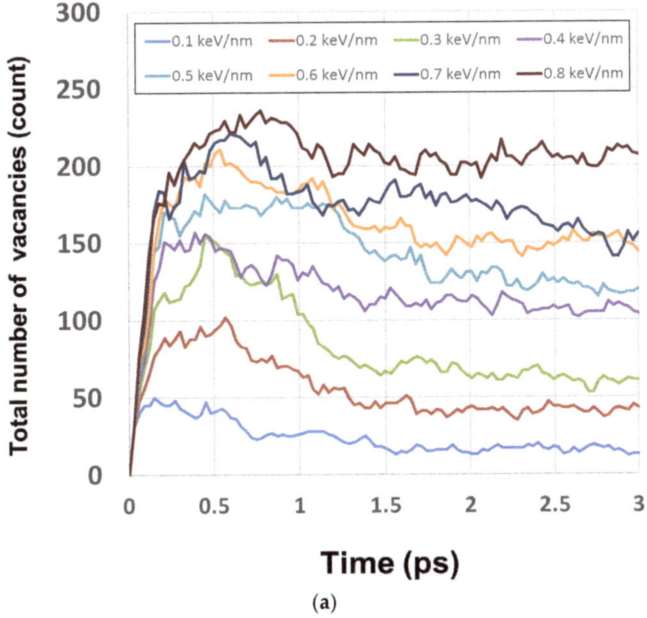

Figure 5. *Cont.*

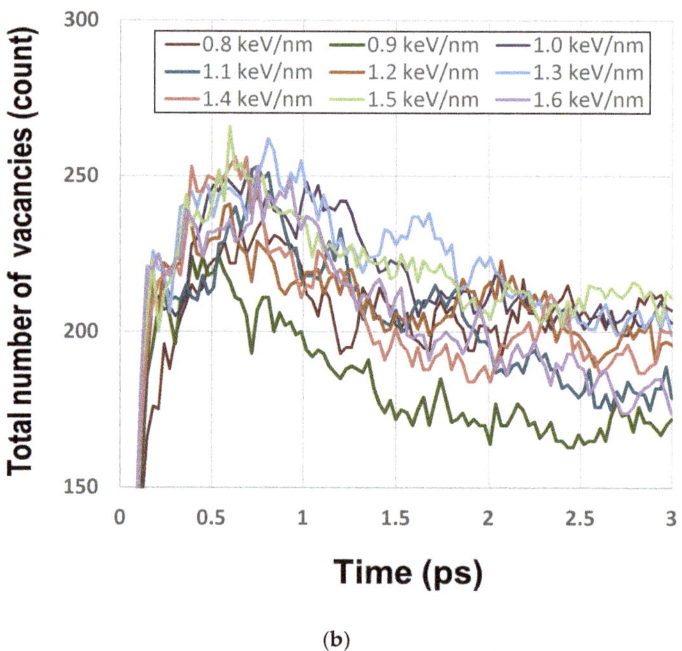

(b)

Figure 5. Time change of the total number of vacancies for the values of (**a**) *gSe* = 0.1–0.8 keV/nm and (**b**) *gSe* = 0.8–1.6 keV/nm.

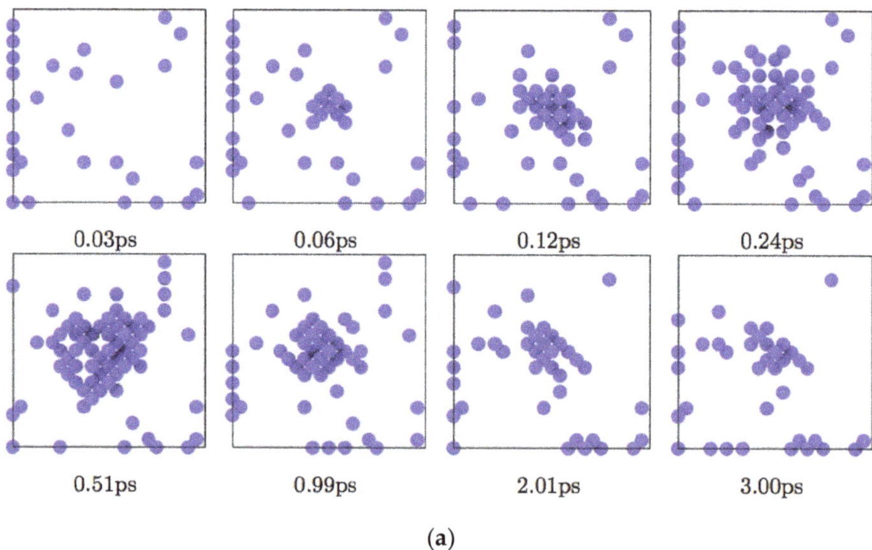

(a)

Figure 6. *Cont.*

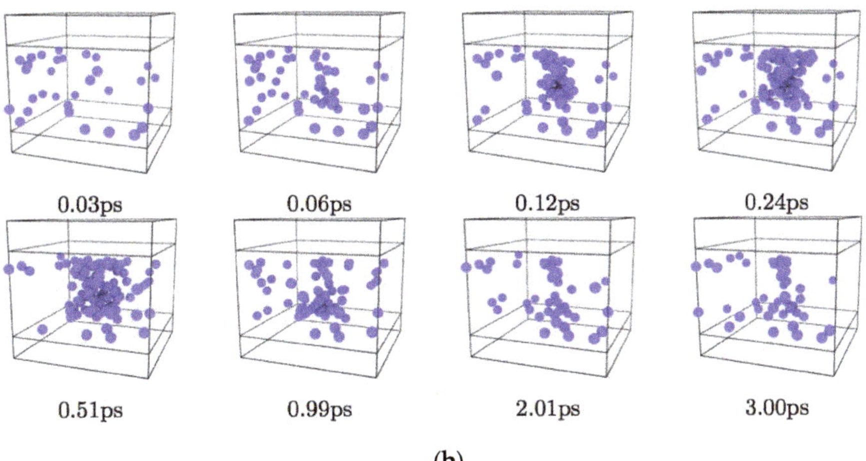

0.03ps   0.06ps   0.12ps   0.24ps

0.51ps   0.99ps   2.01ps   3.00ps

(b)

Figure 6. Distributions of vacancies in the CeO$_2$ systems viewed from (**a**) <001> and (**b**) <100> directions for the values of gSe = 0.2 keV/nm.

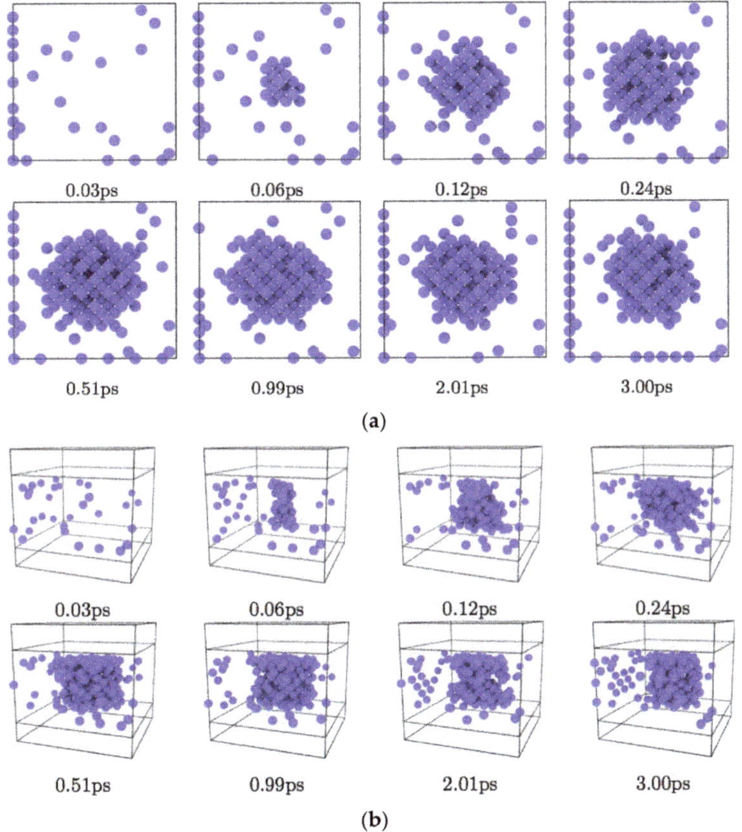

0.03ps   0.06ps   0.12ps   0.24ps

0.51ps   0.99ps   2.01ps   3.00ps

(a)

0.03ps   0.06ps   0.12ps   0.24ps

0.51ps   0.99ps   2.01ps   3.00ps

(b)

Figure 7. Distributions of vacancies in the CeO$_2$ systems viewed from (**a**) <001> and (**b**) <100> directions for the values of gSe = 0.8 keV/nm.

Figure 8 shows the time change of oxygen Frenkel pairs for gSe = 0.4, 0.8, 1.2, and 1.6 keV/nm. The number of oxygen Frenkel pairs could be separated into two categories, higher gSe (0.8, 1.2 and1.6 keV/nm) and lower gSe (0.4 keV/nm). The number of oxygen Frenkel pairs in the higher gSe category was larger than that of the lower gSe category; however, there was no clear correlation between the Frenkel pairs and gSe in the higher category. This behavior was similar to that for the number of vacancies and the radius of the nanopore. It should be noted that the number of oxygen Frenkel pairs is one order of magnitude lower than the total number of vacancies shown in Figure 5. This difference comes from the difference of definition of vacancy and Frenkel pairs. As shown in Table 2, the distance between the vacant site and the escaped oxygen atom should be larger than 0.234 nm. In contrast, the corresponding distances defined for the vacancy are much smaller than that for oxygen Frenkel pairs, 0.097 nm (for O) and 0.138 nm (for Ce). The time averaged distribution of the iNN oxygen Frenkel pairs, where i = 1, 2, … , 7, is shown for the values of gSe = 0.4, 0.8, 1.2, and 1.6 keV/nm in Figure 9a–d, respectively. As seen from these figures, the short-distance Frenkel pairs, 1NN to 4NN, were in the majority for lower gSe (0.4 keV/nm), whereas the long-distance Frenkel pairs were found for the higher values of gSe (0.8, 1.2 and1.6 keV/nm).

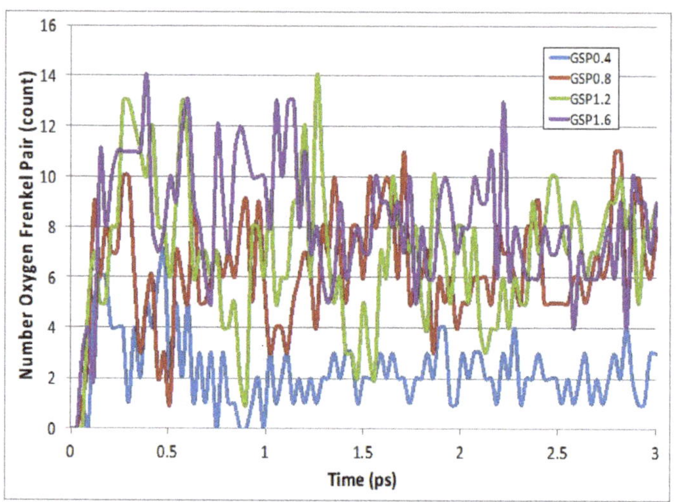

Figure 8. Time change of oxygen Frenkel pairs for gSe = 0.4, 0.8, 1.2, and1.6 keV/nm.

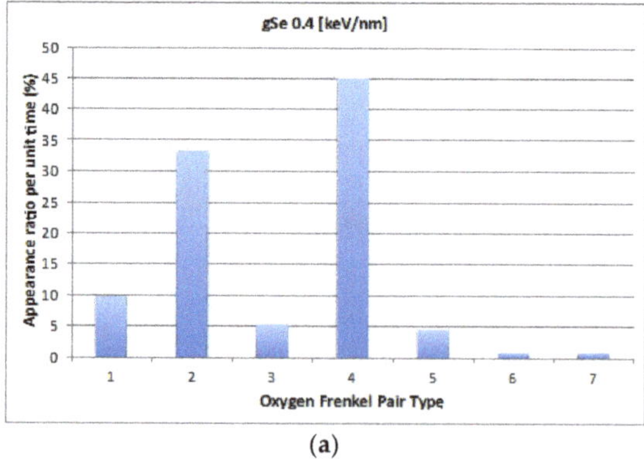

(a)

Figure 9. *Cont.*

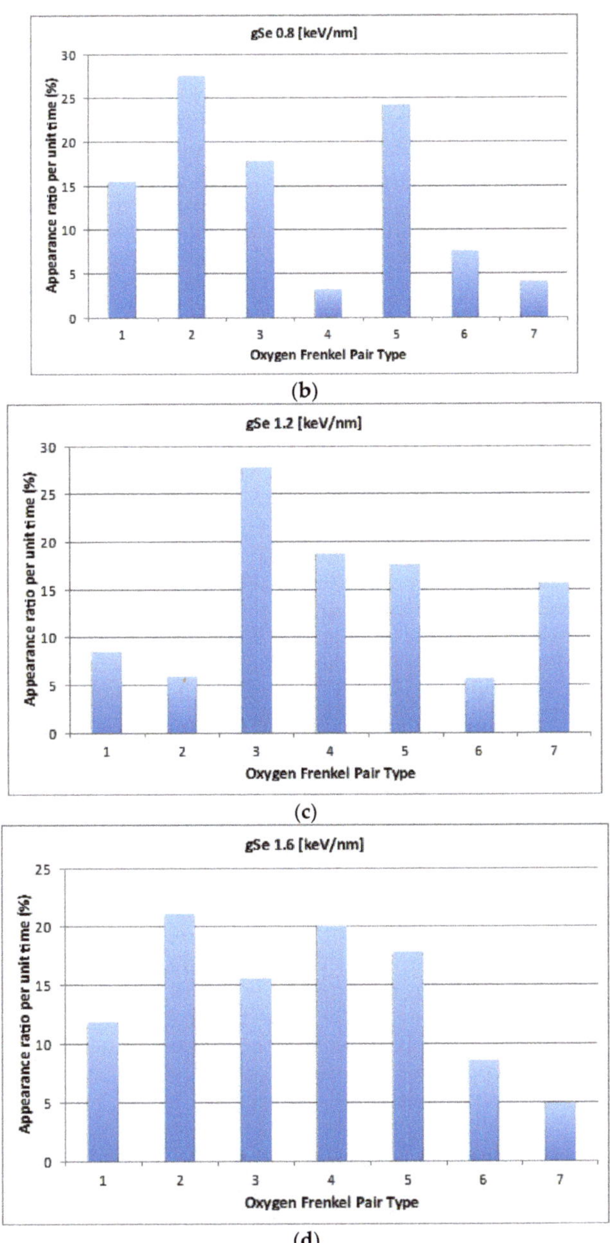

Figure 9. The time averaged distribution of iNN oxygen Frenkel pair, where $i = 1, 2, \ldots , 7$, is shown for the values of gSe = 0.4, 0.8, 1.2, and 1.6 keV/nm in (**a**–**d**), respectively.

In the case of CeO_2, the displacement threshold energy for O is 20–30 eV, much smaller than that for Ce 50–60 eV [30]. Therefore, oxygen-induced defects are a sensitive indicator for the ion irradiation process. Oxygen Frenkel pairs are important because the aggregation of multiple oxygen Frenkel pairs acts as a source of dislocation loops [31]. The dislocation loops create a high-strain energy region, which becomes an initial point of crack formation and propagation in the irradiated specimen.

As can be seen from Figure 4, no hemispherical protrusion was observed on the surfaces; only disordered atoms were seen. In our simulation, the outside of the CeO_2 is a free space, thus cooling by adiabatic extension could not occur. As Schattat et al. [1] pointed out, cooling by adiabatic extension is critical to obtain the hemispherical protrusion, especially for crystallizing as in the CeO_2 case. We considered that the protrusion was in a vapor-like state initially and then was cooled significantly by adiabatic expansion into a hemispherical-shaped crystal (see Figures 2 and 5 in reference [3]). This process is significant and resembles the cluster formation process in the cluster-beam deposition method; the cylindrical region (the nanopore) acts similar to a nozzle in the cluster-beam deposition apparatus [32,33].

The animation of the formation process of the nano-hillock of CaF_2 presented by Rymzhanov et al. [26] suggests that a strong Coulomb interaction plays an important role in forming the nano-hillock structure with single crystalline form. Karlušić et al. [27] simulated the formation process of the nano-hillock of MgO and Al_2O_3 with two different charge states, equilibrium charge state Z_{eff} = +8.47 and fixed charge state Z_{eff} = +6. They showed that the nano-hillock was formed for the equilibrium charge state, whereas a very small nano-hillock (Al_2O_3) or no nano-hillock (MgO) was formed for the fixed charge state. This finding also supports the importance of the strong Coulomb interaction. For the present case of CeO_2, the equilibrium charge state of Ce is Z_{eff} = +2.7 and the fixed charge state is Z_{eff} = +4, indicating a weak Coulomb interaction. In addition, it should be noted that hillock formation experiment/simulation for the system with a free surface was done with 23 MeV I ion giving Se= 8.53 keV/nm (MgO) and 9.1 keV/nm (Al_2O_3), estimated by the SRIM code [27], whereas the irradiation experiment/simulation for the bulk system was done with 167 MeV Xe ion giving Se= 21 keV/nm (MgO), 24.9 keV/nm (Al_2O_3) [25]. This leads to another possible mechanism for nano-hillock formation: moderately strong beam irradiation to a strong Coulomb interaction system with a free surface.

4. Conclusions

Using the MD method, we simulated the nanopore formation process by applying a thermal spike to single crystal CeO_2. The nanopore was formed abruptly at around 0.3 ps after the irradiation and grew to its maximum size at 0.5 ps. Then, it shrank in the time to 1.0 ps and was finally equilibrated. The nanopore size increased with increasing effective stopping power gSe (i.e., the thermal energy deposited per unit length in the specimen), but it saturated when gSe was 0.8 keV/nm or more. This finding will provide useful information for precise control of the size of the nanopores. We classified oxygen Frenkel pairs into seven types using the distance between the vacant site and the corresponding oxygen atom. Irrespective of the value of gSe, the number of interstitial ions became the maximum immediately after irradiation. Subsequently, interstitial ions occupied the vacancies to lower the system energy. When gSe was low, the vast majority of Frenkel pairs were the short-distance type and when gSe was high, both short-distance and long-distance types of Frenkel pairs were produced.

The present study has clarified the essence feature of the nanopore formation process by irradiation; however, there are several issues that should be addressed in further investigations: (i) the MD simulation box is small compared with the radius of cylindrical hole. (ii) The length to width ratio of the track deviates from that of actual experiment. (iii) Several MD simulations per Se are required to confirm quantitative information.

Author Contributions: Conceptualization, Y.S. and A.I.; methodology, Y.S.; software, Y.S.; validation, Y.S. and R.K.; investigation, R.K.; data curation, R.K.; writing—original draft preparation, Y.S.; writing—review and editing, N.I. and A.I.; visualization, R.K.; supervision, A.I.; project administration, Y.S.; funding acquisition, N.I. All authors have read and agreed to the published version of the manuscript.

Funding: This research was funded by JSPS KAKENHI Grant Numbers 16K06963 and 20K05389.

Institutional Review Board Statement: Not applicable.

Informed Consent Statement: Not applicable.

Acknowledgments: Tomoaki Akabane (a former graduate student at Ibaraki University, currently with Fuji Film Medical IT Solutions, Inc. (Tokyo, Japan)) is acknowledged for the development of the program for molecular dynamics simulation.

Conflicts of Interest: The authors declare no conflict of interest. The funders had no role in the design of the study; in the collection, analyses, or interpretation of data; in the writing of the manuscript, or in the decision to publish the results.

References

1. Schattat, B.; Bolse, W.; Klaumuenzer, S.; Zizak, I.; Scholz, R. Cylindrical nanopores in NiO induced by swift heavy ions. *Appl. Phys. Lett.* **2005**, *87*, 173110. [CrossRef]
2. Jensen, J.; Dunlop, A.; Della-Negra, S.; Pascard, H. Tracks in YIG induced by MeV C 60 ions. *Nucl. Instrum. Methods B* **1998**, *135*, 295–301. [CrossRef]
3. Ishikawa, N.; Okubo, N.; Taguchi, T. Experimental evidence of crystalline hillocks created by irradiation of CeO_2 with swift heavy ions: TEM study. *Nanotechnology* **2015**, *26*, 355701. [CrossRef] [PubMed]
4. Ishikawa, N.; Taguchi, T.; Kitamura, A.; Szenes, G.; Toimil-Molares, M.E.; Trautmann, C. TEM analysis of ion tracks and hillocks produced by swift heavy ions of different velocities in $Y_3Fe_5O_{12}$. *J. Appl. Phys.* **2020**, *127*, 055902. [CrossRef]
5. Costantini, J.-M.; Lelong, G.; Guillaumet, M.; Gourier, D.; Takaki, S.; Ishikawa, N.; Watanabe, H.; Yasuda, K. Optical reflectivity of ion-irradiated cerium dioxide sinters. *J. Appl. Phys.* **2019**, *126*, 175902. [CrossRef]
6. Costantini, J.-M.; Gutierrez, G.; Watanabe, H.; Yasuda, K.; Takaki, S.; Lelong, G.; Guillaumet, M.; Weber, W.J. Optical spectroscopy study of modifications induced in cerium dioxide by electron and ion irradiations. *Philos. Mag.* **2019**, *99*, 1695–1714. [CrossRef]
7. Palomares, R.I.; Shamblin, J.; Tracy, C.L.; Neuefeind, J.; Ewing, R.C.; Trautmann, C.; Lang, M. Defect accumulation in swift heavy ion-irradiated CeO_2 and ThO_2. *J. Mater. Chem. A* **2017**, *5*, 12193. [CrossRef]
8. Graham, J.T.; Zhang, Y.; Weber, W.J. Irradiation-induced defect formation and damage accumulation in single crystal CeO_2. *J. Nucl. Mater.* **2018**, *498*, 400–408. [CrossRef]
9. Shelyug, A.; Palomares, R.I.; Lang, M.; Navrotsky, A. Energetics of defect production in fluorite-structured CeO_2 induced by highly ionizing radiation. *Phys. Rev. Mater.* **2018**, *2*, 093607. [CrossRef]
10. Costantini, J.-M.; Miro, S.; Touati, N.; Binet, L.; Wallez, G.; Lelong, G.; Guillaumet, M.; Weber, W.J. Defects induced in cerium dioxide single crystals by electron irradiation. *J. Appl. Phys.* **2018**, *123*, 025901. [CrossRef]
11. Cureton, W.F.; Palomares, R.I.; Walters, J.; Tracy, C.L.; Chen, C.-H.; Ewing, R.C.; Baldinozzi, G.; Lian, J.; Trautmann, C.; Lang, M. Grain size effects on irradiated CeO_2, ThO_2, and UO_2. *Acta Mater.* **2018**, *160*, 47–56. [CrossRef]
12. Maslakov, K.I.; Teterin, Y.A.; Popel, A.J.; Teterin, A.Y.; Ivanov, K.E.; Kalmykov, S.N.; Petrov, V.G.; Petrov, P.K.; Farnan, I. XPS study of ion irradiated and unirradiated CeO_2 bulk and thin film samples. *Appl. Surf. Sci.* **2018**, *448*, 154–162. [CrossRef]
13. Cureton, W.F.; Palomares, R.I.; Tracy, C.L.; O'Quinn, E.C.; Walters, J.; Zdorovets, M.; Ewing, R.C.; Toulemonde, M.; Lang, M. Effects of irradiation temperature on the response of CeO_2, ThO_2, and UO_2 to highly ionizing radiation. *J. Nucl. Mater.* **2019**, *525*, 83–91. [CrossRef]
14. Abanades, S.; Flamant, G. Thermochemical hydrogen production from a two-step solar driven water-splitting cycle based on cerium oxides. *Sol. Energy* **2006**, *80*, 1611–1623. [CrossRef]
15. Sasajima, Y.; Onuki, H.; Ishikawa, N.; Iwase, A. Computer simulation of high-energy-ion irradiation of SiO_2. *Trans. MRS-J.* **2013**, *38*, 497–502.
16. Sasajima, Y.; Ajima, N.; Osada, T.; Ishikawa, N.; Iwase, A. Molecular dynamics simulation of fast particle irradiation to the Gd_2O_3-doped CeO_2. *Nucl. Instrum. Methods B* **2013**, *316*, 176–182. [CrossRef]
17. Sasajima, Y.; Osada, T.; Ishikawa, N.; Iwase, A. Computer simulation of high-energy-ion irradiation of uranium dioxide. *Nucl. Instrum. Methods B* **2013**, *314*, 195–201. [CrossRef]
18. Sasajima, Y.; Ajima, N.; Osada, T.; Ishikawa, N.; Iwase, A. Molecular dynamics simulation of fast particle irradiation to the single crystal CeO_2. *Nucl. Instrum. Methods B* **2013**, *314*, 202–207. [CrossRef]
19. Sasajima, Y.; Ajima, N.; Kaminaga, R.; Ishikawa, N.; Iwase, A. Structure analysis of the defects generated by a thermal spike in single crystal CeO_2: A molecular dynamics study. *Nucl. Instrum. Methods B* **2019**, *440*, 118–125. [CrossRef]
20. Yablinsky, C.A.; Devanathan, R.; Pakarinen, J.; Gan, J.; Severin, D.; Trautmann, C.; Allen, T. Characterization of swift heavy ion irradiation damage in ceria. *J. Mater. Res.* **2015**, *30*, 1473–1484. [CrossRef]
21. Medvedev, N.A.; Rymzhanov, R.A.; Volkov, A.E. Time-resolved electron kinetics in swift heavy ion irradiated solids. *J. Phys. D* **2015**, *48*, 355303. [CrossRef]
22. Rymzhanov, R.A.; Medvedev, N.; Volkov, A.E. Damage threshold and structure of swift heavy ion tracks in Al_2O_3. *J. Phys. D Appl. Phys.* **2017**, *50*, 475301. [CrossRef]

23. Rymzhanova, R.A.; Medvedev, N.; Volkova, A.E.; O'Connell, J.H.; Skuratova, V.A. Overlap of swift heavy ion tracks in Al_2O_3. *Nucl. Instrum. Methods B* **2018**, *435*, 121–125. [CrossRef]
24. Rymzhanova, R.A.; Gorbunovd, S.A.; Medvedeve, N.; Volkova, A.E. Damage along swift heavy ion trajectory. *Nucl. Instrum. Methods B* **2019**, *440*, 25–35. [CrossRef]
25. Rymzhanov, R.A.; Medvedev, N.; O'Connell, J.H.; Janse van Vuuren, A.; Skuratov, V.A.; Volkov, A.E. Recrystallization as the governing mechanism of ion track formation. *Sci. Rep.* **2019**, *9*, 3837. [CrossRef] [PubMed]
26. Rymzhanov, R.A.; O'Connell, J.H.; Janse van Vuuren, A.; Skuatov, V.A.; Medvedev, N.; Volkov, A.E. Insight into picosecond kinetics of insulator surface under ionizing radiation. *J. Appl. Phys.* **2020**, *127*, 015901. [CrossRef]
27. Karlušić, M.; Rymzhanov, R.A.; O'Connell, J.H.; Bröckers, L.; Tomić, L.K.; Siketić, Z.; Fazinić, S.; Dubček, P.; Jakšić, M.; Provatas, G.; et al. Mechanisms of surface nanostructuring of Al_2O_3 and MgO by grazing incidence irradiation with swift heavy ions. *Surf. Interfaces* **2021**, *27*, 101508. [CrossRef]
28. Inaba, H.; Sagawa, R.; Hayashi, H.; Kawamura, K. Molecular dynamics simulation of gadolinia-doped ceria. *Solid State Ion.* **1999**, *122*, 95–103. [CrossRef]
29. Sadus, R.J. *Molecular Simulation of Fuids–Theory, Algorithms and Object-Orientation*; Elsevier: Amsterdam, The Netherlands, 2002; p. 305.
30. Shiiyama, K.; Yamamoto, T.; Takahashi, T.; Guglielmetti, A.; Chartier, A.; Yasuda, K.; Matsumura, S.; Yasunaga, K.; Meis, C. Molecular dynamics simulations of oxygen Frenkel pairs in cerium dioxide. *Nucl. Instrum. Methods B* **2010**, *268*, 2980–2983. [CrossRef]
31. Yasunaga, K.; Yasuda, K.; Matsumura, S.; Sonoda, T. Electron energy-dependent formation of dislocation loops in CeO_2. *Nucl. Instrum. Methods B* **2008**, *266*, 2877–2881. [CrossRef]
32. Takaoka, H. Ionized Cluster Beam Technique for Deposition and Epitaxy. Ph.D. Thesis, Kyoto University, Kyoto, Japan, 23 July 1981; pp. 45–50. [CrossRef]
33. Yamada, I.; Takaoka, H. Ionized Cluster Beams: Physics and Technology. *Jpn. J. Appl. Phys.* **1993**, *32*, 2121. [CrossRef]

quantum beam science

Review

Ion Accelerator Facility of the Wakasa Wan Energy Research Center for the Study of Irradiation Effects on Space Electronics

Satoshi Hatori, Ryoya Ishigami, Kyo Kume and Kohtaku Suzuki *

The Wakasa Wan Energy Research Center (WERC), 64-52-1, Nagatani, Tsuruga 914-0192, Japan; hatori@werc.or.jp (S.H.); rishigami@werc.or.jp (R.I.); kkume@werc.or.jp (K.K.)
* Correspondence: ksuzuki@werc.or.jp

Abstract: The core facility of the Wakasa Wan Energy Research Center (WERC) consists of three ion accelerators: a synchrotron, a tandem accelerator and an ion-implanter. Research on the irradiation effects using these accelerators has been performed on space electronics such as solar cells, radiation detectors, image sensors and LSI circuits. In this report, the accelerator facility and ion-irradiation apparatuses at WERC are introduced, focusing on the research on irradiation effects on space electronics. Then, some recent results are summarized.

Keywords: ion accelerators at WERC; ion beam; irradiation effects on space electronics; single event; total ionization dose; displacement damage

Citation: Hatori, S.; Ishigami, R.; Kume, K.; Suzuki, K. Ion Accelerator Facility of the Wakasa Wan Energy Research Center for the Study of Irradiation Effects on Space Electronics. *Quantum Beam Sci.* **2021**, *5*, 14. https://doi.org/10.3390/qubs5020014

Academic Editor: Akihiro Iwase

Received: 26 March 2021
Accepted: 8 May 2021
Published: 13 May 2021

Publisher's Note: MDPI stays neutral with regard to jurisdictional claims in published maps and institutional affiliations.

1. Introduction

The Wakasa Wan Energy Research Center (WERC) opened in November 1998, and from July 2000, an accelerator facility which forms the core of WERC became available. The accelerator facility consists of three accelerators: a synchrotron, a tandem accelerator and an ion-implanter [1,2]. Using these accelerators, we have been involved in research on the destruction of tumor cells in human bodies (proton beam radiotherapy) [3,4], the transmutation of DNAs in plants and bacteria (ion beam breeding) [5–7], ion beam material modification [8–10] and ion beam analysis for materials [11–13]. The accelerators at WERC have also been utilized to simulate the radiation effects on materials which are related to nuclear power plants [14–16] and those of space electronics [17–25]. Electronic systems are severely affected by the radiation environment in space. There are three main irradiation effects on space electronics: the single event effect (SEE), the total ionizing dose effect (TID) and the displacement damage dose effect (DDD) [26–31]. Before the operation of electronic devices in space, these effects from the radiation environment in space have to be evaluated on earth. Recently, the space industry has been widespread; therefore, the radiation tolerance of space electronics has been actively investigated using electron, gamma-ray, and ions. In this report, we describe the specification of each accelerator, beam lines, and irradiation apparatuses at WERC, focusing on the research of the radiation effects for space electronics. Since the three accelerators can cover a wide energy range of ions from 10 keV to 660 MeV, we can simulate the space radiation field well. Finally, several recent research results of the ion irradiation effects on solar cells, LSI circuits, radiation detectors and image sensors are summarized.

2. Overview of Ion Accelerators at WERC

Figure 1 shows the layout of two ion sources, the tandem accelerator, the synchrotron, the ion-implanter, and their beam transport lines at WERC [1,2]. The tandem accelerator is interfaced to the synchrotron, and used as an injector for the synchrotron. The synchrotron can accelerate protons up to 200 MeV. Such a high energy proton beam has been used mainly for the study of SEE on space electronic devices at irradiation room 4. The tandem accelerator has also been used independently to accelerate protons up to 10 MeV. This medium energy proton beam has been used for the study of TID and DDD on image sensors at irradiation room 2. The ion-implanter provides 10 to 200 keV ions with high beam current. This low energy and intense beam has been used for research of DDD on solar cells at irradiation room 1. The detailed specification of each accelerator will be explained in the following subsections.

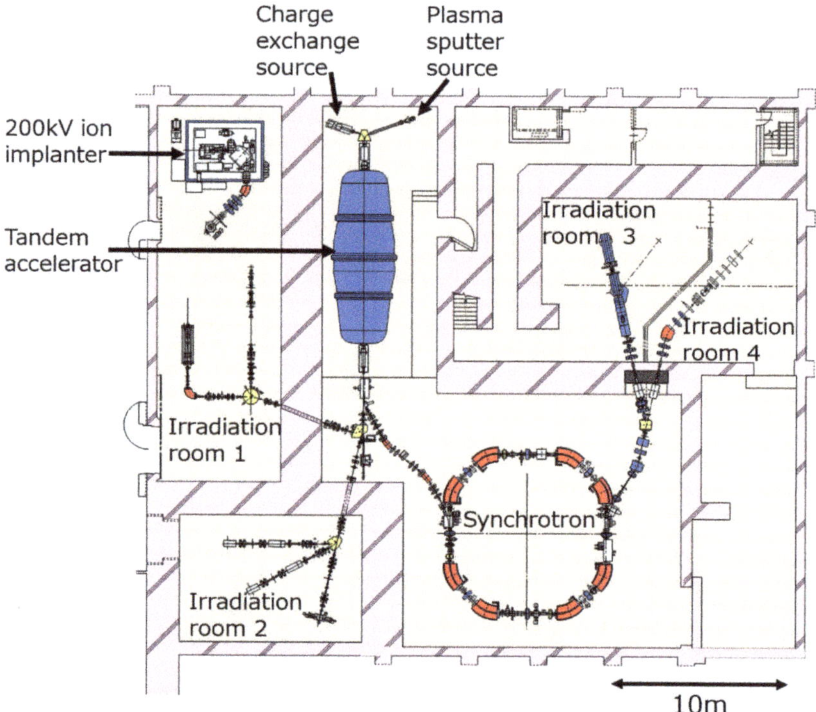

Figure 1. Layout of two ion sources, the tandem accelerator, the synchrotron, the ion-implanter, and their beam transport lines at WERC [1,2].

2.1. Tandem Accelerator at WERC

The tandem accelerator at WERC (Figure 2) accelerates negative ions which are injected from the two ion sources alternatively. One of the ion sources is a plasma sputter type source, and it generates hydrogen ions and heavy ions such as carbon ions. The typical peak intensity of pulsed negative hydrogen ions, which are used for the proton acceleration by the synchrotron, is 4 mA with a 20 Hz period and 25 μs duration. The other ion source is a kind of charge exchange type for the ionization of gas elements. Especially, the highly intense ^{4}He negative ion beam, the current of which amounts to 40 μA, enables α particle irradiation with a high-density energy deposition into target materials.

The acceleration voltage of the tandem accelerator amounts up to 5 MV. The high tension is generated by the Schenkel circuit with a 58-step voltage doubler rectifier, which rectifies 40 kHz RF power by a resonant transformer and an oscillator. The Schenkel rectifier, the acceleration tube, the high-tension terminal, and the resonant transformer are sustained by the insulation column support in the pressure tank. The pressure tank is filled with SF_6 insulation gas at 0.6 MPa gauge. In order to enable the acceleration of the highly intense beam from the ion sources, the conveyor current through the Schenkel rectifier amounts to 1 mA at the terminal high tension of 5 MV. The voltage ripple is 0.4 kV$_{pp}$ at the terminal high tension of 5 MV.

The charge exchange canal in the terminal has an inner diameter of 15 mm and a charge exchange effective length of about 1 m. Although the diameter of 15 mm seems to be large, the geometrical transmission efficiency is about 70% because of the large emittance of the negative ion from the ion sources. Argon gas for the charge exchange (stripper gas) is introduced into the middle of the canal. A differential pumping and the recirculation by four turbo-molecular pumps, each with a pumping speed of 50 L/s, enable the concentration of stripper gas in the canal with a large inner diameter. Such a stripper gas system realizes the 99% conversion of negative ions to positive at a gas thickness of 4 Pa·m. Additionally, the transmission efficiency along the following transport beam line to the synchrotron is achieved to be almost perfect. Therefore, a pulsed proton beam with peak intensity of 3 mA can be injected into the synchrotron. Table 1 summarizes the specifications of the tandem accelerator at WERC.

Figure 2. Tandem accelerator at WERC.

Table 1. Specifications of tandem accelerator.

Categories	Specifications
Generation of high voltage	Cascade of voltage doubler rectifiers (Schenkel Rectifier)
Max terminal voltage	5 MV
Max conveyer current	1 mA
Voltage ripple	$0.4\,\mathrm{kV_{pp}}$ @ 5 MV
Insulation gas	SF_6 0.6 MPa gauge
Accelerator tube	Glass–metal organic bonding
Charge exchange	Ar–gas stripper, recirculation and concentrationby 4 TMP (50L/s/pump)
Injected ion ME (mass·energy value)	6 MeV·amu

2.2. Synchrotron

A proton beam with a maximum energy of 10 MeV is injected from the tandem accelerator into the synchrotron and is accelerated up to 200 MeV. Recently, for the proton acceleration, the injection energy has been set at 7 MeV.

The overview of the synchrotron at WERC is shown in Figure 3. The circumference of the synchrotron is 33.2 m. The super periodicity is 4. Each lattice has a QF-BM-QD-BM permutation, and is operated in separated function style. Here, QF and QD are focusing and defocusing quadrupole magnets in the median plane, respectively, and BM means a bending magnet. Horizontal and vertical tunes are 1.75 and 0.85, respectively. A period of 2 s for the synchrotron acceleration consists of four modes of "injection and capture", "acceleration", "extraction", and "deceleration, then return to injection". The injection is performed in a multi-turn injection style. The RF knock out system slowly extracts the accelerated beam out of the separatrix in the horizontal phase space. The separatrix is generated by the excitation of sextupole magnets. By adjusting the RF strength by time, the variation in the intensity of the extracted beam by time is controlled to be reduced for event-by-event mode experiments such as a counter performance test and SEE investigations on semiconductor devices. The usual range of the beam intensity for the irradiation of space electronic devices is from 0.1 to 3 nA. Table 2 tabulates the principal specifications of the synchrotron at WERC.

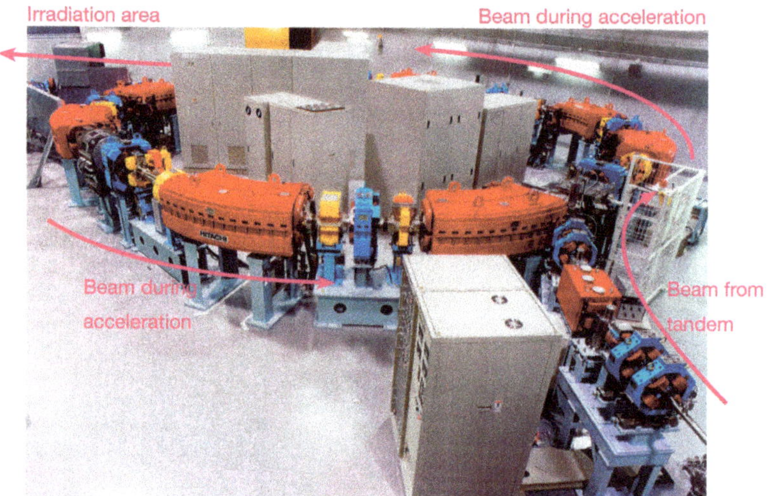

Figure 3. Synchrotron at WERC.

Table 2. Specifications of synchrotron.

Categories		Specifications	
Incident energy	H$^+$	10 MeV	(Bρ = 0.46 Tm)
	Heavy Ions	2.08 MeV/u	(Bρ = 0.42 Tm)
Acceleration energy	H$^+$	200 MeV	(Bρ = 2.15 Tm)
	Heavy Ions	55 MeV/u	(Bρ = 2.15 Tm)
Acceleration period		0.5 Hz	
Lattice		QF-BM-QD-BM	
Injection		Multi-turn injection	
RF cavity		Asynchronous RF cavity	
Extraction		Resonance-RF knockout	
Superperiodicity		4	
Circumference		33.2 m	

Table 2. *Cont.*

Categories		Specifications
Tune	Horizontal ν_x	1.75
	Vertical ν_y	0.85
Bending magnet	Bending angle	45 deg
	Maximum field	1.12 T
	Radius	1.91 m
Momentum compaction		0.31
Natural chromaticity	x	−0.34
	y	−0.36

2.3. Ion-Implanter

A high current ion beam irradiation equipment named the 200 kV ion-implanter has been installed at WERC for research on the radiation effects of materials related to atomic power plants and of space electronics. This accelerator has also been utilized for the modification of semiconductors and metallic alloys. The schematic diagram of the 200 kV ion-implanter are shown in Figure 4.

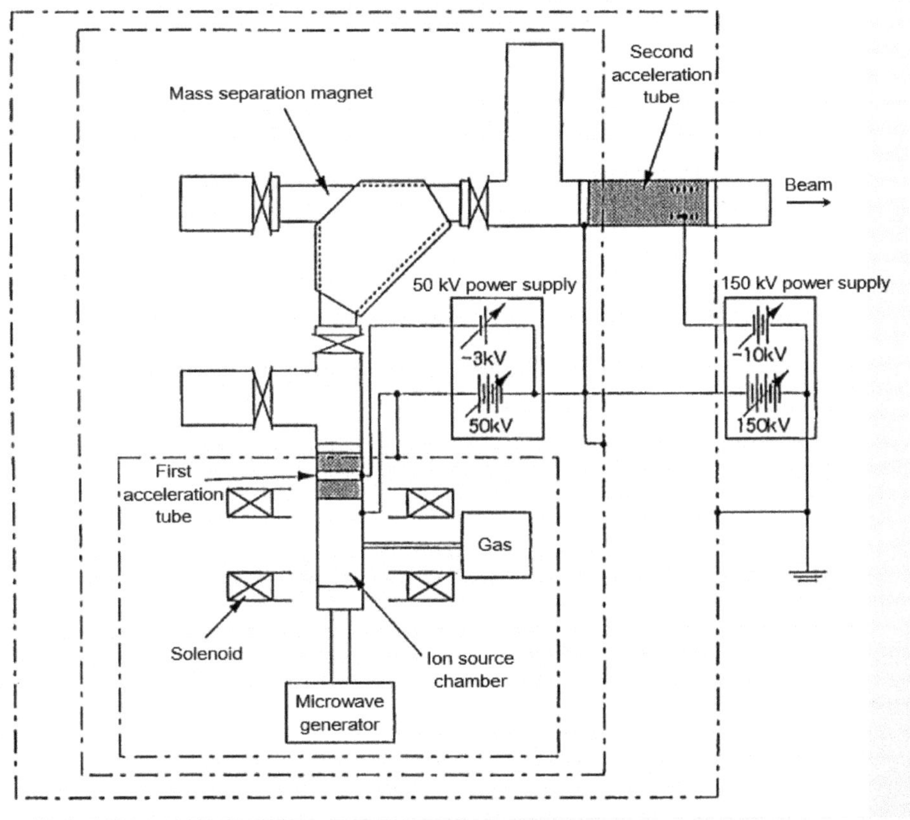

Figure 4. The schematic diagram of 200 kV ion-implanter at WERC.

The 200 kV ion-implanter consists of a microwave ion source, a 50 kV first acceleration tube, a mass separation magnet, and a 150 kV second acceleration tube. The microwave ion source supplies positive ions from gas targets, such as H_2, He, N_2, Ne, Ar, etc. Positive ions generated in the ion source are accelerated up to 50 keV by the first acceleration tube. Then, they are sorted by mass with the mass separation magnet. In the case of the H_2 gas target, we can select H^+ or H_2^+ by changing the magnet field of the mass separator. After the mass separation, the ions are accelerated again up to 200 keV by the second acceleration tube. The ion energy can be changed from 10 to 200 keV. The maximum beam currents available from the first and the second acceleration tubes are 50 and 30 mA, respectively. In the actual irradiation experiments, however, the beam current is reduced to less than several tens of micro A/cm^2 to avoid beam heating of the target materials during the irradiation.

3. Beamlines and Irradiation Apparatuses for the Research on Irradiation Effects on Space Electronics

3.1. Beamline of the Synchrotron in Irradiation Room 4

In irradiation room 4, we can perform irradiation experiments with ion beams extracted from the synchrotron. Figure 5 shows the layout of the beam line in irradiation room 4. After an ion beam is accelerated by the synchrotron and transported to irradiation room 4, the beam is deflected to the direction of the target samples by a bending magnet and is shaped by two sets of quadrupole magnet doublets. Then, the ion beam is emitted into the atmosphere through a copperized polyimide thin film window. In order to form the appropriate irradiation field on the target surface, the ion beam is wobbled by a set of wobbling magnets and/or scattered by a tungsten scatterer and collimated by a brass block collimator.

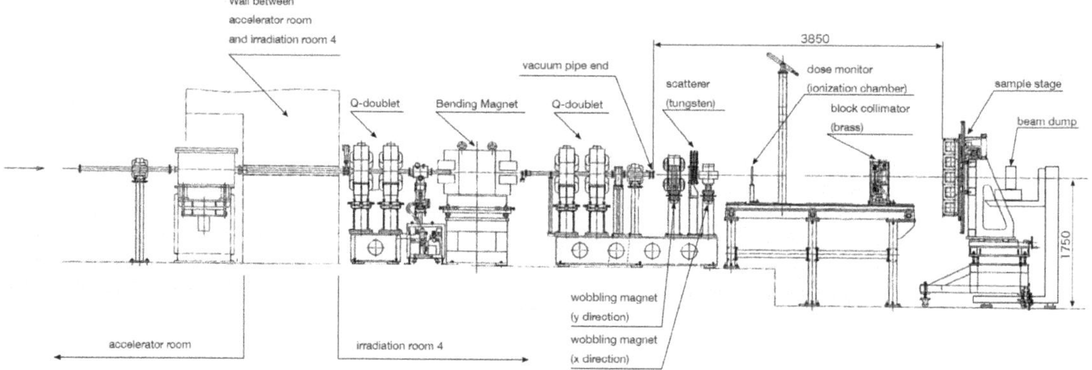

Figure 5. Beam line in irradiation room 4.

The intensity distribution of the cross section of the primary proton beam at 200 MeV kinetic energy has almost an axially symmetrical Gaussian shape with around 1.5 cm standard deviation about beam direction. We call this beam as "pencil beam". The usual irradiation area on the sample is within 2 cm by 2 cm squares; therefore, the pencil beam covers the irradiation area with a flatness (the ratio of intensity at the edge of the irradiation area to that at the center) of more than 80%. Sometimes, much more severe uniformity is required to the intensity distribution. In such case, we use a tungsten scatterer with a thickness of 0.1 mm to make a flatness of more than 94%.

In TID experiments, the beam intensity amounts to around 3 nA, which corresponds to a flux of 2.5×10^8 cm^{-2}s^{-1} in the case of an irradiation area of 2 cm by 2 cm squares and a scatterer thickness of 0.1 mm. In experiments for the single event effect (SEE), the ion beam flux is required to be reduced to less than 1×10^7 cm^{-2}s^{-1}. Thus, the beam intensity is reduced to 0.1 nA by adjusting the strength of the RF kicker. In order to obtain a beam flux of less than 1×10^6 cm^{-2}s^{-1}, a thicker scatterer is used for the reduction in the area density of the beam.

The proton beam can be spread to large targets, such as a 10 cm by 10 cm square for instance. SEE experiments with a flux of less than 10^6 cm^{-2}s^{-1} can be performed by using only the scatterer to form a large irradiation field. In the case of 10^7 cm^{-2}s^{-1} flux experiment, the beam is wobbled by a set of wobbling magnets to improve the beam utilization efficiency.

In several SEE experiments, the cross section for the SEE occurrence is measured as a function of the linear energy transfer (LET) of primary beam in the target material. In order to perform the experiment, the energy of the projectile has to be changed for each LET value. Although the beam energy from the synchrotron is variable, the excitation pattern of all magnets and acceleration RF, and beam transport setup have to be changed for each beam energy. Therefore, an energy degrader is used instead of a conventional method for the variation of the beam energy. The degrader, made of resin which is water-equivalent in the proton energy loss, can change its thickness in units of 1 mm from 2 to 270 mm. The degrader is set at a position immediate upstream of the irradiation target.

An ionization chamber (shown as "dose monitor" in Figure 5) is used for the dose control. As the ionization efficiency relates to a function of the energy deposition by passing particles, the dose measured by the ionization chamber is calibrated for each energy variation by measuring the beam current at the beam dump. The probability of the recombination of ion-electron pairs is a function of density of the pairs produced in the ionization chamber. Therefore, the calibration of the dose monitor is also performed for each variation of beam intensity and irradiation area by the comparison between the averaged current signals from the dose monitor and beam dump for 3~5 times measurements. The doses are controlled with an error (standard deviation) of almost 2% and 1% for the 100 MeV proton irradiation to the area of 10 cm by 10 cm with flux of 3×10^6 and 1×10^7 cm^{-2}s^{-1}, respectively.

The synchrotron beam is used to measure the radiation effects not only on space electronics but also on radiation detectors to be used in space for astronomical phenomena. As usual detectors count each particle and/or gamma/X-ray, the charged particle energy and flux under the circumstances around the space probe have to be simulated. In these experiments, the beam intensity is reduced to less than 10 kHz. The intensity is controlled by the charged particle counting using a counter-telescope with two sets of plastic scintillators read by photomultiplier tubes.

3.2. Beamline of the Tandem Accelerator in Irradiation Room 2

The ion beam extracted from the tandem accelerator is not only injected into the synchrotron system but is also directly transported to irradiation room 1 and 2. Three beam lines are installed in irradiation room 2. One of the beam lines is used for the research on irradiation effects on space electronics. Hydrogen (H), helium (He), carbon (C), nickel (Ni), and copper (Cu) ions are available for the irradiation experiments. The beam line consists of two scanning magnets and two apertures; it ends with an irradiation chamber with high and cold temperature sample stages with three-axis goniometers. Figure 6 shows the beam line and the irradiation chamber. Ion beams are injected into the irradiation chamber from the back to the front in the figure. A sample stage on the goniometer can be moved about 60 mm horizontally and about 45 mm vertically with a precision of less than 0.05 mm and rotated 180° about the horizontal axis with a precision of less than 0.05°. The shape of the beam is checked using a fluorescent plate which is placed on the sample stage. The beam diameter on the sample stage can be reduced to less than 10 mm. To ensure a homogeneous irradiation for large area targets, ion beams are scanned horizontally with a frequency of 50 Hz and vertically with a frequency of 0.1–5.0 Hz by the scanning magnets with the continuous triangle waveform current. The irradiation can be performed at high (20 to 700 °C) or cold (−173 to 20 °C) temperatures. The temperature on the sample stages is measured by a thermocouple. The maximum setting area of samples is 60×60 mm^2 on the sample stage. H and He beams can be scanned over the above area. The scan area for Ni beams is narrower than the light ion cases because Ni is much heavier than H or He. The horizontal and vertical scan widths for a 10 MeV Ni^{3+} beam are 24 and 20 mm at maximum, respectively. The degree of vacuum in the irradiation chamber is less than 9×10^{-4} Pa, even during the irradiation. Electronic device samples can be wired in the chamber via electrical feedthroughs to measure the electronic characteristics of the devices during the irradiation.

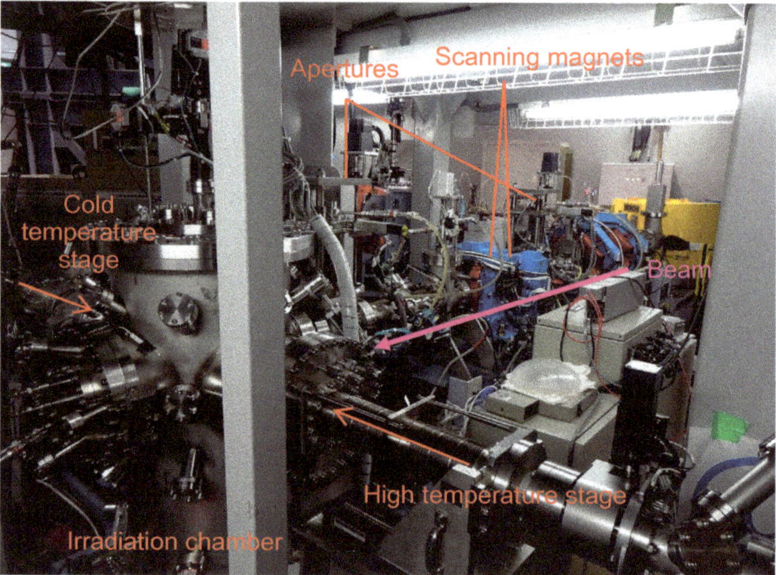

Figure 6. Beamline and irradiation chamber for the irradiation of space electronics in irradiation room 2.

3.3. Beamline of the 200 kV Ion-Implanter in Irradiation Room 1

The ion beam accelerated by the ion-implanter in irradiation room 1 is manipulated using a bending magnet and three quadrupole magnets. The beam size is defined by X-Y slits (shown in Figure 7). The beam can be uniformly spread up to 10 cm squares. The performances of the 200 kV ion-implanter are summarized in Table 3. An irradiation chamber has five sample stages (Figure 8, left). Three of them have a heating system, and the sample temperature can be controlled from room temperature to 1000 °C during the irradiation. The other two sample stages have a water cooling system, which can keep samples around room temperature even during the irradiation. To check the beam uniformly, a fluorescence plate is set at the bottom of the target stages (Figure 8, right). An aperture for defining the beam shape and a −500 V secondary electron suppresser are placed in front of each sample stage. The beam current on the sample stage is measured by a pico-ammeter.

Table 3. Performance of the 200 kV ion-implanter.

Categories	Specifications
Beam elements	H, H_2, He, Ne, N, N_2, O, O_2, Ar, CO, etc.
Beam energy	10 to 200 keV
Typical beam current	<1 mA
Beam current density	~20 $\mu A/cm^2$
Irradiation area	Maximum 100×100 mm^2
Target stage	Heating stage 3, Cooling stage 2
Heating stage control	Room temperature to 1000 °C

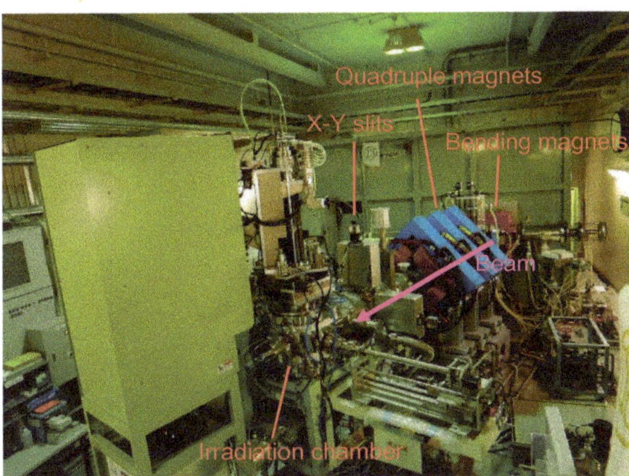

Figure 7. Beamline and irradiation chamber of the ion-implanter.

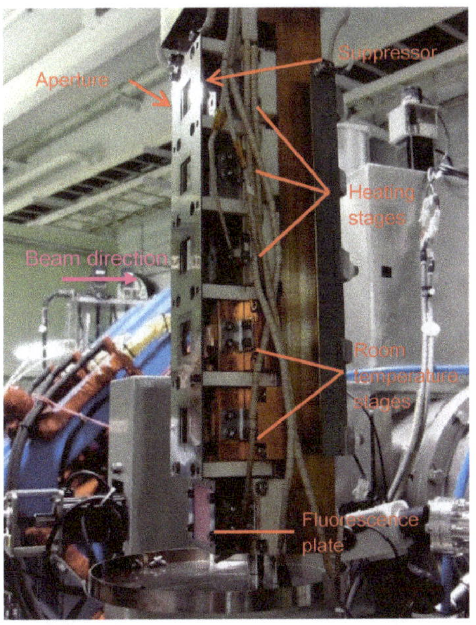

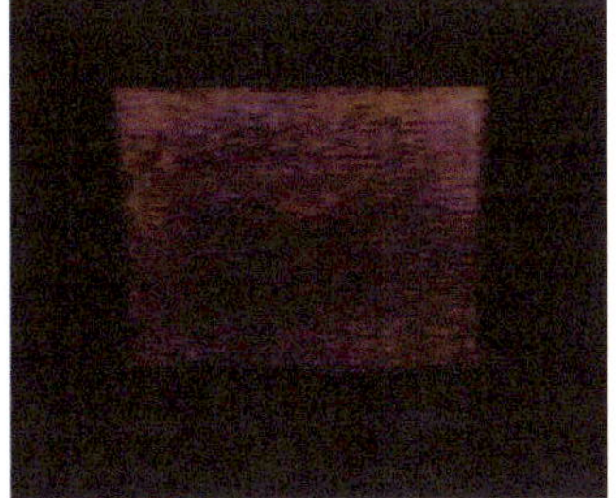

Figure 8. Sample stages (**left**), and captured image of the fluorescence plate for beam irradiation (**right**).

4. Recent Results of Ion Irradiation Effects on Space Electronics at WERC

4.1. Recent Experiments Using Accelerators at WERC

Many radioactive sources exist in space such as electrons, X/γ-ray, and ions. Protons are the main ion component in space; thus, ion irradiation effects are investigated mainly using proton beam. Typical beam currents for irradiation experiments are ~1 nA, ~100 nA, and 1–10 μA, for the synchrotron, the tandem, and the ion-implanter, respectively. The synchrotron is used usually for research on SEE because high energy proton beam can simulate the high-density electronic excitation by the space radiation. On the other hand, the medium energy proton beam from the tandem accelerator is used for research on TID/DDD. The low energy ion beam with a high intensity from the ion-implanter is used exclusively for research on DDD. A lot of irradiation experiments for space electronics have been performed using the three accelerators at WERC. These experiments are summarized in Table 4. We introduce some experiments in the following sections.

Table 4. Summary table of irradiation experiments for space electronics at WERC.

Accelerator	Target	Purpose	Effect	Ion and Energy
Synchrotron	Si-CMOS image sensor	X-ray detector	TID/DDD	p, 100 MeV
	ASIC	Signal processing for X-ray detector	TID	p, 90 MeV
	Silicon Photomultiplier	X-ray detector	SEE	p, 200 MeV
	Electric substrate	Microcomputer	SEE	p, 20–200 MeV
	LaBr$_3$(Ce) scintillator	X-ray detector	Radioactivation	p, 20–140 MeV
	AlGaN/GaN HEMT	Power device	TID	p, 11.1–66.8 MeV
Tandem	AlGaN/GaN HEMT	Power device	DDD	p, 2 MeV
	CCD	Optical imager	TID	p, 8 MeV
Ion-Implanter	InGap/GaAs/Ge solar cell	Solar cell	DDD	p, 50–150 keV
	Perovskite solar cell	Solar cell	DDD	p, 50 keV

4.2. Research Results Using the Synchrotron

In the X/gamma ray astronomy field, various X/gamma ray detectors and electronic circuits dedicated to the detectors, which are durable in space, have been developed. One of the detectors uses a CMOS (complementary metal-oxide semiconductor) imaging sensor. The CMOS sensor can be driven with a small power consumption since the signal transmission is not accompanied by charge transfer and high voltage analog circuits are not needed, such as CCD (charge-coupled device) sensors. The CMOS sensor is chip-integrable together with its peripheral circuit; therefore, it is possible to reduce the size and cost of the detector. For such a reason, the CMOS image sensor has been adopted especially for usage in small space probes.

HiZ-GUNDAM [32] is a future satellite mission for the exploration of the early universe by detecting high-redshift gamma-ray bursts by using the wide-field X-ray detector. The Kanazawa University group will adopt a CMOS image sensor GSENSE6060BSI fabricated by Gpixel Inc. for the detection of soft X-ray [17]. The radiation tolerance test was conducted using GSENSE400BSI-TVISB, which is close in pixel size and has the same resistivity and epilayer thickness as GSENSE6060BSI. The test for the sensor was performed by the irradiation of a 100 MeV proton with a flux of 1×10^7 cm^{-2}s^{-1} and a fluence of 4.8×10^{10} cm^{-2}, which corresponds to the absorbed dose of 5 krad during 6 years in orbit. After the irradiation, the energy spectra of Mn-K$_\alpha$ and K$_\beta$ X-ray were obtained at temperatures of +20 °C and −20 °C (Figure 9). Although the significant damage appears as the increase in the dark noise and the worsening of the energy resolution, the background tail of the low energy side caused by the dark noise is reduced in the operation at −20 °C (Figure 9, right). Although the MOS structure seems to be affected by TID, the increase in dark current noise is considered to be caused by bulk damage from DDD.

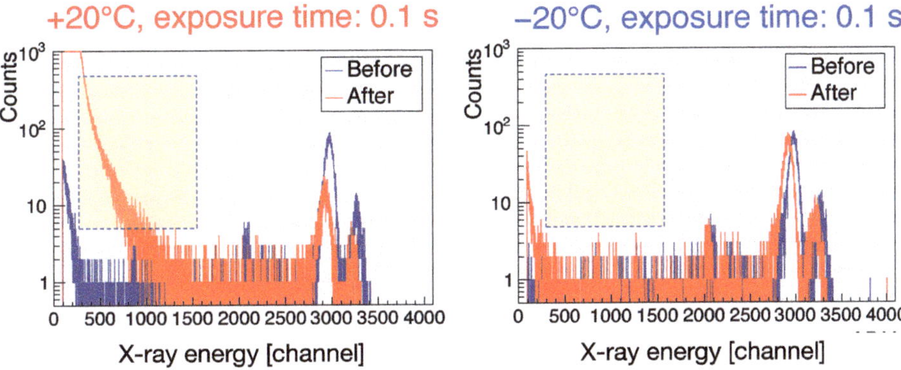

Figure 9. Comparison of the Mn-K$_\alpha$ and K$_\beta$ spectra obtained before and after the 100 MeV proton irradiation. Although the noise tail appears on the low-energy side at +20 °C, after cooling of sample to −20 °C, contribution of dark noise is significantly reduced. (Reprint Figure 10 in p.5 of N. Ogino et al., Nucl. Inst. and Methods in Phys. Research, A987(2021) 164843).

The Hiroshima University group has chosen a system of a CsI(Tl) scintillator and a silicon photomultiplier (Si-PM) for the X/gamma ray detector onboard CubeSats. The Si-PM is a kind of avalanche photodiode (APD) and the multi pixelated Si-PM (Multi Pixel Photon Counter, MPPC) were supplied by Hamamatsu Photonics K.K. (HPK). Two types of MPPCs (S13360-6050CS and S14160-6050HS) were irradiated with a 200 MeV proton beam. Figure 10 shows the comparison between the gamma ray spectra obtained several minutes and 7 months after the 300 rad irradiation. Just after the irradiation, the 22.2 keV photo peak from ^{109}Cd (blue line) disappears in the spectrum and the threshold level rises. In the gamma ray spectrum measured after the MPPC that was left at room temperature for 7 months, the recovery of 22.2 keV photo peak can be found in the spectrum. It is a certain annealing effect against the irradiation for the MPPC [18].

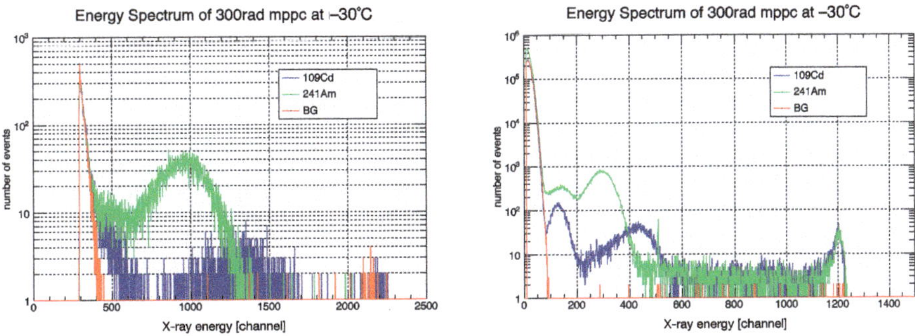

Figure 10. Gamma-ray spectra with MPPC S13360-6050CS at −30 °C. Left figure is acquired at several minutes after 300 rad irradiation and right figure shows the spectra measured at 7 months after irradiation. (Reprint Figure 4 in p.3 of N. Hirade et al., Nucl. Inst. and Meth. in Phys. Research A986(2021) 164673).

Not only radiation detectors but also electronic devices such as a power module, integrated circuits used in an analog amplifier, a logic gate, a microcomputer, etc. have to be evaluated for the radiation tolerance. The gate oxide structure of MOSFET (metal-oxide semiconductor field effect transistor) and IGBT (insulated gate bipolar transistor) used in the power module is subject to the TID effect. SEE may also be observed as the single event upset (SEU) in the switching device by the charged particle-induced ionization in the depletion layer. Especially, SEE in IGBT in the power module and CMOS in the logic gate with a parasitic thyristor in themselves may generate single event latchup (SEL), causing the thermal runaway. Once SEE occurs in the gate region of power MOS–FET or IGBT, the destruction of the gate may cause another thermal runaway, i.e., single event burnout (SEB).

The Tokyo Institute of Technology group has developed APD X-ray detectors and their dedicated ASIC (application-specific integrated circuit). The 0.35 μm CMOS technology is used for the developed ASIC. The ASIC has dual interlock cell (DICE) circuits with a master–slave flip-flop's scheme to prevent malfunction due to the distortion or the noisy clock signal and the accidental activation of CMOS, i.e., an SEU event. The evaluation test of the radiation tolerance was performed for the ASIC with 90 MeV protons because the threshold and saturation energies for the cross section of SEE in the logic device have been reported to be set at around 10 MeV and 100 MeV, respectively. The proton beam was directly irradiated on the ASIC with a flux of 4.8×10^7 s^{-1}cm^{-2} and a fluence of 1.0×10^{11} cm^{-2}, which corresponds to the total dose for 160 years in the International Space Station (ISS) orbit [19]. During the irradiation, the currents from the positive and negative power supplies for the charge sensitive amplifier (CSA) on the ASIC were monitored in order to check whether an SEL event occurred or not. The digital bit data of DICEs were duplicated on a copy register in order to detect SEU events during the irradiation. There was no indication of an SEL event and any SEU events were not obtained either. The ASIC has 1030-bit registers; therefore, the SEU cross-section is estimated to be 2.2×10^{-14} cm^{2}bit^{-1} at most.

The Kyushu Institute of Technology group has tried to obtain the excitation function of the SEE cross section by the variation of kinetic energy of proton [20]. The measurements were conducted for the microchips of PIC16F877, Raspberry Pi Zero and Raspberry Pi 3B. The obtained excitation function shows that the threshold energy for SEE in the integrated circuits may be set at around 10 MeV.

The Aoyama Gakuin University and Nagoya University group have paid attention to the radiation effect on the detectors in space from the radioactivation point of view. A gamma-ray detector using new scintillator LaBr$_3$ (Ce) has been developed for the CALorimetric Electron Telescope (CALET) experiment on ISS. LaBr$_3$ (Ce) detectors were irradiated with 20, 70, and 140 MeV protons with fluences of $7.9 \times 10^{10}, 7.1 \times 10^9$, and 5.0×10^9 cm^{-2}, respectively [21]. After the irradiation, the activation of the scintillators was investigated by measuring gamma-ray energy spectra by Ge detectors.

4.3. Research Results Using the Tandem Accelerator and the Synchrotron

An AlGaN/GaN high-electron mobility transistor (HEMT) has a lot of advantages for space applications, such as having high efficiency, high power and being lightweight. In order to quantify the radiation tolerance of the HEMT in space environment, Sasaki et al. investigated the change in its device characteristics by using the 11.1 MeV, 20.6 MeV and 66.8 MeV proton beams from the synchrotron [22]. The irradiation was performed with the proton flux of 1.7×10^8 cm^{-2}s^{-1} under off-bias ($V_{gs} = -5.0$ V, $V_{ds} = 100$ V) condition. No difference was observed in the breakdown voltage at any irradiation energy (Figure 11).

To study the DDD effect on the HEMT, the change in the output power due to the 2 MeV proton irradiation was also examined by using the tandem accelerator [22]. The 2 MeV proton beam flux was 7×10^{11} cm^{-2}s^{-1}. The output power at last decreased by -4 dB and -10 dB at fluences of 3×10^{14} and 1×10^{15} cm^{-2}, respectively (Figure 12). Sasaki et al. have explained this irradiation effect as being due to the lattice defects produced by the proton irradiation. As the effect by proton fluence of 7×10^{11}/cm^2 roughly corresponds to the radiation damage of 10 years in the geostationary orbit, the experimental result indicates that this device has a sufficient margin for proton irradiation tolerance to operate in the geostationary orbit. They have also found that when the channel temperature was 125 °C, the output power was partially recovered by radio frequency (RF) operation or reverse bias stress ($V_{gs} = -5$ V, $V_{ds} = 30$ V).

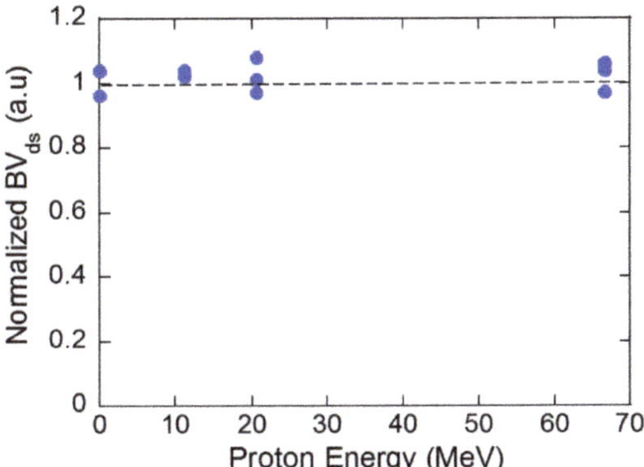

Figure 11. Normalized off-state (V_{gs} = −5.0 V) drain breakdown voltage (BV_{ds}) of HEMT under 11.1–66.8 MeV proton irradiation.

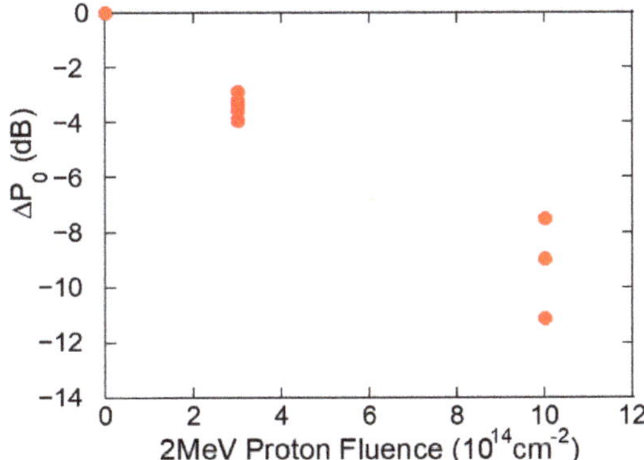

Figure 12. Change in output power of HEMT by 2 MeV proton irradiation.

4.4. Research Results Using the Tandem Accelerator

The Martian Moons eXploration (MMX) is a project of the Japan Aerospace Exploration Agency (JAXA) to explore the two moons (Phobos and Deimos) of Mars with a scheduled launch in the mid-2020s. A spacecraft will enter into the orbit around Mars and will collect scientific data and gather samples from the moons' surfaces. Unlike the Earth, Mars does not have a global magnetic field. Therefore, electronic devices attached to the MMX spacecraft orbiting around Mars and its moons will be exposed to interplanetary high energy protons with the wide energy range. Ozaki et al. have performed the experiments on irradiation resistance evaluation of CCD imagers, which is one of the important electronic devices installed to the MMX spacecraft, by using an 8 MeV proton beam from the tandem accelerator [23]. The irradiation effect has been observed as an increase in dark current. They have performed a similar experiment using a 70 MeV proton beam at another facility. From the proton energy dependence of the irradiation effect on CCD imagers, the experimental data are now being analyzed in terms of the total ionization dose (TID) and the non-ionization energy loss (NIEL).

4.5. Research Results Using the Ion-Implanter

As perovskite solar cells have beneficial features as a high-power conversion efficiency as high as 24%, low-cost, thin-coating, lightweight, and large-area fabrication, they are expected for the application to the space industry. To confirm the irradiation effect on perovskite type solar cells, Kanaya et al. performed the 50 keV proton irradiation with total doses of 1×10^{13}, 1×10^{14}, and 1×10^{15} cm^{-2} using the 200 kV ion-implanter [24]. To investigate the irradiation effect on the grain structure and lattice structure, three types of perovskite solar cells with different grain sizes (~240, ~340, and ~690 nm) were used and the irradiation effects were evaluated by a scanning electron microscope (SEM), X-ray diffraction (XRD), photoluminescence (PL), and time-resolved photoluminescence (TRPL). From the experimental results of SEM and XRD, they have concluded that there is little difference between the initial and irradiated samples in the grain structure or the crystal structure. The result of the TRPL spectrum measurements is shown in Figure 13. Even by the irradiation up to the fluence of 1×10^{14} cm^{-2}, the TRPL spectrum curve does not exhibit a significant decrease. After the irradiation with the fluence of 1×10^{15} cm^{-2}, the lifetime of TRPL spectrum decreases for each grain size. The experimental result shows that perovskite solar cells have a high radiation resistance against the proton irradiation up to the fluence of 10^{14} cm^{-2}.

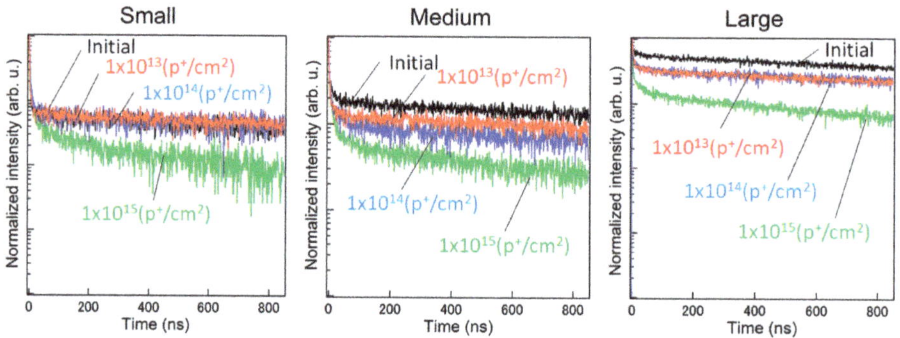

Figure 13. Effect of 50 keV proton irradiation on normalized TRPL decay curves for small grain (~240 nm) sample, medium grain (~340 nm) sample, and large grain (~690 nm) sample. (Reprint Figure 2 in p. 6992 of S. Kanaya et al., J. Phys. Chem. Lett. 10(2019) 6990–6995).

Triple-junction (InGaP/InGaAs/Ge) solar cells are typical successors for the space application and currently widely utilized due to their higher photovoltaic conversion efficiency and better radiation resistance compared to conventional single-junction silicon space solar cells. Imaizumi et al. investigated the effects of the fluence rate and the irradiation mode (defocused beam or scanned beam) of 50, 100 and 150 keV proton irradiation ground tests on the degradation of a triple-junction solar cell and a single-junction silicon solar cell using two accelerators: the ion-implanter at WERC, and that at the National Institutes for Quantum and Radiological Science and Technology (QST)-Takasaki. They have found that the photovoltaic properties of both the solar cells are degraded irrespective of the fluence rate or the irradiation mode. The details are discussed in [25].

Finally, we mention the study of the proton irradiation effect on carbonaceous chondrites by Nakauchi et al. [33]. Although this study is not directly related to the irradiation effect on space electronics, the result obtained by the ion irradiation at WERC is academically quite interesting and important in the field of the astrophysics.

Airless planetary bodies, from which carbonaceous chondrites (CCs) originate, have been mainly exposed to the irradiation with solar wind protons. To investigate the effect of solar wind protons on the surfaces of planetary bodies, Nakauchi et al. irradiated serpentine and saponite samples, which are two major components of CCs, with 10 keV H_2^+ ions up to the fluence of 1.7×10^{18} protons/cm^2, and their reflectance spectra were measured in the wavelength range from 1.5 to 5.5 μm. As the 10 keV H_2^+ molecule ion becomes two 5 keV proton ions at the sample surface by the Coulomb explosion process, the 10 keV H_2^+ ion irradiation corresponds to the 5 keV proton irradiation. Figure 14 shows that the absorption strength for serpentine at the wavelengths of 2.85 μm and 3.0 μm increases with increasing the proton dose. The absorption at 2.85 μm and that at 3.0 μm correspond to Si-OH and the total amount of OH and H_2O, respectively. This experimental result is the first direct evidence of H_2O formation at the surfaces of airless planetary bodies only by solar wind protons.

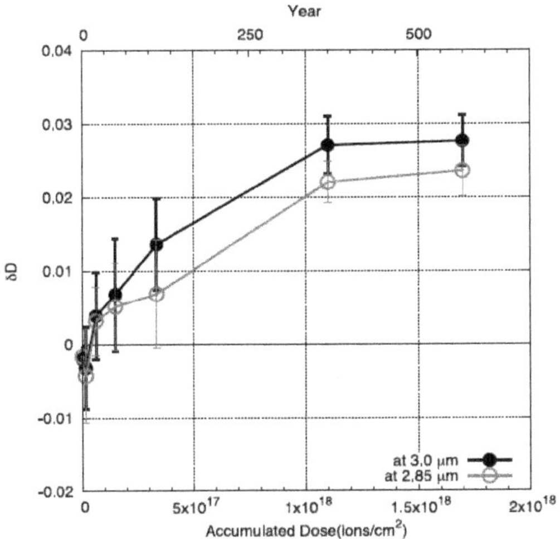

Figure 14. Change in absorption strength at 2.85 μm and 3.0 μm as a function of proton dose for serpentine. (Reprint Figure 6 in p.6 of Y. Nakauchi et al., Icarus 355(2021) 114140).

5. Summary

The ion accelerator facility at WERC, which consists of three kinds of accelerators (the synchrotron, the tandem accelerator, and the ion-implanter), has been utilized for the ground tests of the irradiation-induced degradation of several space electronics such as space solar cells, radiation detectors, image sensors, and LSI circuits. The facility has also contributed to the fundamental study on the astrophysics. Through collaborations with several universities, private companies, and national institutes such as JAXA and QST, further research and development using the ion facility at WERC can be expected for the space industry.

Author Contributions: Original draft preparation, writing and editing, S.H., R.I. and K.S. Final check of draft, K.K. All authors have read and agreed to the published version of the manuscript.

Funding: This research received no external funding.

Data Availability Statement: The data presented in this study are available on request from the authors.

Acknowledgments: The authors thank Yonetoku, Ogino, Takahashi, Yatsu, Arimoto, Yamaoka, Okumaya, Sasaki, Ozaki, Kanaya, and Nakauchi for their great help. The authors also thank T. Ohshima and M. Imaizumi for fruitful discussion.

Conflicts of Interest: The authors declare no conflict of interest.

References

1. Hatori, S.; Kurita, T.; Hayashi, Y.; Yamada, M.; Yamada, H.; Mori, J.; Hamachi, H.; Kimura, S.; Shimoda, T.; Hiroto, M.; et al. Developments and applications of accelerator system at the Wasasa Wan Energy Research Center. *Nucl. Instrum. Methods B* **2005**, *241*, 862–869. [CrossRef]
2. Hatori, S.; Fukumoto, S.; Hayashi, Y.; Kagawa, H.; Kurita, T.; Minehara, E.J.; Nagasaki, S.; Nakata, Y.; Odagiri, T.; Shimada, M.; et al. Tandem accelerator as an injector for the medical-use synchrotron at the wakasa wan energy research center. In Proceedings of the 1st International Particle Accelerator Conference, Kyoto, Japan, 23–28 May 2010; pp. 714–716.

3. Fujinaga, H.; Sakai, Y.; Yamashita, T.; Arai, K.; Terashima, T.; Komura, T.; Seki, A.; Kawaguchi, K.; Nasti, A.; Yoshida, K.; et al. Biological characteristics of gene expression features in pancreatic cancer cells induced by proton and X-ray irradiation. *Int. J. Radiat. Biol.* **2019**, *95*, 571–579. [CrossRef]

4. Lin, C.; Kume, K.; Mori, T.; Martinez, M.E.; Okazawa, H.; Kiyono, Y. Predictive Value of Early-Stage Uptake of 3′-Deoxy-3′-18F-Fluorothymidine in Cancer Cells Treated with Charged Particle Irradiation. *J. Nucl. Med.* **2015**, *56*, 945–950. [CrossRef]

5. Das, S.K.; Masuda, M.; Hatashita, M.; Sakurai, A.; Sakakibara, M. Optimization of culture medium for cordycepin production using Cordyceps militaris mutant obtained by ion beam irradiation. *Process Biochem.* **2010**, *45*, 129–132. [CrossRef]

6. Takagi, K.; Tsukada, T.; Izumi, M.; Kazama, K.; Hayashi, Y.; Abe, T. Localization of γH2AX Fluorescence along Tracks of Argon Beams. *RIKEN Accel. Prog. Rep.* **2010**, *43*, vii–viii.

7. Hatashita, M.; Takagi, K.; Ichida, H.; Abe, T. Effect of ion-beam irradiation on mycelial growth of medicinal mushrooms. *RIKEN Accel. Prog. Rep.* **2019**, *53*, 200.

8. Ito, Y.; Yamauchi, T.; Yamamoto, A.; Sasase, M.; Nishio, S.; Yasuda, K.; Ishigami, R. Ion beam synthesis of 3C-SiC layers in Si and its application in buffer layer for GaN epitaxial growth. *Appl. Surf. Sci.* **2004**, *238*, 159–164. [CrossRef]

9. Ishigami, R.; Batchuluun, C.; Yasuda, K. Increase in coercivity of Fe-Pt thin film permanent magnets by N_2^+ ion implantation and heat treatment in hydrogen gas. *Nucl. Instrum. Methods B* **2012**, *275*, 58–62. [CrossRef]

10. Iwase, A.; Yoneda, K.; Ishigami, R.; Matsui, T. Restoration of ion beam irradiation induced metastable magnetic states and lattice structures of FeRh thin films by heat treatments at elevated temperatures. *J. Magn. Magn. Mater.* **2020**, *515*, 167286. [CrossRef]

11. Yasuda, K.; Kajitori, Y.; Oishi, M.; Nakamura, H.; Haruyama, Y.; Saito, M.; Suzuki, K.; Ishigami, R.; Hibi, S. Upgrades of a time-of-fight elastic recoil detection analysis measurement system for thin flm analysis. *Nucl. Instrum. Methods B* **2019**, *442*, 53–58. [CrossRef]

12. Suzuki, K.; Nakata, Y. Development of an in-air ERDA system for hydrogen analysis. *Nucl. Instrum. Methods B* **2019**, *450*, 135–138. [CrossRef]

13. Suzuki, K.; Tsuchiya, B.; Yasuda, K.; Nakata, Y. Light element analysis of ceramics using in-air ERDA and TOF-ERDA. *Nucl. Instrum. Methods B* **2020**, *478*, 169–173. [CrossRef]

14. Fukumoto, K.; Tone, K.; Onitsuka, T.; Ishigami, R. Effect of Ti addition on microstructural evolution of V–Cr–Ti alloys to balance irradiation hardening with swelling suppression. *Nucl. Mater. Energy* **2018**, *15*, 122–127. [CrossRef]

15. Das, A.K.; Ishigami, R.; Kamal, I. Proton implantation effect on SUS-316 stainless steel. *J. Alloys Compd.* **2015**, *629*, 319–321. [CrossRef]

16. Fukumoto, K.; Kitamura, Y.; Miura, S.; Fujita, K.; Ishigami, R.; Nagasaka, T. Irradiation Hardening Behavior of He-Irradiated V-Cr-Ti Alloys with Low Ti Addition. *Quantum Beam Sci.* **2021**, *5*, 1. [CrossRef]

17. Ogino, N.; Arimoto, M.; Sawano, T.; Yonetoku, D.; Yu, P.; Watanabe, S.; Hiraga, J.S.; Yuhi, D.; Hatori, S.; Kume, K. Performance verification of next-generation Si CMOS soft X-ray detector for space applications. *Nucl. Instrum. Methods A* **2021**, *987*, 164843. [CrossRef]

18. Hirade, N.; Takahashi, H.; Uchida, N.; Ohno, M.; Torigoe, K.; Fukazawa, Y.; Mizuno, T.; Matake, H.; Hirose, K.; Hisadomi, S.; et al. Annealing of proton radiation damages in Si-PM at room temperature. *Nucl. Instrum. Methods A* **2021**, *986*, 164673. [CrossRef]

19. Arimoto, M.; Harita, S.; Sugita, S.; Yatsu, Y.; Kawai, N.; Ikeda, H.; Tomida, H.; Isobe, N.; Ueno, S.; Mihara, T.; et al. Development of a 32-channel ASIC for an X-ray APD detector onboard the ISS. *Nucl. Instrum. Methods A* **2018**, *882*, 138–147. [CrossRef]

20. Okuyama, K.; (Hitachi Ltd., Ibaraki-ken, Japan). Personal communication, 2021.

21. Mizushima, T.; Yoshida, A.; Yamaoka, K.; Hatori, S.; Kume, K. Study on possible protpn-induced background of LaBr$_3$(Ce) scintillator in a low-earth orbit. *Proc. SPIE* **2020**, *11454*, 114542S.

22. Sasaki, H.; Hisaka, T.; Kadoiwa, K.; Kamo, Y.; Ishimura, E.; Hatori, S.; Ishigami, R.; Kume, K. Reliability Study of Proton Irradiation Effects on AlGaN/GaN HEMTs. In Proceedings of the 34th Annual Reliability of Compound Semiconductors (ROCS) Workshop, Minneapolis, MN, USA, 29 April 2019; pp. 83–86.

23. Ozaki, M.; (Kobe University, Kobe, Japan). Personal communication, 2021.

24. Kanaya, S.; Kim, G.M.; Ikegami, M.; Miyasaka, T.; Suzuki, K.; Miyazawa, Y.; Toyota, H.; Osonoe, K.; Yamamoto, T.; Hirose, K. Proton Irradiation Tolerance of High-Efficiency Perovskite Absorbers for Space Applications. *J. Phys. Chem. Lett.* **2019**, *10*, 6990–6995. [CrossRef]

25. Imaizumi, M.; Ohshima, T.; Yuri, Y.; Suzuki, K.; Ito, Y. Effects of beam conditions for ground irradiation tests on degradation of photovoltaic characteristics of space solar cells. *Quantum Beam Sci.* 2021; to be published.

26. Ferlet-Cavros, V.; Massengill, L.W.; Gouker, P. Single Event Transients in Digital CMOS—A Review. *IEEE Trans. Nucl. Sci.* **2013**, *60*, 1767–1790. [CrossRef]

27. Srour, J.R.; Palko, J.W. Displacement Damage Effects in Irradiated Semiconductor Devices. *IEEE Trans. Nucl. Sci.* **2013**, *60*, 1740–1766. [CrossRef]

28. Makino, T.; Onoda, S.; Ohshima, T.; Kobayashi, D.; Ikeda, H.; Hirose, K. A Methodology for Reconstructing DSET Pulses from Heavy-Ion Broad-Beam Measurement. *Quantum Beam Sci.* **2020**, *4*, 15. [CrossRef]

29. Kobayashi, D.; Hirose, K.; Sakamoto, K.; Okamoto, S.; Baba, S.; Shindou, H.; Kawasaki, O.; Makino, T.; Ohshima, T.; Mori, Y.; et al. Data-Retention-Voltage-Based Analysis of Systematic Variations in SRAM SEU Hardness: A Possible Solution to Synergistic Effects of TID. *IEEE Trans. Nucl. Sci.* **2020**, *67*, 328–335. [CrossRef]

30. Imaizumi, M.; Nakamura, T.; Takamoto, T.; Ohshima, T.; Tajima, M. Radiation degradation characteristics of component subcells in inverted metamorphic triple-junction solar cells irradiated with electrons and protons. *Prog. Photovolt. Res. Appl.* **2017**, *25*, 161–174. [CrossRef]

31. Hoeffgen, S.K.; Metzger, S.; Steffens, M. Investigating the Effects of Cosmic Rays on Space Electronics. *Front. Phys.* **2020**, *8*, 318. [CrossRef]

32. Yonetoku, D.; Mihara, T.; Sawano, T.; Ikeda, H.; Harayama, A.; Takata, S.; Yoshida, K.; Seta, H.; Toyanago, A.; Kagawa, Y.; et al. High-z gamma-ray bursts for unraveling the dark ages mission HiZ-GUNDAM. *Proc. SPIE* **2014**, *9144*, 15–26.

33. Nakauchi, Y.; Abe, M.; Ohtake, M.; Matsumoto, T.; Tsuchiyama, A.; Kitazato, K.; Yasuda, K.; Suzuki, K.; Nakata, Y. The formation of H_2O and Si-OH by H_2^+ irradiation in major minerals of carbonaceous chondrites. *Icarus* **2021**, *355*, 114140. [CrossRef]

quantum beam science

Article

Development of Pulsed TEM Equipped with Nitride Semiconductor Photocathode for High-Speed Observation and Material Nanofabrication

Hidehiro Yasuda [1,*], Tomohiro Nishitani [2,3], Shuhei Ichikawa [1], Shuhei Hatanaka [1], Yoshio Honda [3] and Hiroshi Amano [3]

1. Division of Materials and Manufacturing Science, Graduate School of Engineering, Osaka University, Osaka 565-0871, Japan; ichikawa@uhvem.osaka-u.ac.jp (S.I.); hatanaka@uhvem.osaka-u.ac.jp (S.H.)
2. Photo Electron Soul Inc., Tokyo 100-8901, Japan; t.nishitani@photoelectronsoul.com
3. Institute of Materials and Systems for Sustainability Center for Integrated Research of Future Electronics, Nagoya University, Aichi 464-8601, Japan; honda@nuee.nagoya-u.ac.jp (Y.H.); amano@nuee.nagoya-u.ac.jp (H.A.)
* Correspondence: yasuda@uhvem.osaka-u.ac.jp

Abstract: The development of pulsed electron sources is applied to electron microscopes or electron beam lithography and is effective in expanding the functions of such devices. The laser photocathode can generate short pulsed electrons with high emittance, and the emittance can be increased by changing the cathode substrate from a metal to compound semiconductor. Among the substrates, nitride-based semiconductors with a negative electron affinity (NEA) have good advantages in terms of vacuum environment and cathode lifetime. In the present study, we report the development of a photocathode electron gun that utilizes photoelectron emission from a NEA-InGaN substrate by pulsed laser excitation, and the purpose is to apply it to material nanofabrication and high-speed observation using a pulsed transmission electron microscope (TEM) equipped with it.

Keywords: laser photocathode; pulsed electron sources; pulsed transmission electron microscope

check for **updates**

Citation: Yasuda, H.; Nishitani, T.; Ichikawa, S.; Hatanaka, S.; Honda, Y.; Amano, H. Development of Pulsed TEM Equipped with Nitride Semiconductor Photocathode for High-Speed Observation and Material Nanofabrication. *Quantum Beam Sci.* **2021**, *5*, 5. https://doi.org/10.3390/qubs5010005

Received: 21 December 2020
Accepted: 27 January 2021
Published: 1 February 2021

Publisher's Note: MDPI stays neutral with regard to jurisdictional claims in published maps and institutional affiliations.

1. Introduction

For material nanofabrication using an electron beam, a sample is irradiated with a high-intensity electron beam using a field-emission electron gun. The development of a pulsed electron source with high emittance in a short time is effectively used not only for material nanofabrication but also for high-speed observation of electron irradiation-induced material behavior and observation with reduced electron irradiation effect of polymers when used for TEM. For example, the dislocation motion is in the scale from 1 s to 1 ms, the nucleation and growth in the scale from 1 ms to 0.1 ps, and the magnetic switching or melting and solidification in the scale from 10 ns to 0.1 ps. The time scale on which physical phenomena, such as phase transitions, occur is in the scale from 1 ns to 10 fs [1].

The development of a TEM equipped with a laser photocathode electron gun has been in progress since around 2000. As shown in Figure 1, the first machine was manufactured at the Berlin Institute of Technology. Using the pulsed electrons emitted from a Ta cathode excited by an Nd:YAG laser (wavelength, 266 nm; pulse width, 7 ns; current density, 700 A/cm^2; brightness, ~4 × 10^6 A/cm^2 str), a spatial resolution of ~100 nm and a temporal resolution of ~1 ns were achieved by the pulsed electrons [2]. After that, in 2007, at Lawrence Livermore National Laboratory in the United States, a spatial resolution of ~10 nm and a temporal resolution of ~1 ns were achieved [3]. In 2008, at the California Institute of Technology in the United States, the temporal resolution was improved to ~10 ps in the same spatial resolution of ~10 nm [4]. Recently in 2016, at CNRS National Laboratory in France, an atomic resolution of ~0.2 nm and a temporal resolution of ~370 fs were successfully achieved [5].

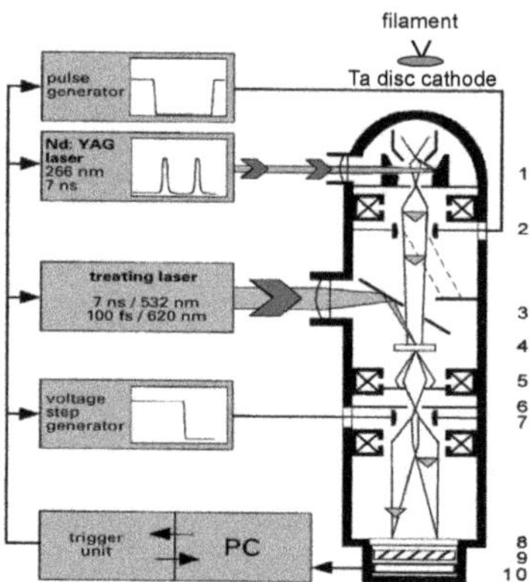

Figure 1. The first machine of pulsed TEM equipped with a photocathode manufactured at the Berlin Institute of Technology [2].

In the developments mentioned above, the cathode material is a metal, but by replacing it with compound semiconductors, it is possible to further increase the emittance. Using III–V compound semiconductor substrates, such as GaAs, with a negative electron affinity (NEA) surface is useful to generate pulsed electrons with small energy dispersion with high emittance. However, it is difficult to handle because the usable vacuum environment is severe, and an extremely high vacuum is required to extend the lifetime [6–11].

The advantage of using the NEA semiconductor surface is it reduces the work function. The role of Cs in NEA surface preparation has not been clarified in detail. The NEA treatment method involves forming a dipole layer of Cs with a positive charge on the semiconductor surface by repeating Cs vapor deposition and oxygen adsorption at a substrate temperature of about 550 °C. This dipole layer is thought to reduce the work function. Lowering of the work function to the energy near the conduction band minimum (CBM) on the semiconductor surface leads the electrons with an extremely narrow energy width excited just above the CBM to be emitted from the surface by the tunnel effect. As a result, it is possible to obtain an electron beam in which the energy dispersion is extremely small. Even if the work function is lowered on the NEA metal surface, the energy dispersion of electrons excited to the unoccupied level cannot be as small as that of a semiconductor with a bandgap. NEA treatment on the semiconductor surface is extremely effective from the viewpoint of electron energy dispersion.

On the other hand, the nitride semiconductor has an advantage in that the cathode lifetime is long because the vacuum environment used is not so severe. However, there are not many examples of the development of TEM using such nitride semiconductor photocathodes with a NEA surface.

In the present study, a pulsed TEM equipped with a photocathode electron gun, in which photoelectron emission from a NEA-InGaN substrate by pulsed laser excitation was utilized, was developed for the purpose of high-speed observation and material nanofabrication. In the paper, we will introduce some of the performances of the apparatus.

2. Development of the Apparatus

Figure 2 shows the appearance of a pulse TEM equipped with the developed photocathode electron gun. The pulsed electron gun consists of an ultrahigh vacuum chamber capable of NEA surface processing, an acceleration tube, an ion pump, two non-evaporable getter (NEG) pumps, a Cs vapor deposition source for NEA surface processing, a semiconductor pulse laser for pulse electron generation, and an auxiliary laser for photocurrent measurement during NEA surface processing. The accelerating voltage is 50 kV, and the ultimate vacuum of the chamber is ~5×10^{-9} Pa. The wavelength of the semiconductor pulse laser for pulse electron generation is 404 nm, and it is possible to generate pulses from continuous wave (CW) to nanoseconds. The pulse electron gun was assembled by installing it to the Hitachi HF-2000 TEM (Hitachinaka, Ibaraki, Japanc) from which the cold field-emission electron gun and the acceleration tube were removed.

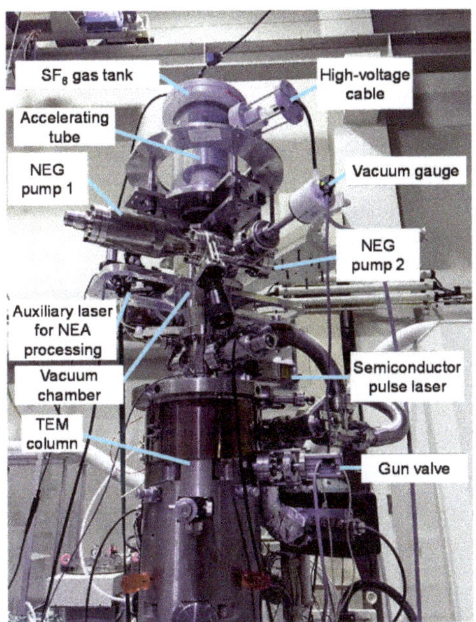

Figure 2. The appearance of a pulse TEM equipped with the developed photocathode electron gun.

Figure 3 shows schematic illustrations of the operation. The cathode and acceleration tube have a structure that slides up and down. During the pulsed electron generation, the cathode and the anode are in a close state, and the pulsed electrons excited by the semiconductor pulse laser are introduced into the TEM column. On the other hand, during NEA surface treatment, the cathode moves to the upper part of the chamber, and the cathode substrate is heated at approximately 550 °C in an O_2 atmosphere of about 10^{-6} Pa and under Cs vapor deposition. At the same time, the activation process is performed while monitoring the state with the photocurrent generated by the auxiliary laser excitation.

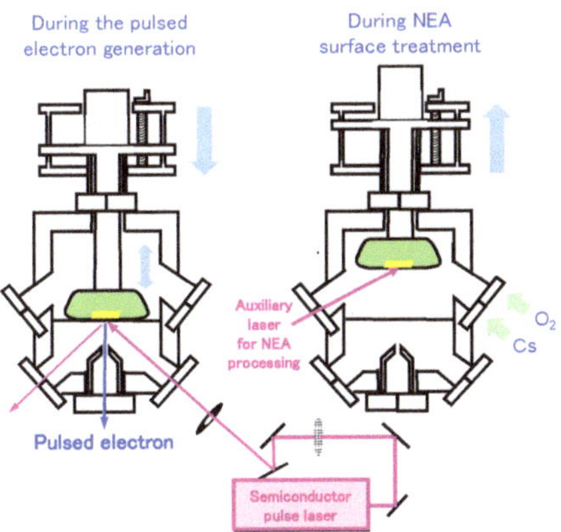

Figure 3. Schematic illustrations of the operation. The cathode and acceleration tube have a structure that slides up and down.

The cathode substrate is a p-type InGaN thin film. An $In_{0.12}Ga_{0.88}N$ thin film doped with Mg with a thickness of about 250 nm is deposited on the sapphire substrate via a GaN buffer layer with a thickness of about 2.2 mm. In order to evaluate the performance of the apparatus, a cathode substrate passed about 2 years or more from the NEA surface treatment was used. The electron emission current immediately after the NEA surface treatment is about 10~100 mA, but it remains 1 mA even after about 2 years or more. The imaging conditions were a laser pulse width of 100 ms and a repetition frequency of 5 kHz. An imaging plate was used for image recording.

3. Performance Characterization

In order to evaluate the performance of the apparatus, the exposure time required for pulse imaging was measured. Figure 4 shows a bright-field image that was taken by conventional TEM with a thermal emitter and size dispersion of Au nanoparticles on an amorphous carbon film used for pulse imaging as a standard sample. The size distribution was fitted by the lognormal distribution function, and the average particle size was approximately 13 nm.

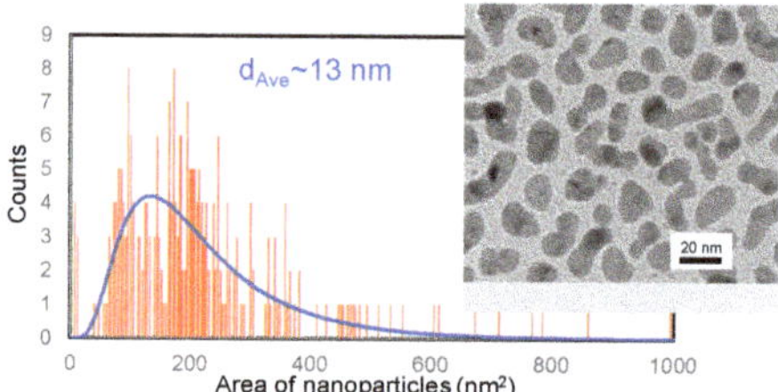

Figure 4. A bright-field image and size dispersion of Au nanoparticles on an amorphous carbon film used for pulse imaging as a standard sample. The size distribution was fitted by the lognormal distribution function.

Figure 5 shows the results of the pulse imaging of Au nanoparticles mentioned above by bright-field images. The contrast becomes clearer as the number of integrated pulses increases. It is evident that the image can be identified at about 4–8 pulses under the conditions. A factor in which laser frequency affects image quality is the amount of specimen drift within the total exposure time. If the total exposure time is shortened, the effect of the specimen drift will be reduced, which will lead to an improvement in image quality. In Figure 5, even with a total of 2048 pulses, the total exposure time is about 0.2 s, and even if an interval of 100 ms is included, it is 0.4 s, and the effect of the specimen drift is very small.

Figure 6 shows an example of the contrast profile of Au nanoparticles on an amorphous carbon support film. The contrast change of the part indicated by a red line in Au nanoparticles in Figure 6a is shown in Figure 6b. The dark contrast of the Au nanoparticle is shown as an image signal by normalization through the bright contrast of an amorphous carbon support film. An enlargement of the contrast in the amorphous carbon support film is shown in Figure 6c. The deviation from the average value of the contrast is defined as noise. Figure 6d shows the signal-to-noise (S/N) ratio as a function of the number of pulses (exposure time).

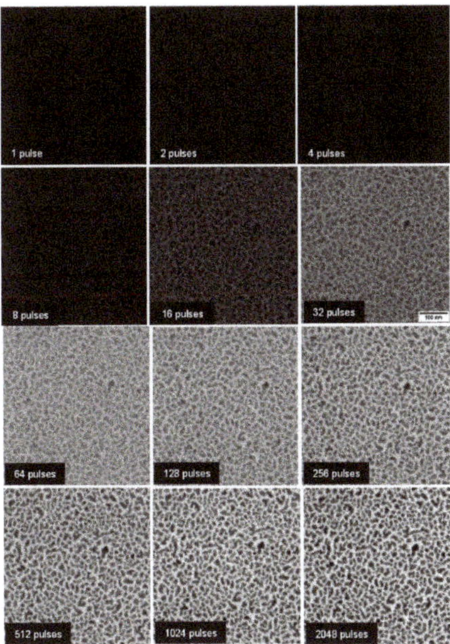

Figure 5. The results of pulse imaging of Au nanoparticles.

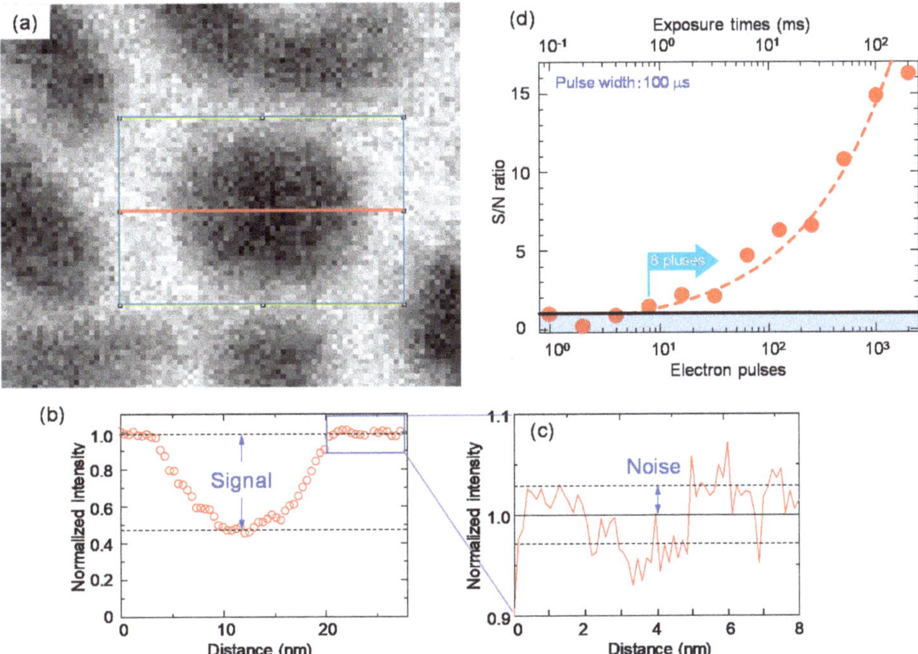

Figure 6. An example of the contrast profile of Au nanoparticles on an amorphous carbon support film. (**a**) A bright-field image, (**b**) the contrast change of the part indicated by a red line in Au nanoparticles, (**c**) an enlargement of the contrast in the amorphous carbon support film, (**d**) S/N ratio as a function of the number of pulses (exposure time).

Based on this result, if the minimum exposure time is determined assuming that Au nanoparticles can be identified when the S/N ratio is 1 or more, it is revealed that an image can be recorded with an exposure time of 800 ms or more under the present condition. Immediately after NEA surface processing, which has a high quantum efficiency of the electron emission, it is possible to take an emission current that is one to two orders of magnitude larger, so it is considered that the minimum exposure time will be shorter. This performance suggests that it is sufficient for high-speed in situ observation as a pulsed TEM.

The performance of pulsed TEM mentioned above can also be applied to material nanofabrication. When a semiconductor photocathode is utilized as the electron gun of a semiconductor inspection device, the stability of the electron beam is extremely important. The required performance is a small current fluctuation, which is a standard deviation from the average current during a constant current generation for a long time. In particular, if the current fluctuation over a long time is large, it affects the performance, and the same accuracy as making the TEM image contrast negligibly small is required. It is considered that the fact that sufficient performance can be obtained by applying a semiconductor photocathode to a pulsed TEM is consistent with satisfying the necessary and sufficient conditions for utilizing a semiconductor photocathode in a semiconductor inspection device.

Recently, in the semiconductor manufacturing process, in order to solve the problems of low inspection throughput and drawing throughput, a new demand for higher brightness and a multi-electron beam source in which an individual source has high brightness and uniformity is searched. In order to obtain high brightness, an electron source that generates an electron beam with a large current and low-emittance characteristics is required. It is suggested that semiconductor photocathodes that enable the generation of a high-current, low-emittance, and high-speed pulse multi-electron beam are also useful as next-generation electron beam sources that respond to new demands of material nanofabrication.

4. Conclusions

We developed a TEM equipped with a NEA-InGaN photocathode electron gun with an accelerating voltage of 50 kV and evaluated its performance by imaging with a pulsed electron beam. The cathode has a long lifetime and can be imaged even after about 2 years of NEA surface treatment. Under the conditions, a bright-field image can provide an S/N ratio of 1 or more with an exposure time of 800 ms. By increasing the emittance further, the temporal and spatial resolutions can be improved.

Author Contributions: Conceptualization, T.N.; Data curation, S.I. and S.H.; Investigation, Y.H.; Writing—original draft, H.Y.; Writing—review & editing, H.A. All authors have read and agreed to the published version of the manuscript.

Funding: This work was supported by MEXT/JSPS KAKENHI, Grant Number JP19H00666, and by the joint research of "Research on observations by semiconductor photocathode electron beams".

Conflicts of Interest: The funders had no role in the design of the study; in the collection, analyses, or interpretation of data; in the writing of the manuscript, or in the decision to publish the results.

References

1. Chergui, M.; Zewail, A.H. Electron and X-ray methods of ultrafast structural dynamics: Advances and applications. *ChemPhysChem* **2009**, *10*, 28–43. [CrossRef] [PubMed]
2. Bostanjoglo, O.; Elschner, R.; Mao, Z.; Nink, T.; WeingaKrtner, M. Nanosecond electron microscopes. *Ultramicroscopy* **2000**, *81*, 141–147. [CrossRef]
3. Armstrong, M.R.; Reed, B.W.; Torralva, B.R.; Browning, N.D. Prospects for electron imaging with ultrafast time resolution. *Appl. Phys. Lett.* **2007**, *90*, 114101. [CrossRef]
4. Barwick, B.; Park, H.S.; Kwon, O.-H.; Baskin, J.S.; Zewail, A.H. 4D Imaging of Transient Structures and Morphologies in Ultrafast Electron Microscopy. *Science* **2008**, *322*, 1227–1231. [CrossRef]
5. Bücker, K.; Picher, M.; Crégut, O.; LaGrange, T.; Reed, B.W.; Park, S.T.; Masiel, D.J.; Banhart, F. Electron beam dynamics in an ultrafast transmission electron microscope with wehnelt electrode. *Ultramicroscopy* **2016**, *171*, 8–18.

6. Sato, D.; Nishitani, T.; Honda, Y.; Amano, H. Observation of relaxation time of surface charge limit for InGaN photocathodes with negative electron affinity. *Jpn. J. Appl. Phys.* **2016**, *55*, 05FH05. [CrossRef]
7. Kashima, M.; Sato, D.; Koizumi, A.; Nishitani, T.; Honda, Y.; Amano, H.; Iijima, H.; Meguro, T. Analysis of negative electron affinity InGaN photocathode by temperature-programed desorption method. *J. Vac. Sci. Technol. B* **2018**, *36*, 06JK02. [CrossRef]
8. Sato, D.; Nishitani, T.; Honda, Y.; Amano, H. Recovery of quantum efficiency on Cs/O-activated GaN and GaAs photocathodes by thermal annealing in vacuum. *J. Vac. Sci. Technol. B* **2020**, *38*, 012603. [CrossRef]
9. Sato, D.; Honda, A.; Koizumi, A.; Nishitani, T.; Honda, Y.; Amano, H. Optimization of InGaN thickness for high-quantum-efficiency Cs/O-activated InGaN photocathode. *Microelectron. Eng.* **2020**, *223*, 111229. [CrossRef]
10. Kuwahara, M.; Kusunoki, S.; Jin, X.G.; Nakanishi, T.; Takeda, Y.; Saitoh, K.; Ujihara, T.; Asano, H.; Tanaka, N. 30-kV spin-polarized transmission electron microscope with GaAs–GaAsP strained superlattice photocathode. *Appl. Phys. Lett.* **2012**, *101*, 033102. [CrossRef]
11. Kuwahara, M.; Kusunoki, S.; Nambo, Y.; Saitoh, K.; Jin, X.G.; Ujihara, T.; Asano, H.; Takeda, Y.; Tanaka, N. Coherence of a spin-polarized electron beam emitted from a semiconductor photocathode in a transmission electron microscope. *Appl. Phys. Lett.* **2014**, *105*, 193101. [CrossRef]

MDPI

St. Alban-Anlage 66

4052 Basel

Switzerland

Tel. +41 61 683 77 34

Fax +41 61 302 89 18

www.mdpi.com

Quantum Beam Science Editorial Office

E-mail: qubs@mdpi.com

www.mdpi.com/journal/qubs